Regional Trends 35

2000 edition

Editors : Jil Matheson
Gwyneth Edwards

Production team : Martin Smith
David Harper
Nina Mill
Anne-Marie Manners
Victoria Jackson
Nigel King
Mike Short
Sunita Dedi
Steve Whyman
Christine Lillistone

Nicola Amaranayake
Liza Murray
Max Bonini
Conor Shipsey
Alex Clifton-Fearnside
Dave Vincent
Aubrey Stoll
Philip Papaiah
Jan Kiernan
Mario Alemanno

David Penny
David Ham
Mayur Patel
Keith Tyrrell
Betty Ankamah
Ronnie Amako
Brian Yin
Ramona Insalaco
Matt Richardson
Sam Welch

Charts: Richard Birch

Maps: Gaynor Tizzard

London: The Stationery Office

Contact points

For enquiries about this publication, contact
Kevin Laverty:
Tel: 020 7533 5739
E-mail: kevin.laverty@ons.gov.uk

To order this publication, call The Stationery Office
on **0870 600 5522**. See also back cover.

For general enquiries, contact the National Statistics
Public Enquiry Service on **0845 601 3034**
(minicom: 01633 812399)
E-mail: info@statistics.gov.uk
Fax: 01633 652747
Letters: Room DG/18, 1 Drummond Gate,
London SW1V 2QQ

You can also find National Statistics on the internet —
go to **www.statistics.gov.uk**

About the Office for National Statistics

The Office for National Statistics (ONS) is the government agency responsible for compiling, analysing and disseminating many of the United Kingdom's economic, social and demographic statistics, including the retail prices index, trade figures and labour market data, as well as the periodic census of the population and health statistics. The Director of ONS is also the National Statistician and the Registrar General for England and Wales, and the agency administers the statutory registration of births, marriages and deaths there.

A National Statistics publication

National Statistics are produced to high professional standards set out in the National Statistics Code of Practice. They undergo regular quality assurance reviews to ensure that they meet customer needs. They are produced free from any political interference.

Contents

Contents

6 Housing

7 Health

Contents

12 Regional accounts

13 Industry and Agriculture

14 Sub-regions of England

Contents

Introduction

Regional Trends seeks to contribute to decision making at national, local and European levels and to inform debate about the current state of the nation. It brings together data from a wide range of sources, both from within government and outside to paint a comprehensive picture of the regions of the United Kingdom.

In recent years there has been much interest in regional diversity across the United Kingdom. While it is true that differences in income, housing costs, health and unemployment exist, in many other aspects of daily life, the regions are similar.

One often-quoted difference is that income is higher in the south east and lower in the north. Over the period 1996-97 to 1998-99, average gross weekly household income in London and the South East was almost one and a half times that of households in the North East and the difference has been similar since 1995-96. Personal income was also higher: London and the South East have the highest proportion of people with a personal taxable income of £50,000 or more. However, this does not give a full picture; weekly household expenditure and housing costs for those living in the south are higher than the national average and house prices continue to rise at a faster rate. The increase in house prices in London between 1998 and 1999 was almost 17 per cent compared with increases of between 6 and 7 per cent in the North East, North West, Yorkshire and the Humber and the East Midlands over the same period.

While these overall regional comparisons give a broad picture, they also mask considerable variability within regions. For example, although average weekly earnings for people in London in 1999 were the highest in the UK; £520 compared with the UK average of £399, within the region, average weekly earnings varied considerably: 10 per cent of men earned more than £1,008 but 10 per cent also earned less than £242.

One area where there is a strong north-south difference is in people's health. In particular, deaths from circulatory diseases are much higher in the north, although the differences have narrowed slightly in recent years. The contrast is particularly striking for women; in 1998, the age-adjusted mortality rates from ischaemic heart disease among women in the four northern NHS regions and Scotland were all above 200 people per 100,000 population, whereas in the southern English regions the rates were all below 180. There are also differences in people's behaviour; for example, the proportion of men who drank more than eight units of alcohol per day and women who drank more than 6 units per day in the week before interview were higher in the North West than in the East of England, London, the South East and the South West while people in Northern Ireland are least likely to drink. Almost half of men and three fifths of women in Northern Ireland had not had an alcoholic drink in the week prior to interview, far higher than in any other region.

Comprehensive and up-to-date statistics about regions and local areas are increasingly in demand. In response to this, we have produced the *Region in Figures* series to complement *Regional Trends*. The set of nine publications presents a wide range of sub-regional data at the lowest possible geographical level for each Government Office Region in England and will be as valuable as the information in *Regional Trends* is to regional decision-making.

Overview

Regional Trends provides a unique description of the regions of the United Kingdom. In 17 chapters it covers a wide range of demographic, social, industrial and economic statistics, taking a look at most aspects of life. The chapters fall broadly into four sections: regional profiles (Chapter 1), the European Union (Chapter 2), the main topic areas (Chapters 3 to 13) and sub-regional statistics (Chapters 14 to 17). To make comparison between regions easy, information is given in clear tables, maps and charts.

Regional statistics are essential for a wide range of people: for example, policy-makers and planners in both the public and private sectors; marketing professionals; researchers; students and teachers; journalists; and anyone with general regional interests. *Regional Trends* brings together data from diverse sources and, for some topics, is the only publication where data for the whole of the United Kingdom are available in one place. Wherever systems/data sources for the four countries of the United Kingdom are sufficiently comparable, figures have been aggregated to give a national average/total.

Coverage and definitions

Due to variations in coverage and definitions, some care may be needed when comparing data from more than one source. Readers should consult the Notes and Definitions towards the back of the book as well as reading the footnotes relevant to each table, map and chart for help in analysing trends or comparing different sources

Availability of Electronic data

The contents of *Regional Trends 35* will be available free-of-charge via StatBase, the Government's online statistics service. This can be accessed through the National Statistics website (http://www.statistics.gov.uk).

Further Information

Regional and sub-regional statistics can be found in a range of other GSS publications, statistical bulletins and regular press releases. Much of the information included in the Population and Households and the Labour market chapters of *Regional Trends* can be found on NOMIS, the on-line database run by Durham University under contract to the Office for National Statistics (ONS). It contains government statistics down to the smallest available geographic area, which may be unpublished elsewhere. In addition, sub-regional data for the Government Office Regions in England can be found in the series *Region in Figures*; data for Wales are published in the *Digest of Welsh Statistics* and the *Digest of Welsh Local Area Statistics*; data for Scotland are published in the *Scottish Abstract of Statistics* and the *Scottish Economic Report;* data for Northern Ireland are published in the *Northern Ireland Annual Abstract of Statistics*. Details of these publications and others can be found by using the StatBase 'text search' facility on the National Statistics website (http://www.statistics.gov.uk).

Contributors

The Editors wish to thank all their colleagues in the ONS and the rest of the Government Statistical Service and all contributors in other organisations without whose help this publication would not be possible.

Information

Regional boundaries

The United Kingdom comprises Great Britain and Northern Ireland; Great Britain consists of England plus Wales and Scotland. The Isle of Man and the Channel Isles are not part of the United Kingdom. The Scilly Isles are included as part of Cornwall throughout.

The statistical regions of the United Kingdom comprise the nine Government Office Regions (GORs) for England plus Wales, Scotland and Northern Ireland. The local government administrative structure provides the framework for breaking down the regions into smaller areas. Maps of the statistical regions of the United Kingdom and the sub-regions in each of the four countries are given in Chapters 1 and 14 to 17. Apart from the GORs which are used as far as is possible throughout, there are a number of other regional classifications used in *Regional Trends 35*. Maps of these non-standard regions are given on pages 239 and 240 of the Notes and Definitions.

Nomenclature for Territorial Units (NUTS)

Some data is presented using the European Nomenclature for Territorial Units (NUTS) area classification, primarily economic data in chapters 12 and 13. Further information on the NUTS classification is contained in the Notes and Definitions and maps showing the NUTS areas for the four constituent countries are shown in the sub-regional chapters on pages 214, 224, 232 and 237.

Sub-regional geography

The sub-regional information presented in Chapters 14 to 17 reflect the complete implementation of the local government reorganisation that happened between 1 April 1995 and 1 April 1998. Data for England in Chapter 14 are presented firstly by region. Within each region Unitary Authorities (UAs) are listed first in alphabetical order. Counties are listed next in alphabetical order. Within each County the Local Authority Districts (LAD) are listed alphabetically. Figures for former counties are shown at the end of the region. Chapter 15 for Wales and Chapter 16 for Scotland present data for the UAs and the New Councils respectively which replaced the former two-tier systems on 1 April 1996. Chapter 17 for Northern Ireland continues to give figures at Board or district level as available. Maps on pages 183, 218, 226 and 234 show the boundaries of the Counties/UAs, Districts and Boards as at 1 April 1998.

Full details of the local government reorganisation and the NUTS area classification are given in the *Gazetteer of old and new geographies of the United Kingdom* available from NS Direct Tel: 01633 812078, price £20.

Symbols and conventions

Reference years. Where a choice of years has to be made, the most recent year or a run of recent years is shown together with the past population census years (1991, 1981 etc) and sometimes the mid-points between census years (1996 etc). Other years may be added if they represent a peak or trough in the series.

Rounding of figures. In tables where figures have been rounded to the nearest final digit, there may be an apparent discrepancy between the sum of the constituent items and the total as shown.

Billion. This term is used to represent a thousand million.

Provisional and estimated data. Some data for the latest year (and occasionally for earlier years) are provisional or estimated. To keep footnotes to a minimum, these have not been indicated; source departments will be able to advise if revised data are available.

Survey data. Many of the tables and charts in *Regional Trends* present the results of household surveys which can be subject to large sampling error. Care should therefore be taken in drawing conclusions about regional differences.

Non-calendar years.
Financial year - eg 1 April 1998 - 31 March 1999 would be shown as 1998-99
Academic year - eg September 1998/July 1999 would be shown as 1998/99
Data covering more than one year - eg 1997, 1998 and 1999 would be shown as 1997-1999

Units. Figures are shown in italics when they represent percentages.

Symbols. The following symbols have been used throughout *Regional Trends*:
..	*not available*
.	*not applicable*
-	*negligible (less than half the final digit shown)*
0	*nil*

1 Regional Profiles

Statistical Regions of the United Kingdom

SCOTLAND

NORTHERN IRELAND

NORTH EAST

NORTH WEST

YORKSHIRE AND THE HUMBER

ENGLAND

———— Government Office Region boundary

EAST MIDLANDS

WEST MIDLANDS

WALES

EAST

LONDON

SOUTH WEST

SOUTH EAST

1.1 Key statistics for the North East

	North East	United Kingdom		North East	United Kingdom
Population, 1998 (thousands)	2589.6	59,236.5	Gross domestic product, 1998 (£ million)	25,496	747,544
Percentage under 16	*20.3*	*20.4*	Gross domestic product per head index, 1998 (UK=100)	78.8	100.0
Percentage pension age or over[1]	*18.7*	*18.1*	Total business sites, 1999 (thousands)	75.1	2,508.0
Standardised mortality ratio (UK=100), 1998	114	100	Average dwelling price[4], 1999 (£)	59,442	94,581
Infant mortality rate[2], 1997-1999	5.4	5.8			
Percentage of pupils achieving 5 or more grades A* to C			Motor cars currently licensed, 1998 (thousands)	824	23,878
at GCSE level, 1998/99	*40.8*	*49.1*	Fatal and serious accidents on roads[5], 1998		
			(rate per 100,000 population)	43	66
Economic activity rate[3], Spring 1999 (percentages)	*72.3*	*78.4*	Recorded crime rate[4], 1998-99		
Employment rate[3], Spring 1999 (percentages)	*64.9*	*73.6*	(notifiable offences per 100,000 population)	10,359	9,785
ILO unemployment rate, Spring 1999 (percentages)	*10.1*	*6.0*			
Average gross weekly earnings: males in full-time			Average weekly household income, 1996-1999 (£)[6]	357	430
employment, April 1999(£)	384.60	440.70	Average weekly household expenditure, 1996-1999 (£)[6]	283	333
Average gross weekly earnings: females in full-time			Households in receipt of Income Support/Family Credit[4],		
employment, April 1999 (£)	289.80	325.60	1998-99 (percentages)	*20*	*15*

1 Males aged 65 or over, females aged 60 or over.
2 Deaths of infants under 1 year of age per 1,000 live births.
3 For people of working age, males aged 16 to 64, females aged 16 to 59.
4 Figure for the United Kingdom relates to England and Wales.
5 Figure for the United Kingdom relates to Great Britain.
6 Combined years 1996-97, 1997-98 and 1998-99.

1.2 Population density: by local authority, 1998

Population density, 1998
(persons per sq km)

- 2,500 or over
- 1,000 - 2,499
- 500 - 999
- 250 - 499
- 100 - 249
- 99 or under

1 Wansbeck
2 Newcastle-upon-Tyne
3 Chester-le-Street
4 Hartlepool UA
5 Stockton-on-Tees UA
6 Middlesbrough UA

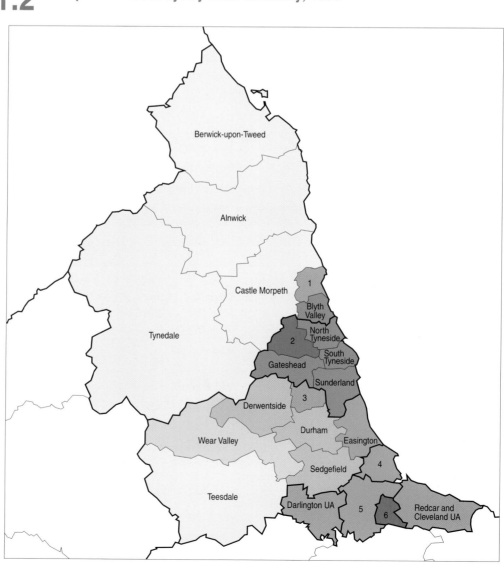

North East

Population

In 1998 the North East of England had a population of 2.6 million. Within the region the population density was highest in Middlesbrough UA at over 2,600 people per sq km and lowest in the district of Tynedale in Northumberland at only 26 people per sq km.

(Tables 3.1 and 14.1)

Mortality

Overall the region has a higher Standardised Mortality Ratio (SMR) than the UK as a whole at 114 in 1998 (UK=100). Within the region the SMR ranged from 90 in Berwick-upon-Tweed and Teesdale to 130 in Blyth Valley.

(Table 14.1)

The Infant Mortality rate for 1997-1999 for the North East was marginally lower than the UK rate (5.4 and 5.8 deaths of infants under 1 year old per 1,000 live births, respectively); within the region it ranged between 4.1 for Redcar and Cleveland UA and 6.4 for Hartlepool UA.

(Table 14.2)

Education

The North East had a higher participation rate for children under 5 in schools than the UK as a whole, 86 per cent compared with 64 per cent for the UK in January 2000.

(Tables 4.3 and 14.3)

In 1998/99, all areas in the region had lower proportions of pupils achieving 5 or more A* to C graded results than the UK average and higher proportions of pupils with no graded results than in the UK as a whole.

(Table 14.3)

Among the population of working age, the proportion of people who's highest qualification was GCE A level or equivalent in the North East was 24.6 per cent in Spring 2000; higher than the UK average of 24.0 per cent.

(Table 4.13)

Labour market

The employment rate for people of working age in the region in Spring 1999, was the lowest in the UK at 64.9 per cent. Within the region the employment rate varied between 74.3 per cent in Darlington UA to 60.9 per cent in Middlesbrough UA in 1998-1999.

(Tables 5.1 and 14.5)

In 1999, average weekly earnings for people in the region were amongst the lowest in the UK, only Northern Ireland had lower average earnings. Within the region average weekly earnings varied considerably; in Middlesbrough UA 10 per cent of men earned more than £678.90 but 10 per cent also earned less than £198.80.

(Tables 5.11 and 14.5)

Economy

Within the region manufacturing industry accounted for some 28.2 per cent of the region's GDP in 1997, compared to 21.3 per cent for the UK as a whole. Agriculture, forestry and fishing accounted for 0.8 per cent of the region's GDP in 1997, compared to 1.5 per cent for the UK.

(Table 12.4)

Almost 34 per cent of the region's 75 thousand business sites in 1999 were in distribution, hotels and catering and repairs industries, the highest rate in the UK; this compares with a UK average of 29.6 per cent.

(Table 13.3)

Over 80 per cent of the value of direct export trade from the region in 1999 was to the EU, the highest rate in the UK, and well above the average of nearly 61 per cent. Over half of the value of direct import trade to the region was from the EU, similar to the UK average of 51.5 per cent.

(Table 13.7)

Environment

Over a sixth of the total area in the North East is designated as an Area of Outstanding Natural Beauty and six per cent is Green Belt land.

(Table 11.12)

1.3 Key statistics for the North West

	North West	United Kingdom		North West	United Kingdom
Population, 1998 (thousands)	6,890.8	59,236.5	Gross domestic product, 1998 (£ million)	75,834	747,544
Percentage under 16	20.9	20.4	Gross domestic product per head index, 1998 (UK=100)	88.2	100.0
Percentage pension age or over[1]	18.2	18.1	Total business sites, 1999 (thousands)	253.7	2,508.0
Standardised mortality ratio (UK=100), 1998	109	100	Average dwelling price[4], 1999 (£)	65,543	94,581
Infant mortality rate[2], 1997-1999	6.5	5.8			
Percentage of pupils achieving 5 or more grades A* to C at GCSE level, 1998/99	46.0	49.1	Motor cars currently licensed, 1998 (thousands)	2,647	23,878
			Fatal and serious accidents on roads[5], 1998 (rate per 100,000 population)	58	66
Economic activity rate[3], Spring 1999 (percentages)	75.7	78.4	Recorded crime rate[4], 1998-99 (notifiable offences per 100,000 population)	10,556	9,785
Employment rate[3], Spring 1999 (percentages)	70.9	73.6			
ILO unemployment rate, Spring 1999 (percentages)	6.2	6.0			
Average gross weekly earnings: males in full-time employment, April 1999(£)	415.10	440.70	Average weekly household income, 1996-1999 (£)[6]	401	430
			Average weekly household expenditure, 1996-1999 (£)[6]	320	333
Average gross weekly earnings: females in full-time employment, April 1999 (£)	299.40	325.60	Households in receipt of Income Support/Family Credit[4], 1998-99 (percentages)	18	15

1 Males aged 65 or over, females aged 60 or over.
2 Deaths of infants under 1 year of age per 1,000 live births.
3 For people of working age, males aged 16 to 64, females aged 16 to 59.
4 Figure for the United Kingdom relates to England and Wales.
5 Figure for the United Kingdom relates to Great Britain.
6 Combined years 1996-97, 1997-98 and 1998-99.

1.4 Population density: by local authority, 1998

Population density, 1998
(persons per sq km)

2,500 or over

1,000 - 2,499

500 - 999

250 - 499

100 - 249

99 or under

1 Barrow-in-Furness
2 Blackpool UA
3 Preston
4 Hyndburn
5 South Ribble
6 Blackburn with Darwen UA
7 Rossendale
8 Sefton
9 Bury
10 Rochdale
11 Salford
12 Manchester
13 Tameside
14 Trafford
15 Liverpool
16 Knowsley
17 St Helens
18 Warrington UA
19 Halton UA
20 Stockport
21 Ellesmere Port and Neston
22 Congleton

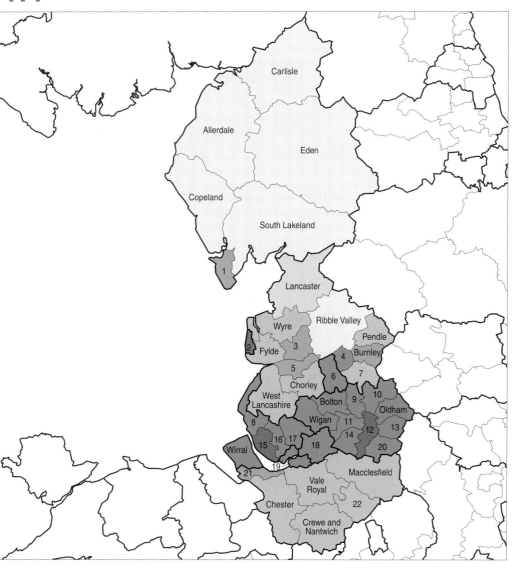

North West

Population

In 1998, the North West of England had a population of 6.9 million. Within the region the population density was highest in Blackpool UA at over 4,300 people per sq km and lowest in the district of Eden in Cumbria at only 23 people per sq km.

(Tables 3.1 and 14.1)

Mortality

Overall the region has a higher Standardised Mortality Ratio (SMR) than the UK as a whole at 109 in 1998 (UK=100). Within the region the SMR ranged from 90 in South Lakeland and Fylde to 129 in Halton UA.

(Table 14.1)

The Infant Mortality rate for 1997-1999 for the North West was higher than the UK rate (6.5 and 5.8 deaths of infants under 1 year old per 1,000 live births, respectively); within the region it ranged between 5.0 for Warrington UA and 8.4 for Blackburn with Darwen UA.

(Table 14.2)

Education

The North West had a higher participation rate for children under 5 in schools than the UK as a whole, 71 per cent compared with 64 per cent for the UK in January 2000.

(Tables 4.3 and 14.3)

In 1998/99 over half of all the areas in the region had lower proportions of pupils achieving 5 or more A* to C graded results than the UK average and had, with the exception of Warrington UA, Cheshire County, Cumbria and Lancashire County, higher proportions of pupils with no graded results than the UK as a whole.

(Table 14.3)

Among the population of working age, the proportion of people who's highest qualification was GCE A level or equivalent in the North West was 25.4 per cent in Spring 2000; higher than the UK average of 24.0 per cent.

(Table 4.13)

Labour market

The employment rate for people of working age in the region in Spring 1999, was among the lowest in the UK at 70.9 per cent. Within the region the employment rate varied between 80.0 per cent in Warrington UA to 62.0 per cent in Blackburn with Darwen UA in 1998-1999.

(Tables 5.1 and 14.5)

In 1999, average weekly earnings for people in the region were £372.60; lower than the UK average of £398.70. Within the region average weekly earnings varied considerably; in Cheshire County 10 per cent of men earned more than £722.80 but 10 per cent also earned less than £215.30.

(Tables 5.11 and 14.5)

Economy

Within the region manufacturing industry accounted for some 26.2 per cent of the region's GDP in 1997, compared to 21.3 per cent for the UK as a whole. Agriculture, forestry and fishing accounted for 1.1 per cent of the region's GDP in 1997, compared to 1.5 per cent for the UK.

(Table 12.4)

One third of the region's 254 thousand business sites in 1999 were in distribution, hotels and catering and repairs industries, higher than the UK average of 29.6 per cent.

(Table 13.3)

Almost 68 per cent of the value of direct export trade from the region in 1999 was to the EU, higher than the UK average of nearly 61 per cent. Imports from the EU accounted for 43 per cent of the value of direct import trade to the region, lower than the UK average of 51.5 per cent.

(Table 13.7)

Environment

Over a sixth of the total area in the North West is Green Belt land and over a tenth is designated as an Area of Outstanding Natural Beauty.

(Table 11.12)

1.5 Key statistics for Yorkshire and the Humber

	Yorkshire and the Humber	United Kingdom		Yorkshire and the Humber	United Kingdom
Population, 1998 (thousands)	5,042.9	59,236.5	Gross domestic product, 1998 (£ million)	55,232	747,544
Percentage under 16	20.6	20.4	Gross domestic product per head index, 1998 (UK=100)	87.8	100.0
Percentage pension age or over[1]	18.3	18.1	Total business sites, 1999 (thousands)	187.9	2,508.0
Standardised mortality ratio (UK=100), 1998	103	100	Average dwelling price[4], 1999 (£)	63,524	94,581
Infant mortality rate[2], 1997-1999	6.5	5.8			
Percentage of pupils achieving 5 or more grades A* to C at GCSE level, 1998/99	41.9	49.1	Motor cars currently licensed, 1998 (thousands)	1,808	23,878
			Fatal and serious accidents on roads[5], 1998 (rate per 100,000 population)	65	66
Economic activity rate[3], Spring 1999 (percentages)	77.6	78.4	Recorded crime rate[4], 1998-99 (notifiable offences per 100,000 population)	11,770	9,785
Employment rate[3], Spring 1999 (percentages)	72.5	73.6			
ILO unemployment rate, Spring 1999 (percentages)	6.5	6.0			
Average gross weekly earnings: males in full-time employment, April 1999(£)	395.80	440.70	Average weekly household income, 1996-1999 (£)[6]	390	430
			Average weekly household expenditure, 1996-1999 (£)[6]	320	333
Average gross weekly earnings: females in full-time employment, April 1999 (£)	297.90	325.60	Households in receipt of Income Support/Family Credit[4], 1998-99 (percentages)	18	15

1 Males aged 65 or over, females aged 60 or over.
2 Deaths of infants under 1 year of age per 1,000 live births.
3 For people of working age, males aged 16 to 64, females aged 16 to 59.
4 Figure for the United Kingdom relates to England and Wales.
5 Figure for the United Kingdom relates to Great Britain.
6 Combined years 1996-97, 1997-98 and 1998-99.

1.6 Population density: by local authority, 1998

Population density, 1998
(persons per sq km)

2,500 or over
1,000 - 2,499
500 - 999
250 - 499
100 - 249
99 or under

1 City of Kingston upon Hull UA
2 North Lincolnshire UA
3 North East Lincolnshire UA

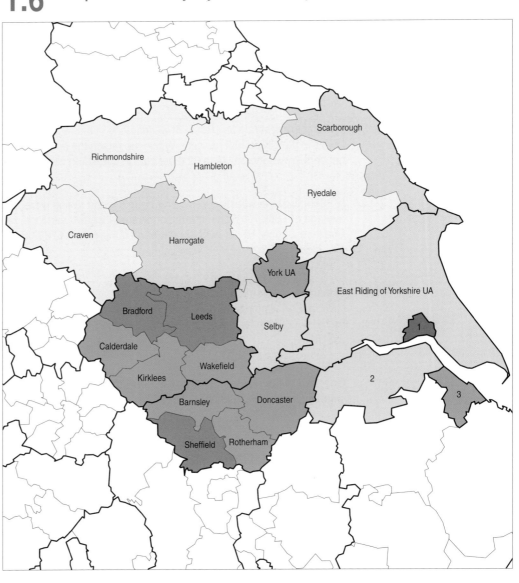

Yorkshire and the Humber

Population

In 1998 Yorkshire and the Humber had a population of 5.0 million. Within the region the population density was highest in City of Kingston upon Hull UA at over 3,687 people per sq km and lowest in the district of Ryedale in North Yorkshire County at only 32 people per sq km.

(Tables 3.1 and 14.1)

Mortality

Overall the region has a slightly higher Standardised Mortality Ratio (SMR) than the UK as a whole at 103 in 1998 (UK=100). Within the region the SMR ranged from 89 in Hambleton and Richmondshire to 116 in Doncaster.

(Table 14.1)

The Infant Mortality rate for 1997-1999 for Yorkshire and the Humber was higher than the UK rate (6.5 and 5.8 deaths of infants under 1 year old per 1,000 live births, respectively); within the region it ranged between 4.7 for North Yorkshire County and 7.3 for the City of Kingston upon Hull UA.

(Table 14.2)

Education

Yorkshire and the Humber had a higher participation rate for children under 5 in schools than the UK as a whole, 73 per cent compared with 64 per cent for the UK in January 2000. The participation rate was also higher than the UK average in all areas within the region, with the exception of East Riding of Yorkshire UA and North Yorkshire County.

(Tables 4.3 and 14.3)

In 1998/99 over half of all areas in the region had lower proportions of pupils achieving 5 or more A* to C graded results than the UK average and with the exception of East Riding of Yorkshire UA, North Lincolnshire UA, York UA and North Yorkshire County, had higher proportions of pupils with no graded results.

(Table 14.3)

Among the population of working age, the proportion of people who's highest qualification was GCE A level or equivalent in Yorkshire and the Humber was 25.2 per cent in Spring 2000; higher than the UK average of 24.0 per cent.

(Table 4.13)

Labour market

The employment rate for people of working age in the region in Spring 1999, was among the lowest in the UK at 72.5 per cent. Within the region the employment rate varied between 80.0 per cent in North Yorkshire County to 64.5 per cent in City of Kingston upon Hull UA in 1998-1999.

(Tables 5.1 and 14.5)

In 1999, average weekly earnings for people in the region were £361.00; lower than the UK average of £398.70. Within the region average weekly earnings varied considerably; in North Yorkshire County 10 per cent of men earned more than £629.60 but 10 per cent also earned less than £192.50.

(Tables 5.11 and 14.5)

Economy

Within the region manufacturing industry accounted for some 27.4 per cent of the region's GDP in 1997 compared to 21.3 per cent for the UK as a whole. Agriculture, forestry and fishing accounted for 1.9 per cent of the region's GDP in 1997, compared to 1.5 per cent for the UK.

(Table 12.4)

One third of the region's 188 thousand business sites in 1999 were in distribution, hotels and catering and repairs industries, higher than the UK average of 29.6 per cent.

(Table 13.3)

Almost 65 per cent of the value of direct export trade from the region in 1999 was to the EU, higher than the UK average of nearly 61 per cent. Imports from the EU accounted for just over 52 per cent of the value of direct import trade to the region, similar to the UK average of 51.5 per cent.

(Table 13.7)

Environment

Over a sixth of the total area in Yorkshire and Humber is Green Belt land and six per cent is desig-nated as an Area of Outstanding Natural Beauty.

(Table 11.12)

1.7 Key statistics for the East Midlands

	East Midlands	United Kingdom		East Midlands	United Kingdom
Population, 1998 (thousands)	4,169.3	59,236.5	Gross domestic product, 1998 (£ million)	49,260	747,544
Percentage under 16	20.2	20.4	Gross domestic product per head index, 1998 (UK=100)	94.8	100.0
Percentage pension age or over[1]	18.4	18.1	Total business sites, 1999 (thousands)	169.5	2,508.0
Standardised mortality ratio (UK=100), 1998	100	100	Average dwelling price[4], 1999 (£)	69,500	94,581
Infant mortality rate[2], 1997-1999	5.8	5.8			
Percentage of pupils achieving 5 or more grades A* to C at GCSE level, 1998/99	47.1	49.1	Motor cars currently licensed, 1998 (thousands)	1,698	23,878
			Fatal and serious accidents on roads[5], 1998 (rate per 100,000 population)	77	66
Economic activity rate[3], Spring 1999 (percentages)	80.1	78.4	Recorded crime rate[4], 1998-99 (notifiable offences per 100,000 population)	10,231	9,785
Employment rate[3], Spring 1999 (percentages)	75.9	73.6			
ILO unemployment rate, Spring 1999 (percentages)	5.2	6.0			
Average gross weekly earnings: males in full-time employment, April 1999(£)	398.30	440.70	Average weekly household income, 1996-1999 (£)[6]	405	430
			Average weekly household expenditure, 1996-1999 (£)[6]	320	333
Average gross weekly earnings: females in full-time employment, April 1999 (£)	286.70	325.60	Households in receipt of Income Support/Family Credit[4], 1998-99 (percentages)	16	15

1 Males aged 65 or over, females aged 60 or over.
2 Deaths of infants under 1 year of age per 1,000 live births.
3 For people of working age, males aged 16 to 64, females aged 16 to 59.
4 Figure for the United Kingdom relates to England and Wales.
5 Figure for the United Kingdom relates to Great Britain.
6 Combined years 1996-97, 1997-98 and 1998-99.

1.8 Population density: by local authority, 1998

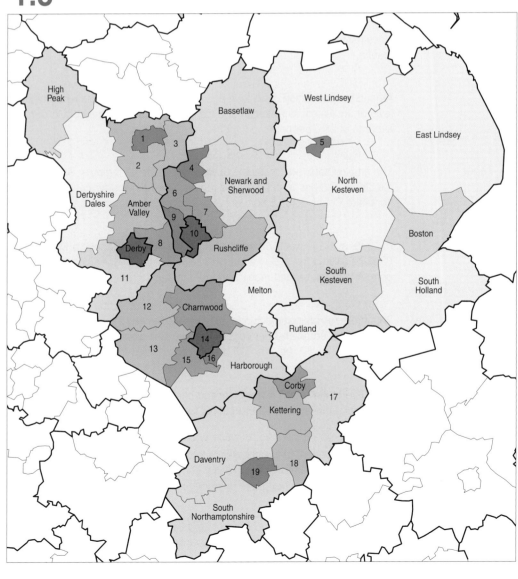

Population density, 1998
(persons per sq km)

	2,500 or over
	1,000 - 2,499
	500 - 999
	250 - 499
	100 - 249
	99 or under

1 Chesterfield
2 North East Derbyshire
3 Bolsover
4 Mansfield
5 Lincoln
6 Ashfield
7 Gedling
8 Erewash
9 Broxtowe
10 Nottingham UA
11 South Derbyshire
12 North West Leicestershire
13 Hinckley and Bosworth
14 Leicester UA
15 Blaby
16 Oadby and Wigston
17 East Northamptonshire
18 Wellingborough
19 Northampton

East Midlands

Population

In 1998 the East Midlands had a population of 4.2 million. Within the region the population density was highest in Leicester UA at over 4,000 people per sq km and lowest in the district of West Lindsey in Lincolnshire at 67 people per sq km.

(Tables 3.1 and 14.1)

Mortality

Overall the region had the same Standardised Mortality Ratio (SMR) as the UK as a whole in 1998 (UK=100). Within the region the SMR ranged from 75 in Rutland UA to 119 in Chesterfield.

(Table 14.1)

The Infant Mortality rate for 1997-1999 for the East Midlands was the same as the UK rate (5.8 deaths of infants under 1 year old per 1,000 live births); within the region it ranged between 5.0 for Derbyshire County and Nottinghamshire County and 7.8 for Leicester UA.

(Table 14.2)

Education

The East Midlands had a similar participation rate for children under 5 in schools to the UK as a whole, 63 per cent compared with 64 per cent for the UK in January 2000.

(Tables 4.3 and 14.3)

In 1998/99 all areas in the region, with the exception of Rutland UA, Derbyshire County, Leicestershire County and Lincolnshire had lower proportions of pupils achieving 5 or more A* to C graded results than the UK average and had, with the exception of Derby UA, Leicester UA, Nottingham UA and Nottinghamshire County, lower proportions of pupils with no graded results.

(Table 14.3)

Among the population of working age, the proportion of people who's highest qualification was GCE A level or equivalent in the East Midlands was 24.1 per cent in Spring 2000; similar to the UK average of 24.0 per cent.

(Table 4.13)

Labour market

The employment rate for people of working age in Spring 1999, at 75.9 per cent, was among the highest in the UK. Within the region the employment rate varied between 83.0 per cent in Leicestershire County to 65.7 per cent in Nottingham UA in 1998-1999.

(Tables 5.1 and 14.5)

In 1999, average weekly earnings for people in the region were £361.70; lower than the UK average of £398.70. Within the region average weekly earnings varied considerably; in Nottingham UA 10 per cent of men earned more than £671.80 but 10 per cent also earned less than £201.50.

(Tables 5.11 and 14.5)

Economy

Within the region manufacturing industry accounted for some 29.9 per cent of the region's GDP in 1997, compared to 21.3 per cent for the UK as a whole. Agriculture, forestry and fishing accounted for 2.2 per cent of the region's GDP in 1997, compared to 1.5 per cent for the UK.

(Table 12.4)

Just over 30 per cent of the region's 170 thousand business sites in 1999 were in distribution, hotels and catering and repairs industries, slightly higher than the UK average of 29.6 per cent.

(Table 13.3)

Over 54 per cent of the value of direct export trade from the region in 1999 was to the EU, lower than the UK average of nearly 61 per cent. Imports from the EU accounted for 50.5 per cent of the value of direct import trade to the region, slightly lower than the UK average of 51.5 per cent.

(Table 13.7)

Environment

Five per cent of the total area in the East Midlands is Green Belt land and three per cent is designated as an Area of Outstanding Natural Beauty.

(Table 11.12)

1.9 Key statistics for the West Midlands

	West Midlands	United Kingdom		West Midlands	United Kingdom
Population, 1998 (thousands)	5,332.5	59,236.5	Gross domestic product, 1998 (£ million)	60,927	747,544
Percentage under 16	20.9	20.4	Gross domestic product per head index, 1998 (UK=100)	91.7	100.0
Percentage pension age or over[1]	18.1	18.1	Total business sites, 1999 (thousands)	208.6	2,508.0
Standardised mortality ratio (UK=100), 1998	101	100	Average dwelling price[4], 1999 (£)	76,633	94,581
Infant mortality rate[2], 1997-1999	6.8	5.8			
Percentage of pupils achieving 5 or more grades A* to C			Motor cars currently licensed, 1998 (thousands)	2,290	23,878
at GCSE level, 1998/99	45.1	49.1	Fatal and serious accidents on roads[5], 1998		
			(rate per 100,000 population)	69	66
Economic activity rate for[3], Spring 1999 (percentages)	79.1	78.4	Recorded crime rate[4], 1998-99		
Employment rate[3], Spring 1999 (percentages)	73.6	73.6	(notifiable offences per 100,000 population)	9,901	9,785
ILO unemployment rate, Spring 1999 (percentages)	6.8	6.0			
Average gross weekly earnings: males in full-time			Average weekly household income, 1996-1999 (£)[6]	408	430
employment, April 1999(£)	414.60	440.70	Average weekly household expenditure, 1996-1999 (£)[6]	316	333
Average gross weekly earnings: females in full-time			Households in receipt of Income Support/Family Credit[4],		
employment, April 1999 (£)	301.00	325.60	1998-99 (percentages)	17	15

1 Males aged 65 or over, females aged 60 or over.
2 Deaths of infants under 1 year of age per 1,000 live births.
3 For people of working age, males aged 16 to 64, females aged 16 to 59.
4 Figure for the United Kingdom relates to England and Wales.
5 Figure for the United Kingdom relates to Great Britain.
6 Combined years 1996-97, 1997-98 and 1998-99.

1.10 Population density: by local authority, 1998

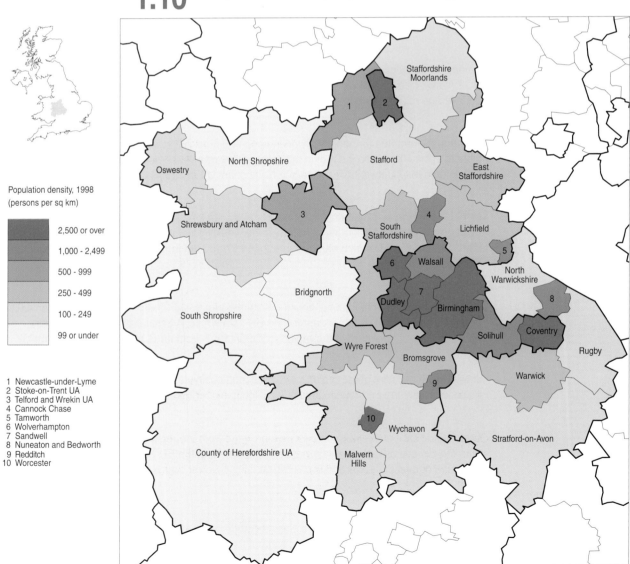

Population density, 1998
(persons per sq km)

- 2,500 or over
- 1,000 - 2,499
- 500 - 999
- 250 - 499
- 100 - 249
- 99 or under

1 Newcastle-under-Lyme
2 Stoke-on-Trent UA
3 Telford and Wrekin UA
4 Cannock Chase
5 Tamworth
6 Wolverhampton
7 Sandwell
8 Nuneaton and Bedworth
9 Redditch
10 Worcester

West Midlands

Population

In 1998 the West Midlands had a population of 5.3 million. Within the region the population density was highest in Birmingham at over 3,800 people per sq km and lowest in South Shropshire at 40 people per sq km.

(Tables 3.1 and 14.1)

Mortality

Overall the region has a slightly higher Standardised Mortality Ratio (SMR) than the UK as a whole at 101 in 1998 (UK=100). Within the region this ranged from 83 in Solihull to 118 in Sandwell.

(Table 14.1)

The Infant Mortality rate for 1997-1999 for the West Midlands was higher than the UK rate (6.8 and 5.8 deaths of infants under 1 year old per 1,000 live births, respectively); within the region it ranged between 4.9 for Shropshire County and 8.1 for Stoke-on-Trent UA.

(Table 14.2)

Education

The West Midlands had a higher participation rate for children under 5 in schools than the UK as a whole, 70 per cent compared with 64 per cent for the UK in January 2000.

(Tables 4.3 and 14.3)

In 1998/99, half of the areas in the region had lower proportions of pupils achieving 5 or more A* to C graded results than the UK average and had, with the exception of Stoke-on-Trent UA, Telford and Wrekin UA, West Midlands (Met. County) and Worcestershire County, lower proportions of pupils with no graded results.

(Table 14.3)

Among the population of working age, the proportion of people who's highest qualification was GCE A level or equivalent in the West Midlands was 21.9 per cent in Spring 2000; lower than the UK average of 24.0 per cent.

(Table 4.13)

Labour market

The employment rate for people of working age in Spring 1999, was among the highest in the UK at 73.6 per cent. Within the region the employment rate varied between 81.1 per cent in Worcestershire County to 68.0 per cent in Stoke-on-Trent UA in 1998-1999.

(Tables 5.1 and 14.5)

In 1999 average weekly earnings for people in the region were £375.60; lower than the UK average of £398.70. Within the region average weekly earnings varied considerably; in Warwickshire 10 per cent of men earned more than £673.70 but 10 per cent also earned less than £215.00.

(Tables 5.11 and 14.5)

Economy

Within the region manufacturing industry accounted for some 29.5 per cent of the region's GDP in 1997, compared to 21.3 per cent for the UK as a whole. Agriculture, forestry and fishing accounted for 1.7 per cent of the region's GDP in 1997, compared to 1.5 per cent for the UK.

(Table 12.4)

Of the region's 209 thousand business sites in 1999, 31 per cent were in distribution, hotels and catering and repairs industries, slightly higher than the UK average of 29.6 per cent.

(Table 13.3)

Over 59 per cent of the value of direct export trade from the region in 1999 was to the EU, slightly below the UK average of nearly 61 per cent. Imports from the EU accounted for over 64 per cent of the value of direct import trade to the region, the highest percentage in the UK, and well above the UK average of 51.5 per cent.

(Table 13.7)

Environment

Around a fifth of the total area in the West Midlands is Green Belt land and a tenth is designated as an Area of Outstanding Natural Beauty.

(Table 11.12)

1.11 Key statistics for the East of England

	East	United Kingdom		East	United Kingdom
Population, 1998 (thousands)	5,377.0	59,236.5	Gross domestic product, 1998 (£ million)	76,308	747,544
Percentage under 16	20.1	20.4	Gross domestic product per head index, 1998 (UK=100)	114.2	100.0
Percentage pension age or over[1]	18.5	18.1	Total business sites, 1999 (thousands)	241.1	2,508.0
Standardised mortality ratio (UK=100), 1998	93	100	Average dwelling price[4], 1999 (£)	94,679	94,581
Infant mortality rate[2], 1997-1999	4.8	5.8			
Percentage of pupils achieving 5 or more grades A* to C			Motor cars currently licensed, 1998 (thousands)	2,429	23,878
at GCSE level, 1998/99	52.2	49.1	Fatal and serious accidents on roads[5], 1998		
			(rate per 100,000 population)	74	66
Economic activity rate[3], Spring 1999 (percentages)	81.5	78.4	Recorded crime rate[4], 1998-99		
Employment rate[3], Spring 1999 (percentages)	78.1	73.6	(notifiable offences per 100,000 population)	7,017	9,785
ILO unemployment rate, Spring 1999 (percentages)	4.1	6.0			
Average gross weekly earnings: males in full-time			Average weekly household income, 1996-1999 (£)[6]	451	430
employment, April 1999(£)	436.00	440.70	Average weekly household expenditure, 1996-1999 (£)[6]	346	333
Average gross weekly earnings: females in full-time			Households in receipt of Income Support/Family Credit[4],		
employment, April 1999 (£)	323.90	325.60	1998-99 (percentages)	11	15

1 Males aged 65 or over, females aged 60 or over.
2 Deaths of infants under 1 year of age per 1,000 live births.
3 For people of working age, males aged 16 to 64, females aged 16 to 59.
4 Figure for the United Kingdom relates to England and Wales.
5 Figure for the United Kingdom relates to Great Britain.
6 Combined years 1996-97, 1997-98 and 1998-99.

1.12 Population density: by local authority, 1998

Population density, 1998
(persons per sq km)

- 2,500 or over
- 1,000 - 2,499
- 500 - 999
- 250 - 499
- 100 - 249
- 99 or under

1 Norwich
2 Cambridge
3 Ipswich
4 South Bedfordshire
5 Luton UA
6 North Hertfordshire
7 Stevenage
8 St. Albans
9 Welwyn Hatfield
10 Broxbourne
11 Harlow
12 Three Rivers
13 Watford
14 Hertsmere
15 Brentwood
16 Castle Point
17 Southend-on-Sea UA

East of England

Population

In 1998 the East of England had a population of 5.4 million. Within the region the population density was highest in Luton at over 4,200 people per sq km and lowest in the district of Breckland in Norfolk at 90 people per sq km.

(Tables 3.1 and 14.1)

Mortality

Overall the region has a lower Standardised Mortality Ratio (SMR) than the UK as a whole at 93 in 1998 (UK=100). Within the region this ranged from 75 in Three Rivers to 119 in Watford.

(Table 14.1)

The Infant Mortality rate for 1997-1999 for the East of England was lower than the UK average (4.8 and 5.8 deaths of infants under 1 year old per 1,000 live births, respectively); within the region it ranged between 4.2 for Southend-on-Sea UA, Essex County and Hertfordshire and 7.7 for Luton UA.

(Table 14.2)

Education

The East of England had a lower participation rate for children under 5 in schools than the UK as a whole, 54 per cent compared with 64 per cent for the UK in January 2000. The participation rate is also lower in all areas within the region, with the exception of Hertfordshire.

(Tables 4.3 and 14.3)

In 1998/99 all areas in the region, with the exception of Luton UA, Peterborough UA, Thurrock UA and Bedfordshire County had higher proportions of pupils achieving 5 or more A* to C graded results than the UK average and with the exception of Peterborough UA, Southend-on-Sea UA, Thurrock UA and Cambridgeshire County had lower proportions of pupils with no graded results.

(Table 14.3)

Among the population of working age, the proportion of people who's highest qualification was GCE A level or equivalent in the East of England was 23.6 per cent in Spring 2000; lower than the UK average of 24.0 per cent.

(Table 4.13)

Labour market

The employment rate for people of working age in Spring 1999 was the second highest in the UK at 78.1 per cent. Within the region the employment rate varied between 81.5 per cent in Hertfordshire to 70.5 per cent in Southend-on-Sea UA in 1998-1999.

(Tables 5.1 and 14.5)

In 1999 average weekly earnings for people in the region were £396.60; lower than the UK average of £398.70. Within the region average weekly earnings varied considerably; in Luton UA 10 per cent of men earned more than £838.60 but 10 per cent also earned less than £243.80.

(Tables 5.11 and 14.5)

Economy

Within the region manufacturing industry accounted for some 17.8 per cent of the region's GDP in 1997, compared to 21.3 per cent for the UK as a whole. Agriculture, forestry and fishing accounted for 2.0 per cent of the region's GDP in 1997, compared to 1.5 per cent for the UK.

(Table 12.4)

Over 27 per cent of the region's 241 thousand business sites in 1999 were in distribution, hotels and catering and repairs industries, slightly lower than the UK average of 29.6 per cent.

(Table 13.3)

Over 53 per cent of the value of direct export trade from the region in 1999 was to the EU, lower than the UK average of nearly 61 per cent. Imports from the EU accounted for 62.5 per cent of the value of direct import trade to the region, higher than the UK average of 51.5 per cent.

(Table 13.7)

Environment

Almost an eighth of the total area in the East of England is Green Belt land and six per cent is designated as an Area of Outstanding Natural Beauty.

(Table 11.12)

1.13   Key statistics for London

	London	United Kingdom		London	United Kingdom
Population, 1998 (thousands)	7,187.2	59,236.5	Gross domestic product, 1998 (£ million)	116,444	747,544
Percentage under 16	*20.7*	*20.4*	Gross domestic product per head index, 1998 (UK=100)	130.4	100.0
Percentage pension age or over[1]	*14.9*	*18.1*	Total business sites, 1999 (thousands)	378.4	2,508.0
Standardised mortality ratio (UK=100), 1998	95	100	Average dwelling price[4], 1999 (£)	150,094	94,581
Infant mortality rate[2], 1997-1999	5.9	5.8			
Percentage of pupils achieving 5 or more grades A* to C			Motor cars currently licensed, 1998 (thousands)	2,369	23,878
at GCSE level, 1998/99	*46.7*	*49.1*	Fatal and serious accidents on roads[5], 1998		
			(rate per 100,000 population)	87	66
Economic activity rate[3], Spring 1999 (percentages)	*77.3*	*78.4*	Recorded crime rate[4], 1998-99		
Employment rate[3], Spring 1999 (percentages)	*71.4*	*73.6*	(notifiable offences per 100,000 population)	12,354	9,785
ILO unemployment rate, Spring 1999 (percentages)	*7.6*	*6.0*			
Average gross weekly earnings: males in full-time			Average weekly household income, 1996-1999 (£)[6]	523	430
employment, April 1999(£)	584.40	440.70	Average weekly household expenditure, 1996-1999 (£)[6]	376	333
Average gross weekly earnings: females in full-time			Households in receipt of Income Support/Family Credit[4],		
employment, April 1999 (£)	422.80	325.60	1998-99 (percentages)	*17*	*15*

1 Males aged 65 or over, females aged 60 or over.
2 Deaths of infants under 1 year of age per 1,000 live births.
3 For people of working age, males aged 16 to 64, females aged 16 to 59.
4 Figure for the United Kingdom relates to England and Wales.
5 Figure for the United Kingdom relates to Great Britain.
6 Combined years 1996-97, 1997-98 and 1998-99.

1.14   Population density: by London Borough, 1998

Population density, 1998
(persons per sq km)

- 10,000 or over
- 7,500 - 9,999
- 5,000 - 7,499
- 2,500 - 4,999
- 2,499 or under

1 Waltham Forest
2 Camden
3 Islington
4 Hackney
5 Tower Hamlets
6 Newham
7 Barking and Dagenham
8 Hammersmith and Fulham
9 Kensington and Chelsea
10 Westminster
11 City of London
12 Richmond upon Thames
13 Wandsworth
14 Lambeth
15 Southwark
16 Lewisham
17 Kingston upon Thames

London

Population

In 1998 London had a population of 7.2 million. Within the region the population density was highest in Kensington and Chelsea at over 14,161 people per sq km and lowest in the City of London at only 1,737 people per sq km.

(Tables 3.1 and 14.1)

Mortality

Overall the region has a lower Standardised Mortality Ratio (SMR) than the UK as a whole at 95 in 1998 (UK=100). Within the region this ranged from 79 in Kensington and Chelsea, Westminster and Richmond-upon-Thames to 112 in Newham.

(Table 14.1)

The Infant Mortality rate for 1997-1999 for London was in line with the UK rate (5.9 and 5.8 deaths of infants under 1 year old per 1,000 live births, respectively).

(Table 14.2)

Education

London had a higher participation rate for children under 5 in schools than the UK as a whole, 69 per cent compared with 64 per cent for the UK in January 2000.

(Table 4.3)

In 1998/99 Outer London had a lower proportion of pupils achieving 5 or more A* to C graded results than the UK average while Inner London had higher proportions of pupils with no graded results.

(Table 14.3)

Among the population of working age, the proportion of people who's highest qualification was GCE A level or equivalent in London was 18.8 per cent in Spring 2000, lower than the UK average of 24.0 per cent.

(Table 4.13)

Labour market

The employment rate for people of working age in Spring 1999 in the region was among the lowest in the UK at 71.4 per cent.

(Table 5.1)

In 1999 average weekly earnings for people in the region were the highest in the UK; £520.00 compared with the UK average of £398.70. Within the region average weekly earnings varied considerably; 10 per cent of men earned more than £1,008.60 but 10 per cent also earned less than £242.30.

(Tables 5.11 and 14.5)

Economy

Within the region manufacturing industry accounted for some 11.5 per cent of the region's GDP in 1997, compared to 21.3 per cent for the UK as a whole. Agriculture, forestry and fishing accounted for virtually none of the region's GDP in 1997, compared to 1.5 per cent for the UK.

(Table 12.4)

Almost 39 per cent of the region's 378 thousand business sites in 1999 were in financial intermediation, real estate, renting and business activities, the highest rate in the UK; this compares with a UK average of 25 per cent.

(Table 13.3)

Almost 45 per cent of the value of direct export trade from the region in 1999 was to the EU, the lowest rate in the UK. Imports from the EU accounted for under 42 per cent of the value of direct import trade to the region, below the UK average of 51.5 per cent.

(Table 13.7)

Environment

Over a fifth of the total area in London is Green Belt land.

(Table 11.12)

1.15 Key statistics for the South East

	South East	United Kingdom		South East	United Kingdom
Population, 1998 (thousands)	8,003.8	59,236.5	Gross domestic product, 1998 (£ million)	116,176	747,544
Percentage under 16	20.0	20.4	Gross domestic product per head index, 1998 (UK=100)	116.7	100.0
Percentage pension age or over[1]	18.5	18.1	Total business sites, 1999 (thousands)	373.3	2,508.0
Standardised mortality ratio (UK=100), 1998	91	100	Average dwelling price[4], 1999 (£)	118,385	94,581
Infant mortality rate[2], 1997-1999	4.7	5.8			
Percentage of pupils achieving 5 or more grades A* to C at GCSE level, 1998/99	53.8	49.1	Motor cars currently licensed, 1998 (thousands)	3,709	23,878
			Fatal and serious accidents on roads[5], 1998 (rate per 100,000 population)	60	66
Economic activity rate[3], Spring 1999 (percentages)	82.8	78.4	Recorded crime rate[4], 1998-99		
Employment rate[3], Spring 1999 (percentages)	79.7	73.6	(notifiable offences per 100,000 population)	7,897	9,785
ILO unemployment rate, Spring 1999 (percentages)	3.6	6.0			
Average gross weekly earnings: males in full-time employment, April 1999(£)	471.20	440.70	Average weekly household income, 1996-1999 (£)[6]	511	430
			Average weekly household expenditure, 1996-1999 (£)[6]	376	333
Average gross weekly earnings: females in full-time employment, April 1999 (£)	341.00	325.60	Households in receipt of Income Support/Family Credit[4], 1998-99 (percentages)	10	15

1 Males aged 65 or over, females aged 60 or over.
2 Deaths of infants under 1 year of age per 1,000 live births.
3 For people of working age, males aged 16 to 64, females aged 16 to 59.
4 Figure for the United Kingdom relates to England and Wales.
5 Figure for the United Kingdom relates to Great Britain.
6 Combined years 1996-97, 1997-98 and 1998-99.

1.16 Population density: by local authority, 1998

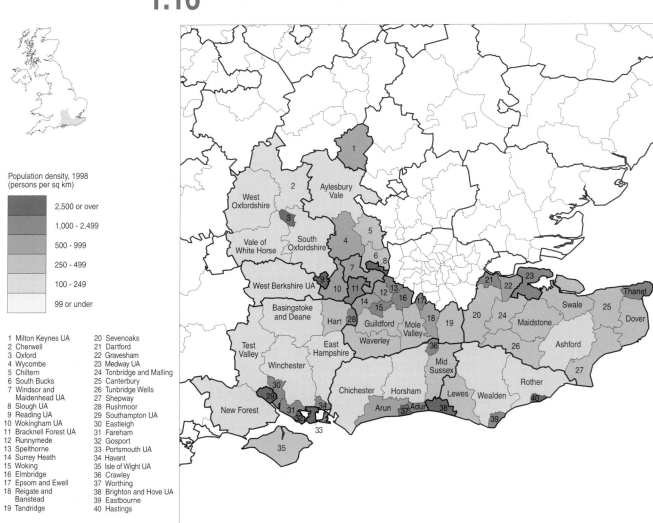

Population density, 1998
(persons per sq km)

- 2,500 or over
- 1,000 - 2,499
- 500 - 999
- 250 - 499
- 100 - 249
- 99 or under

1 Milton Keynes UA
2 Cherwell
3 Oxford
4 Wycombe
5 Chiltern
6 South Bucks
7 Windsor and Maidenhead UA
8 Slough UA
9 Reading UA
10 Wokingham UA
11 Bracknell Forest UA
12 Runnymede
13 Spelthorne
14 Surrey Heath
15 Woking
16 Elmbridge
17 Epsom and Ewell
18 Reigate and Banstead
19 Tandridge

20 Sevenoaks
21 Dartford
22 Gravesham
23 Medway UA
24 Tonbridge and Malling
25 Canterbury
26 Tunbridge Wells
27 Shepway
28 Rushmoor
29 Southampton UA
30 Eastleigh
31 Fareham
32 Gosport
33 Portsmouth UA
34 Havant
35 Isle of Wight UA
36 Crawley
37 Worthing
38 Brighton and Hove UA
39 Eastbourne
40 Hastings

South East

Population	In 1998 the South East had a population of 8.0 million. Within the region the population density was highest in Portsmouth UA at 4,750 people per sq km and lowest in Chichester in West Sussex at 137 people per sq km.

(Tables 3.1 and 14.1)

Mortality	Overall the region has a lower Standardised Mortality Ratio (SMR) than the UK as a whole at 91 in 1998 (UK=100). Within the region this ranged from 78 in Epsom and Ewell to 109 in Dartford.

(Table 14.1)

The Infant Mortality rate for 1997-1999 was lower for the South East than in the UK as a whole (4.7 and 5.8 deaths of infants under 1 year old per 1,000 live births, respectively); within the region it ranged between 2.6 for Bracknell Forest UA and 7.2 for Portsmouth UA.

(Table 14.2)

Education	The South East had a lower participation rate for children under 5 in schools than the UK as a whole, 47 per cent compared with 64 per cent for the UK in January 2000. The participation rate was lower in all areas within the region, with the exception of Slough UA.

(Tables 4.3 and 14.3)

In 1998/99 over half of all areas in the region had higher proportions of pupils achieving 5 or more A* to C graded results than the UK average and with the exception of Brighton and Hove, Isle of Wight, Milton Keynes, Portsmouth, Reading and Southampton UAs, had lower proportions of pupils with no graded results.

(Table 14.3)

Among the population of working age, the proportion of people who's highest qualification was GCE A level or equivalent in the South East was 24.8 per cent in Spring 2000; higher than the UK average of 24.0 per cent.

(Table 4.13)

Labour market	The employment rate for people of working age in Spring 1999 was the highest in the UK at 79.7 per cent. Within the region the employment rate varied between 85.0 per cent in West Berkshire UA (Newbury), Windsor and Maidenhead UA and Wokingham UA to 70.8 per cent in Southampton UA in 1998-1999.

(Tables 5.1 and 14.5)

In 1999, average weekly earnings for people in the region were amongst the highest in the UK, only London had higher average earnings. Within the region average weekly earnings varied considerably. For example in Bracknell Forest UA 10 per cent of men earned more than £1,150.60 but 10 per cent also earned less than £260.20.

(Tables 5.11 and 14.5)

Economy	Within the region manufacturing industry accounted for some 16.2 per cent of the region's GDP in 1997, compared to 21.3 per cent for the UK as a whole. Agriculture, forestry and fishing accounted for 0.9 per cent of the region's GDP in 1997, compared to 1.5 per cent for the UK.

(Table 12.4)

Thirty one per cent of the region's 373 thousand business sites in 1999 were in financial intermediation, real estate, renting and business activities, higher than the UK average of 25 per cent.

(Table 13.3)

In 1999, 60 per cent of the value of direct export trade from the region was to the EU, similar to the UK rate of nearly 61 per cent. Imports from the EU accounted for over 61 per cent of the value of direct import trade to the region, higher than the UK average of 51.5 per cent.

(Table 13.7)

Environment	Almost a third of the total area in the South East (including London) is designated as an Area of Outstanding Natural Beauty and around a fifth is Green Belt land.

(Table 11.12)

1.17 Key statistics for the South West

	South West	United Kingdom		South West	United Kingdom
Population, 1998 (thousands)	4,901.3	59,236.5	Gross domestic product, 1998 (£ million)	56,068	747,544
Percentage under 16	19.3	20.4	Gross domestic product per head index, 1998 (UK=100)	91.9	100.0
Percentage pension age or over[1]	21.1	18.1	Total business sites, 1999 (thousands)	230.7	2,508.0
Standardised mortality ratio (UK=100), 1998	89	100	Average dwelling price[4], 1999 (£)	90,274	94,581
Infant mortality rate[2], 1997-1999	5.1	5.8			
Percentage of pupils achieving 5 or more grades A* to C at GCSE level, 1998/99	52.8	49.1	Motor cars currently licensed, 1998 (thousands)	2,230	23,878
			Fatal and serious accidents on roads[5], 1998 (rate per 100,000 population)	50	66
Economic activity rate[3], Spring 1999 (percentages)	82.1	78.4	Recorded crime rate[4], 1998-99 (notifiable offences per 100,000 population)	8,201	9,785
Employment rate[3], Spring 1999 (percentages)	78.1	73.6			
ILO unemployment rate, Spring 1999 (percentages)	4.7	6.0			
Average gross weekly earnings: males in full-time employment, April 1999(£)	402.90	440.70	Average weekly household income, 1996-1999 (£)[6]	411	430
			Average weekly household expenditure, 1996-1999 (£)[6]	316	333
Average gross weekly earnings: females in full-time employment, April 1999 (£)	297.80	325.60	Households in receipt of Income Support/Family Credit[4], 1998-99 (percentages)	12	15

1 Males aged 65 or over, females aged 60 or over.
2 Deaths of infants under 1 year of age per 1,000 live births.
3 For people of working age, males aged 16 to 64, females aged 16 to 59.
4 Figure for the United Kingdom relates to England and Wales.
5 Figure for the United Kingdom relates to Great Britain.
6 Combined years 1996-97, 1997-98 and 1998-99.

1.18 Population density: by local authority, 1998

Population density, 1998
(persons per sq km)

- 2,500 or over
- 1,000 - 2,499
- 500 - 999
- 250 - 499
- 100 - 249
- 99 or under

1 Forest of Dean
2 Tewkesbury
3 Gloucester
4 Cheltenham
5 South Gloucestershire UA
6 Swindon UA
7 City of Bristol UA
8 North Somerset UA
9 Bath and North East Somerset UA
10 West Wiltshire
11 Sedgemoor
12 Poole UA
13 Bournemouth UA
14 Christchurch
15 Exeter
16 Restormel
17 Plymouth UA
18 Torbay UA
19 Weymouth and Portland
20 Penwith

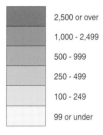

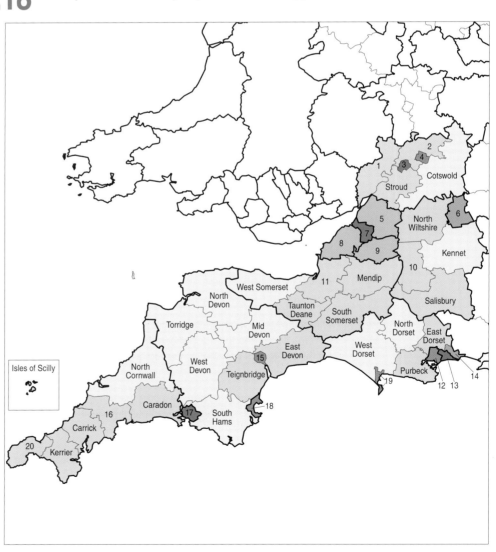

South West

Population

In 1998 the South West had a population of 4.9 million. Within the region the population density was highest in City of Bristol UA at over 3,600 people per sq km and lowest in the district of West Devon at only 41 people per sq km.

(Tables 3.1 and 14.1)

Mortality

Overall the region has a lower Standardised Mortality Ratio (SMR) than the UK at 89 in 1998 (UK=100). Within the region the SMR ranged from 74 in East Dorset to 99 in Swindon UA and the Forest of Dean.

(Table 14.1)

The Infant Mortality rate for 1997-1999 for the South West was lower than the UK rate (5.1 and 5.8 deaths of infants under 1 year old per 1,000 live births, respectively); within the region it ranged between 3.6 for Bath and North East Somerset UA and 6.4 for North Somerset UA.

(Table 14.2)

Education

The South West had a lower participation rate for children under 5 in schools than the UK as a whole, 49 per cent compared with 64 per cent for the UK in January 2000. Participation rates were lower than the UK average for all areas within the region, with the exception of City of Bristol and Torbay UAs.

(Tables 4.3 and 14.3)

In 1998/99 all areas in the region, with the exception of City of Bristol and Swindon UAs had higher proportions of pupils achieving 5 or more A* to C graded results than the UK average and with the exception of the City of Bristol UA, North Somerset UA and Torbay UA, had lower proportions of pupils with no graded results.

(Table 14.3)

Within the population of working age, the proportion of people who's highest qualification was GCE A level or equivalent in the South West was 24.8 per cent in Spring 2000; higher than the UK average of 24.0 per cent.

(Table 4.13)

Labour market

The employment rate for people of working age in Spring 1999 was the second highest in the UK at 78.1 per cent. Within the region the employment rate varied between 85.2 per cent in Swindon to 70.0 per cent in Plymouth UA in 1998-1999.

(Tables 5.1 and 14.5)

In 1999 average weekly earnings for people in the region were £364.90; lower than the UK average of £398.70. Within the region average weekly earnings varied considerably. For example in Swindon UA 10 per cent of men earned more than £724.80 but 10 per cent also earned less than £245.00.

(Tables 5.11 and 14.5)

Economy

Within the region manufacturing industry accounted for some 19.8 per cent of the region's GDP in 1997, compared to 21.3 per cent for the UK as a whole. Agriculture, forestry and fishing accounted for 2.9 per cent of the region's GDP in 1997, compared to 1.5 per cent for the UK.

(Table 12.4)

Over 29 per cent of the region's 231 thousand business sites in 1999 were in distribution, hotels and catering and repairs industries, slightly lower than the UK average of 29.6 per cent.

(Table 13.3)

Almost 69 per cent of the value of direct export trade from the region in 1999 was to the EU, higher than the UK average of nearly 61 per cent. Imports from the EU accounted for just over 40 per cent of the value of direct import trade to the region, below the UK average of 51.5 per cent.

(Table 13.7)

Environment

Almost a third of the total area in the South West is designated as an Area of Outstanding Natural Beauty and four per cent is Green Belt land.

(Table 11.12)

1.19 Key statistics for Wales

	Wales	United Kingdom		Wales	United Kingdom
Population, 1998 (thousands)	2,933.3	59,236.5	Gross domestic product, 1998 (£ million)	29,027	747,544
Percentage under 16	20.4	20.4	Gross domestic product per head index, 1998 (UK=100)	79.4	100.0
Percentage pension age or over[1]	19.9	18.1	Total business sites, 1999 (thousands)	115.1	2,508.0
Standardised mortality ratio (UK=100), 1998	101	100	Average dwelling price[4], 1999 (£)	62,424	94,581
Infant mortality rate[2], 1997-1999	6.0	5.8			
Percentage of pupils achieving 5 or more grades A* to C			Motor cars currently licensed, 1998 (thousands)	1,129	23,878
at GCSE level, 1998/99	47.5	49.1	Fatal and serious accidents on roads[5], 1998		
			(rate per 100,000 population)	47	66
Economic activity rate[3], Spring 1999 (percentages)	73.7	78.4	Recorded crime rate[4], 1998-99		
Employment rate[3], Spring 1999 (percentages)	68.4	73.6	(notifiable offences per 100,000 population)	8,951	9,785
ILO unemployment rate, Spring 1999 (percentages)	7.0	6.0			
Average gross weekly earnings: males in full-time			Average weekly household income, 1996-1999 (£)[6]	359	430
employment, April 1999(£)	384.00	440.70	Average weekly household expenditure, 1996-1999 (£)[6]	308	333
Average gross weekly earnings: females in full-time			Households in receipt of Income Support/Family Credit[4],		
employment, April 1999 (£)	298.30	325.60	1998-99 (percentages)	18	15

1 Males aged 65 or over, females aged 60 or over.
2 Deaths of infants under 1 year of age per 1,000 live births.
3 For people of working age, males aged 16 to 64, females aged 16 to 59.
4 Figure for the United Kingdom relates to England and Wales.
5 Figure for the United Kingdom relates to Great Britain.
6 Combined years 1996-97, 1997-98 and 1998-99.

1.20 Population density: by local authority, 1998

Population density, 1998
(persons per sq km)

- 2,500 or over
- 1,000 - 2,499
- 500 - 999
- 250 - 499
- 100 - 249
- 99 or under

1 Swansea UA
2 Neath Port Talbot UA
3 Bridgend UA
4 Rhondda, Cynon, Taff UA
5 Merthyr Tydfil UA
6 Caerphilly UA
7 Blaenau Gwent UA
8 Torfaen UA
9 The Vale of Glamorgan UA
10 Newport UA

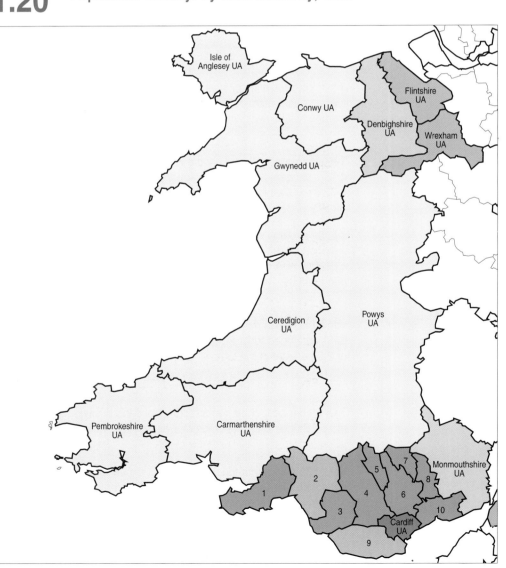

Wales

Population

In 1998 Wales had a population of 2.9 million. Within the region the population density was highest in Cardiff at over 2,290 people per sq km and lowest in Powys at only 24 people per sq km.

(Tables 3.1 and 15.1)

Mortality

Overall Wales has a slightly higher Standardised Mortality Ratio (SMR) than the UK as a whole at 101 in 1998 (UK =100). Within Wales this ranged from 82 in Ceredigion to 127 in Merthyr Tydfil.

(Table 15.1)

The Infant Mortality rate for 1997-1999 for Wales was marginally higher than the UK rate (6.0 and 5.8 deaths of infants under 1 year old per 1,000 live births, respectively); within the country it ranged between 3.8 for Powys and Merthyr Tydfil and 8.4 for Blaenau Gwent.

(Table 15.2)

Education

Wales had a higher participation rate for children under 5 in schools than the UK as a whole, 78 per cent compared with 64 per cent in the UK in January 2000. The participation rate was also higher than the UK average for all areas within the country, with the exception of Powys and Monmouthshire.

(Tables 4.3 and 15.3)

In 1998/99 half the areas in the country had lower proportions of pupils achieving 5 or more A* to C graded results than the UK average and with the exception of Gwynedd, Conwy, Flintshire, Powys and Pembrokeshire, had higher proportions of pupils with no graded results.

(Table 15.3)

Among the population of working age, the proportion of people who's highest qualification was GCE A level or equivalent in Wales was 20.9 per cent in Spring 2000, lower than the UK average of 24.0 per cent.

(Table 4.13)

Labour market

The employment rate for people of working age in Spring 1999, was among the lowest in the UK at 68.4 per cent. Within the country the employment rate varied in 1998-1999 between 76.9 per cent in Powys to 54.1 per cent in Merthyr Tydfil.

(Tables 5.1 and 15.5)

In 1999 average weekly earnings for people in Wales were amongst the lowest in the UK, only the North East of England and Northern Ireland had lower average earnings.

(Table 5.11)

Economy

In Wales, manufacturing industry accounted for some 27.9 per cent of the country's GDP in 1997, compared to 21.3 per cent for the UK as a whole. Agriculture, forestry and fishing accounted for 1.8 per cent of the country's GDP in 1997, compared to 1.5 per cent for the UK.

(Table 12.4)

Nearly 31 per cent of the country's 115 thousand business sites in 1999 were in distribution, hotels and catering and repairs industries, slightly higher than the UK average of 29.6 per cent.

(Table 13.3)

Almost 72 per cent of the value of direct export trade from Wales in 1999 was to the EU, higher than the UK average of nearly 61 per cent. Imports from the EU accounted for just under 33 per cent of the value of direct import trade to Wales, lower than the UK average of 51.5 per cent.

(Table 13.7)

Environment

A fifth of the total area of Wales is within a National Park.

(Table 11.12)

1.21 Key statistics for Scotland

	Scotland	United Kingdom		Scotland	United Kingdom
Population, 1998 (thousands)	5,120.0	59,236.5	Gross domestic product, 1998 (£ million)	61,052	747,544
Percentage under 16	19.8	20.4	Gross domestic product per head index, 1998 (UK=100)	95.6	100.0
Percentage pension age or over[1]	18.0	18.1	Total business sites, 1999 (thousands)	198.4	2,508.0
Standardised mortality ratio (UK=100), 1998	116	100			
Infant mortality rate[2], 1997-1999	5.3	5.8			
Percentage of pupils achieving 5 or more grades A* to C			Motor cars currently licensed, 1998 (thousands)	1,775	23,878
at GCSE level, 1998/99	57.8	49.1	Fatal and serious accidents on roads[4], 1998		
			(rate per 100,000 population)	71	66
Economic activity rate[3], Spring 1999 (percentages)	76.6	78.4	Recorded crime rate, 1998-99		
Employment rate[3], Spring 1999 (percentages)	70.8	73.6	(notifiable offences per 100,000 population)	8,510	..
ILO unemployment rate, Spring 1999 (percentages)	7.4	6.0			
Average gross weekly earnings: males in full-time			Average weekly household income, 1996-1999 (£)[5]	387	430
employment, April 1999(£)	406.00	440.70	Average weekly household expenditure, 1996-1999 (£)[5]	306	333
Average gross weekly earnings: females in full-time			Households in receipt of Income Support/Family Credit[4],		
employment, April 1999 (£)	297.70	325.60	1998-99 (percentages)	17	15

1 Males aged 65 or over, females aged 60 or over.
2 Deaths of infants under 1 year of age per 1,000 live births.
3 For people of working age, males aged 16 to 64, females aged 16 to 59.
4 Figure for the United Kingdom relates to Great Britain.
5 Combined years 1996-97, 1997-98 and 1998-99.

1.22 Population density: by local authority, 1998

Population density, 1998
(persons per sq km)

- 2,500 or over
- 1,000 - 2,499
- 500 - 999
- 250 - 499
- 100 - 249
- 99 or under

1 Aberdeen City
2 Dundee City
3 Clackmannanshire
4 West Dunbartonshire
5 East Dunbartonshire
6 Falkirk
7 Inverclyde
8 Renfrewshire
9 Glasgow City
10 North Lanarkshire
11 West Lothian
12 City of Edinburgh
13 Midlothian
14 East Lothian
15 North Ayrshire
16 East Renfrewshire
17 East Ayrshire
18 South Lanarkshire
19 South Ayrshire

Scotland

Population

In 1998 Scotland had a population of 5.1 million. Within the country the population density was highest in Glasgow City at 3,540 people per sq km and lowest in the Highland council at only 8 people per sq km.

(Tables 3.1 and 16.1)

Mortality

Overall the country has a higher Standardised Mortality Ratio (SMR) than the UK as a whole at 116 in 1998 (UK =100). Within Scotland this ranged from 97 in East Dunbartonshire and East Renfrewshire to 139 in Glasgow City.

(Table 16.1)

The Infant Mortality rate for 1997-1999 for Scotland was lower than the UK rate (5.3 and 5.8 deaths of infants under 1 year old per 1,000 live births, respectively); within Scotland it ranged between 2.6 for East Lothian and 8.1 for West Dunbartonshire.

(Table 16.2)

Education

Scotland had a higher participation rate for children under 5 in schools than the UK as a whole, 66 per cent compared with 64 per cent for the UK in January 2000.

(Table 4.3)

In 1998/99 all areas in the country with the exception of Dundee City and Glasgow City, had higher proportions of pupils achieving 5 or more 1 to 3 graded results (equivalent to 5 or more A* to C graded results) than the UK average and with the exception of Dundee City, East Lothian, Eilean Siar, Glasgow City, Moray and Perth and Kinross had lower proportions of pupils with no graded results.

(Table 16.3)

Among the population of working age, the proportion of people who's highest qualification was SCE H (Higher) level or equivalent in Scotland was 30.4 per cent in Spring 2000; higher than the UK average of 24.0 per cent.

(Table 4.13)

Labour market

The employment rate for people of working age in Spring 1999, was among the lowest in the UK at 70.8 per cent. Within the country the employment rate varied between 83.3 per cent in the Shetland Islands to 57.0 per cent in Glasgow City in 1998-1999.

(Tables 5.1 and 16.5)

In 1999 average weekly earnings for people in Scotland were £364.90; lower than the UK average of £398.70.

(Table 5.11)

Economy

In Scotland, manufacturing industry accounted for some 21.8 per cent of the country's GDP in 1997, compared to 21.3 per cent for the UK as a whole. Agriculture, forestry and fishing accounted for 2.3 per cent of the country's GDP in 1997, compared to 1.5 per cent for the UK.

(Table 12.4)

Just over 31 per cent of the country's 198 thousand business sites in 1999 were in distribution, hotels and catering and repairs industries, slightly higher than the UK average of 29.6 per cent.

(Table 13.3)

In 1999, 72 per cent of the value of direct export trade from Scotland was to the EU, higher than the UK average of nearly 61 per cent. Imports from the EU accounted for 30.5 per cent of the value of direct import trade to Scotland, the lowest percentage in the UK, much lower than the UK average of 51.5 per cent.

(Table 13.7)

Environment

Over an eighth of the total area in Scotland is designated as a National Scenic Area and two per cent is Green Belt land.

(Table 11.12)

1.23   Key statistics for Northern Ireland

	Northern Ireland	United Kingdom		Northern Ireland	United Kingdom
Population, 1998 (thousands)	1,688.6	59,236.5	Gross domestic product, 1998 (£ million)	15,966	747,544
Percentage under 16	24.5	20.4	Gross domestic product per head index, 1998 (UK=100)	75.8	100.0
Percentage pension age or over[1]	15.2	18.1	Total business sites, 1999 (thousands)	76.2	2,508.0
Standardised mortality ratio (UK=100), 1998	101	100			
Infant mortality rate[2], 1998	5.6	5.8			
Percentage of pupils achieving 5 or more grades A* to C			Motor cars currently licensed, 1998 (thousands)	585	23,878
at GCSE level, 1998/99	56.0	49.1	Fatal and serious accidents on roads, 1998		
			(rate per 100,000 population)	74	66
Economic activity rate[3], Spring 1999 (percentages)	71.9	78.4			
Employment rate[3], Spring 1999 (percentages)	66.6	73.6			
ILO unemployment rate, Spring 1999 (percentages)	7.2	6.0	Average weekly household income, 1996-1999 (£)[4]	347	430
Average gross weekly earnings: males in full-time			Average weekly household expenditure, 1996-1999 (£)[4]	301	333
employment, April 1999(£)	376.80	440.70			
Average gross weekly earnings: females in full-time					
employment, April 1999 (£)	295.10	325.60			

1 Males aged 65 or over, females aged 60 or over.
2 Deaths of infants under 1 year of age per 1,000 live births.
3 For people of working age, males aged 16 to 64, females aged 16 to 59.
4 Combined years 1996-97, 1997-98 and 1998-99.

1.24   Population density: by local authority, 1998

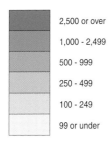

Population density, 1998
(persons per sq km)

- 2,500 or over
- 1,000 - 2,499
- 500 - 999
- 250 - 499
- 100 - 249
- 99 or under

1 Newtownabbey
2 Carrickfergus
3 Belfast
4 North Down
5 Castlereagh

Northern Ireland

Population

In 1998 Northern Ireland had a population of 1.7 million. Within the region the population density was highest in Belfast district at over 2,600 people per sq km and lowest in Moyle at only 31 people per sq km.

(Tables 3.1 and 17.1)

Mortality

Overall the region has a slightly higher Standardised Mortality Ratio (SMR) than the UK as a whole at 101 in 1998 (UK =100). Within Northern Ireland this ranged from 86 in Cookstown to 117 in Derry.

(Table 17.1)

The Infant Mortality rate for 1998 for Northern Ireland was around the same as the UK rate for 1997-99 (5.6 and 5.8 deaths of infants under 1 year old per 1,000 live births, respectively); within the region it ranged between 4.2 for the Eastern Health and Social Services Board area and 8.1 for the Western area.

(Table 17.2)

Education

Northern Ireland had a lower participation rate for children under 5 in schools than the UK as a whole, 57 per cent compared with 64 per cent for the UK in 1999/00.

(Table 4.3)

In 1998/99 all Education Boards in Northern Ireland had higher proportions of pupils achieving 5 or more A* to C graded results than the UK average and had lower proportions of pupils with no graded results.

(Table 17.3)

Among the population of working age, the proportion of people who's highest qualification was GCE A level or equivalent in Northern Ireland was 24.4 per cent in Spring 2000; higher than the UK average of 24.0 per cent.

(Table 4.13)

Labour market

The employment rate for people of working age in Spring 1999 was among the lowest in the UK at 66.6 per cent. Within the region the employment rate varied between 76.2 per cent in Ards to 48.6 per cent in Strabane in 1998-1999.

(Tables 5.1 and 17.4)

In 1999 average weekly earnings for people in the region were the lowest in the UK; £344.90 compared with the UK average of £398.70.

(Table 5.11)

Economy

In Northern Ireland, manufacturing industry accounted for some 20.1 per cent of the region's GDP in 1997, compared to 21.3 per cent for the UK as a whole. Agriculture, forestry and fishing accounted for 4.8 per cent of the region's GDP in 1997, compared to 1.5 per cent for the UK.

(Table 12.4)

Just over 26 per cent of the region's 76 thousand business sites in 1999 were in distribution, hotels and catering and repairs industries, compared with the national average of 29.6 per cent. This was the lowest rate in the UK.

(Table 13.3)

In 1999, 65.5 per cent of the value of direct export trade from Northern Ireland was to the EU, higher than the UK rate of nearly 61 per cent. Imports from the EU accounted for over 43 per cent of the value of direct import trade to Northern Ireland, below the UK average of 51.5 per cent.

(Table 13.7)

Environment

A fifth of the total area in Northern Ireland is designated as an Area of Outstanding Natural Beauty and just under a sixth is Green Belt land.

(Table 11.12)

2 European Union

Population

The most densely populated region in the European Union is Brussels, with almost 5,900 people per square kilometre; London is second with a density of nearly 4,500.

(Table 2.1)

All regions of the United Kingdom have a higher proportion of their population aged under 15 than the EU average. The highest rate of all, at 23.5 per cent in 1997, was in Northern Ireland, equal to the rate in the Republic of Ireland.

(Table 2.1)

Dependency

Dependency ratios – those economically inactive as a percentage of economically active people – in the United Kingdom were lower than the EU average in 1999 except in the North East and Northern Ireland. The South East had one of the lowest rates of all, only Denmark was lower.

(Table 2.2)

Transport

Rates of car ownership in the North East and in Scotland were among the lowest in the European Union; only the Irish Republic and several regions of Greece were lower.

(Table 2.2)

Labour market

At 1.6 per cent, the United Kingdom had the lowest percentage of workers employed in agriculture of all EU countries in 1999.

(Table 2.3)

Only in Luxembourg was there a higher percentage of the workforce employed in the services sector than in the United Kingdom in 1999.

(Table 2.3)

Gross domestic product

Only in London and in the South East was GDP per head in 1997 higher than the overall EU average.

(Table 2.3 and Map 2.5)

The lowest GDP per head in the United Kingdom in 1997 was in Northern Ireland, 18 per cent below the EU average.

(Table 2.3)

Agriculture

The United Kingdom had the highest percentage of land used for agricultural purposes in the European Union.

(Table 2.4)

The United Kingdom had the lowest proportion of agricultural land used for arable purposes within the European Union.

(Table 2.4)

Wales had by far the highest density of sheep and lambs in the European Union, at over 4,800 per 1,000 hectares of utilised agricultural land.

(Table 2.4)

NUTS level 1 areas in the European Union

NETHERLANDS
1 Noord-Nederland
2 Oost-Nederland
3 Zuid-Nederland
4 West-Nederland

BELGIUM
5 Vlaams Gewest
6 Région Wallonne
7 Bruxelles-Brussels

8 LUXEMBOURG

GERMANY
9 Saarland
10 Rheinland-Pfalz
11 Baden-Württemberg
12 Mecklenburg-Vorpommern
13 Hamburg
14 Schleswig-Holstein

non-EU countries

1 NUTS (Nomenclature of Units for Territorial Statistics) is a hierachical classfication of areas that provides a breakdown of the EU's economic territory. See Notes and Definitions.

2.1 Population and vital statistics, 1997

| | Area (sq km) | Popu-lation (thousands) | Persons per sq km | Percentage of population | | Births (per 1,000 population) | Deaths[1] (per 1,000 population) | Infant mortality (per 1,000 births) |
				Aged under 15	Aged 65 or over			
EUR 15	**3,191,120**	**374,094**	**117**	*17.3*	*15.7*	10	9	..
Austria	**83,859**	**8,072**	**96**	*17.7*	*15.1*	10	9	4.7
Ostösterreich	23,554	3,406	145	*16.4*	*16.1*	9	11	4.5
Südösterreich	25,921	1,770	68	*17.4*	*16.0*	9	9	4.6
Westösterreich	34,384	2,896	84	*19.3*	*13.3*	11	8	5.1
Belgium	**30,518**	**10,181**	**334**	*17.8*	*16.2*	11	10	5.4
Bruxelles-Brussels	161	952	5,898	*17.6*	*17.1*	13	11	5.8
Vlaams Gewest	13,512	5,906	437	*17.4*	*15.9*	10	9	5.1
Région Wallonne	16,844	3,324	197	*18.7*	*16.4*	11	11	5.7
Denmark	**43,094**	**5,285**	**123**	*17.5*	*15.5*	12	11	5.3
Finland	**304,529**	**5,140**	**17**	*18.9*	*14.6*	11	9	4.9
Manner-Suomi	303,003	5,115	17	*18.9*	*14.5*	11	9	4.9
Åland	1,527	25	17	*21.9*	*27.2*	11	9	7.0
France	**543,965**	**58,609**	**108**	*19.5*	*15.2*	12	9	4.1
Île de France	12,012	11,073	922	*20.5*	*11.6*	14	6	4.5
Bassin Parisien	145,645	10,516	72	*19.9*	*15.9*	12	9	4.1
Nord-Pas-de-Calais	12,414	4,010	323	*22.5*	*12.9*	13	9	4.3
Est	48,030	5,148	107	*19.2*	*14.7*	12	8	3.6
Ouest	85,099	7,702	91	*19.0*	*17.1*	11	9	4.1
Sud-Ouest	103,599	6,143	59	*17.0*	*18.2*	10	10	3.5
Centre-Est	69,711	6,980	100	*19.5*	*14.0*	12	8	3.6
Méditerranée	67,455	7,036	104	*18.5*	*17.9*	11	9	3.8
Germany	**357,021**	**81,979**	**230**	*15.7*	*16.1*	9	10	4.9
Baden-Württemberg	35,752	10,428	292	*16.7*	*15.3*	11	9	4.2
Bayern	70,548	12,087	171	*16.3*	*15.5*	10	10	4.5
Berlin	891	3,399	3,815	*14.9*	*13.4*	8	10	4.4
Brandenburg	29,476	2,590	88	*15.8*	*14.1*	6	10	4.9
Bremen	404	668	1,653	*14.4*	*17.0*	9	12	6.2
Hamburg	755	1,700	2,251	*13.6*	*16.1*	9	11	5.8
Hessen	21,115	6,035	286	*15.3*	*16.3*	10	10	4.5
Mecklenburg-Vorpommern	23,170	1,799	78	*16.4*	*13.6*	6	9	5.6
Niedersachsen	47,613	7,866	165	*15.6*	*16.7*	10	10	4.8
Nordrhein-Westfalen	34,079	17,915	526	*15.7*	*16.7*	10	10	5.5
Rheinland-Pfalz	19,847	4,025	203	*15.8*	*17.3*	10	10	5.2
Saarland	2,570	1,074	418	*14.1*	*18.9*	9	11	6.3
Sachsen	18,412	4,489	244	*14.2*	*18.2*	6	11	4.4
Sachsen-Anhalt	20,447	2,675	131	*15.2*	*16.4*	6	11	5.0
Schleswig-Holstein	15,771	2,766	175	*16.3*	*16.0*	10	10	4.8
Thüringen	16,172	2,463	152	*15.4*	*15.5*	7	12	4.9
Greece	**131,626**	**10,499**	**80**	*15.0*	*18.8*	9	9	6.4
Voreia Ellada	56,457	3,394	60	*15.6*	*17.3*	10	9	6.5
Kentriki Ellada	53,902	2,641	49	*15.2*	*22.4*	7	9	6.7
Attiki	3,808	3,449	906	*14.1*	*17.9*	10	9	6.5
Nisia Aigaiou, Kriti	17,458	1,015	58	*16.7*	*20.7*	10	9	5.7
Ireland	**70,273**	**3,661**	**52**	*23.5*	*10.5*	14	8	6.2

2.1 *(continued)*

	Area (sq km)	Popu-lation (thousands)	Persons per sq km	Percentage of population		Births (per 1,000 population)	Deaths[1] (per 1,000 population)	Infant mortality (per 1,000 births)
				Aged under 15	Aged 65 or over			
Italy	**301,316**	**57,512**	**191**	*15.4*	*15.6*	9	9	**5.4**
Nord Ovest	34,081	6,059	178	*11.7*	*18.9*	7	12	5.9
Lombardia	23,872	8,974	376	*13.4*	*15.3*	9	9	3.3
Nord Est	39,816	6,568	165	*13.7*	*16.0*	9	9	3.8
Emilia-Romagna	22,124	3,943	178	*11.0*	*19.9*	7	11	5.8
Centro	41,142	5,806	141	*12.5*	*19.3*	7	11	5.1
Lazio	17,227	5,230	304	*15.0*	*14.7*	9	9	5.1
Abruzzo-Molise	15,232	1,605	105	*16.1*	*16.8*	8	10	4.6
Campania	13,595	5,791	426	*21.1*	*11.6*	12	8	6.5
Sud	44,430	6,771	152	*19.6*	*13.2*	10	8	5.8
Sicilia	25,707	5,104	199	*19.8*	*13.6*	11	9	7.8
Sardegna	24,090	1,662	69	*16.9*	*12.7*	8	8	5.2
Luxembourg	**2,586**	**421**	**163**	*18.5*	*14.1*	13	9	**4.2**
Netherlands[2]	**41,526**	**15,611**	**461**	*18.6*	*12.6*	12	8	**4.6**
Noord-Nederland	11,389	1,637	196	*18.3*	*13.6*	11	9	5.2
Oost-Nederland	10,975	3,239	332	*19.7*	*12.1*	12	8	5.0
West-Nederland	11,871	7,286	839	*18.4*	*12.9*	12	8	4.2
Zuid-Nederland	7,291	3,449	486	*18.3*	*11.9*	10	7	4.8
Portugal	**91,906**	**9,946**	**108**	*14.9*	*17.1*	11	10	**6.4**
Continente	88,797	9,444	106	*14.6*	*17.2*	11	10	6.2
Açores	2,330	243	104	*21.0*	*14.2*	14	11	10.6
Madeira	779	259	332	*19.5*	*14.3*	12	9	6.7
Spain	**504,790**	**39,323**	**78**	*15.9*	*16.6*	8	8	**5.2**
Noroeste	45,297	4,313	95	*13.3*	*19.8*	6	10	5.4
Noreste	70,366	4,027	57	*13.2*	*17.9*	5	9	6.8
Madrid	7,995	5,022	628	*15.3*	*13.7*	9	7	4.8
Centro	215,025	5,282	25	*15.6*	*19.9*	8	9	5.1
Este	60,249	10,718	178	*15.3*	*17.3*	9	9	4.6
Sur	98,616	8,384	85	*19.2*	*14.0*	10	8	5.6
Canarias	7,242	1,577	218	*18.1*	*11.8*	10	7	6.2
Sweden	**410,934**	**8,846**	**22**	*18.7*	*17.4*	10	10	**3.1**
United Kingdom[3]	**243,820**	**59,009**	**242**	*19.8*	*15.2*	12	10	**5.8**
North East	8,612	2,602	302	*19.8*	*15.5*	11	11	5.8
North West	14,165	6,900	487	*20.5*	*15.1*	11	11	6.7
Yorkshire and the Humber	15,566	5,045	324	*19.8*	*15.3*	11	10	6.5
East Midlands	15,627	4,156	266	*19.6*	*15.2*	11	10	5.7
West Midlands	13,004	5,328	410	*20.2*	*15.1*	12	10	7.0
East	19,120	5,323	278	*19.6*	*15.3*	12	10	4.8
London	1,584	7,110	4,490	*20.0*	*13.0*	14	8	5.8
South East	19,111	7,941	416	*19.5*	*16.0*	12	10	5.0
South West	23,971	4,867	203	*18.6*	*18.0*	11	11	5.8
Wales	20,768	2,929	141	*19.7*	*16.9*	11	11	5.9
Scotland	78,132	5,131	66	*19.3*	*14.8*	11	11	5.3
Northern Ireland	14,160	1,678	119	*23.5*	*12.3*	14	8	5.6

1 Deaths are by date of occurrences and not date of registration.
2 Including 'central persons register'.
3 Government Office Regions for the United Kingdom equal NUTS-1 regions for the European Union. See Notes and Definitions.

Source: Eurostat

2.2 Social statistics

	Dependency rate[1] 1999	Proportion of 16-18 year olds in education or training (percentages) 1995/96[2]	Causes of death 1994[3] (rate per 100,000 population)				Transport	
			Circulatory system	Cancer (all neoplasms)	All accidents	Motor vehicle accidents	Length of motorways (km) per 1,000 sq km 1998[4]	Private cars per 1,000 population 1997[5]
EUR 15	**131**	..	..	..	..	..	**14**	**429**
Austria	**110**	*82*	**544**	**244**	**40**	**15**	**19**	**469**
Ostösterreich	107	..	660	271	44	15	18	451
Südösterreich	119	..	516	250	37	15	21	502
Westösterreich	109	..	426	210	36	15	19	469
Belgium	**147**	*97*	**399**	**270**	**42**	**18**	**55**	**434**
Bruxelles-Brussels	163	..	409	302	44	12	68	474
Vlaams Gewest	136	..	386	263	38	17	61	442
Région Wallonne	164	..	419	273	48	23	50	408
Denmark	**90**	*83*	**514**	**276**	**47**	**11**	**20**	**337**
Finland	**107**	*90*	**449**	**197**	**51**	**9**	**2**	**379**
Manner-Suomi	..	*90*	449	197	51	9	2	378
Åland	..	*81*	450	307	36	8	0	506
France	**139**	*91*	**301**	**247**	**53**	**15**	**17**	**435**
Île de France	116	..	210	201	39	9	49	376
Bassin Parisien	141	..	312	260	59	18	18	447
Nord-Pas-de-Calais	182	..	304	258	46	10	48	363
Est	132	..	307	243	51	15	20	429
Ouest	138	..	318	263	57	15	9	468
Sud-Ouest	145	..	381	269	63	18	11	493
Centre-Est	132	..	291	238	56	14	21	457
Méditerranée	166	..	340	266	60	17	18	445
Germany	**114**	*92*	**543**	**263**	**33**	**12**	**32**	**500**
Baden-Württemberg	108	*91*	444	236	30	10	29	528
Bayern	101	*91*	508	248	33	14	31	531
Berlin	111	*92*	576	253	24	8	66	350
Brandenburg	105	*91*	583	245	54	25	26	486
Bremen	129	*107*	571	327	35	8	119	429
Hamburg	105	*97*	502	302	45	8	107	418
Hessen	114	*93*	479	270	37	10	45	540
Mecklenburg-Vorpommern	111	*89*	504	223	60	24	11	449
Niedersachsen	123	*91*	542	267	35	14	28	516
Nordrhein-Westfalen	128	*95*	553	281	21	8	64	495
Rheinland-Pfalz	119	*87*	558	267	23	12	42	535
Saarland	132	*94*	626	283	24	8	92	540
Sachsen	110	*93*	702	281	56	15	24	469
Sachsen-Anhalt	118	*88*	648	277	47	18	12	451
Schleswig-Holstein	114	*92*	580	273	32	10	30	518
Thüringen	103	*88*	650	237	34	15	17	474
Greece	**147**	*69*	**460**	**202**	**36**	**19**	**2**	**238**
Voreia Ellada	148	..	465	208	33	18	1	190
Kentriki Ellada	153	..	483	191	35	19	3	120
Attiki	147	..	428	209	40	21	18	389
Nisia Aigaiou, Kriti	132	..	496	189	34	17	0	192
Ireland	**129**	*83*	**402**	**212**	**28**	**11**	**1**	**310**

2.2 *(continued)*

	Dependency rate[1] 1999	Proportion of 16-18 year olds in education or training (percentages) 1995/96[2]	Causes of death 1994[3] (rate per 100,000 population)				Transport	
			Circulatory system	Cancer (all neoplasms)	All accidents	Motor vehicle accidents	Length of motorways (km) per 1,000 sq km 1998[4]	Private cars per 1,000 population 1997[5]
Italy	163	..	422	264	38	14	21	535
Nord Ovest	143	..	532	326	50	15	36	554
Lombardia	128	..	385	307	38	16	23	557
Nord Est	131	..	416	298	43	18	22	544
Emilia-Romagna	122	..	483	341	45	21	28	585
Centro	146	..	501	313	45	16	16	561
Lazio	162	..	373	250	39	14	28	593
Abruzzo-Molise	181	..	470	236	43	12	24	499
Campania	238	..	368	185	24	9	33	521
Sud	227	..	359	179	31	12	14	442
Sicilia	251	..	431	194	29	9	22	498
Sardegna	190	..	339	209	41	14	0	477
Luxembourg	138	..	417	255	49	20	44	562
Netherlands	101	91	336	237	22	8	65	392
Noord-Nederland	110	..	369	257	25	10	37	379
Oost-Nederland	102	..	333	229	23	10	60	395
West-Nederland	99	..	340	244	22	6	85	379
Zuid-Nederland	99	..	315	222	21	10	81	422
Portugal	102	70	430	193	38	22	14	378
Continente	100	73	430	194	38	22	14	398
Açores	151	..	509	208	30	13	0	..
Madeira	120	21	355	157	39	19	0	..
Spain	165	74	342	212	32	15	16	389
Noroeste	179	79	408	250	40	22	18	365
Noreste	151	86	326	231	32	16	16	360
Madrid	143	82	252	182	27	10	61	481
Centro	184	73	408	235	33	16	10	331
Este	145	70	352	222	35	16	31	433
Sur	204	69	331	183	26	15	17	330
Canarias	155	71	264	170	31	8	26	445
Sweden	110	96	515	229	33	6	3	482
United Kingdom[6]	108	71	..	280	21	..	14	379
North East	137	..	553	312	20	7	7	300
North West	120	..	548	296	22	8	41	363
Yorkshire and the Humber	114	64	514	279	20	8	20	340
East Midlands	103	69	495	273	21	9	12	390
West Midlands	109	76	486	278	22	8	29	411
East	97	..	451	263	20	9	13	435
London	104	..	..	252	19	..	45	..
South East	93	..	480	278	18	7	34	441
South West	99	68	540	303	20	8	12	436
Wales	130	64	547	301	23	8	6	365
Scotland	114	82	..	296	31	..	4	327
Northern Ireland	134	77	439	220	23	..	8	..

1 Dependency rates are calculated as the number of non-active persons (total population less labour force) expressed as a percentage of those economically active. 1998 for EUR15 and Greece.

2 Participation rates are calculated by dividing the number of pupils enrolled in a region by the resident population in that region. As some young people may be resident in one region and in education in another, this inter-regional movement may influence the results. Data for Belgium and Portugal are for 1994/95. The UK data exclude Open University, independent and special schools in Wales, and Youth Training with employers, all of which are not available by region and age. For all countries, age is taken at 1 January except for the UK where it is on 31 August (ie the start of the academic year).

3 Unadjusted death rates using 1994 population estimates. 1990 for Belgium. 1992 for United Kingdom. 1993 for Denmark, Germany, Greece, Spain, France, Ireland, Italy, Luxembourg and Austria.

4 1994 for EUR15 and Italy. 1996 for Greece, Ireland, Sweden and United Kingdom. 1997 for Denmark and the Netherlands.

5 1995 for EUR15 and Portugal. 1996 for Germany, Sweden and United Kingdom.

6 Government Office Regions for the United Kingdom equal NUTS-1 regions for the European Union. See Notes and Definitions.

Source: Eurostat

2.3 Economic statistics

	Persons in employ-ment[1], 1999[2] (thousands)	Employment[1], 1999[2] percentage in			Unem-ployment rate[2] (percent-ages) 1998	Long-term unemployed[2] as a percentage of the unem-ployed, 1998[3]	Gross domestic product per head (PPS)[4] EUR 15=100 1997	Estimates [5,6] of the percentage of GDP in 1996 derived from		
		Agriculture	Industry	Services				Agriculture	Industry	Services
EUR 15	**152,494**	*4.7*	*29.6*	*65.5*	*10.1*	*49.1*	*100*	*..*	*..*	*..*
Austria	**3,678**	*6.2*	*29.8*	*64.0*	*4.8*	*34.1*	112	*1.5*	*31.6*	*66.9*
Ostösterreich	1,572	5.3	26.6	68.1	5.2	49.9	126	1.2	27.5	71.3
Südösterreich	778	8.3	32.8	58.9	5.2	30.4	91	2.4	35.0	62.5
Westösterreich	1,327	6.1	31.7	62.2	3.9	11.7	108	1.4	35.4	63.2
Belgium	**3,987**	*2.4*	*25.8*	*71.8*	*9.3*	*62.1*	111	*1.4*	*27.3*	*71.3*
Bruxelles-Brussels	338	0.2	13.4	86.4	14.3	64.2	169	0.0	13.2	86.8
Vlaams Gewest	2,450	2.5	28.0	69.5	6.2	57.4	115	1.6	31.5	66.9
Région Wallonne	1,199	2.8	24.7	72.5	13.5	65.5	88	1.7	25.8	72.5
Denmark	**2,708**	*3.3*	*26.8*	*69.5*	*5.4*	*24.8*	120	*4.1*	*27.6*	*68.3*
Finland	**2,333**	*6.4*	*27.6*	*65.7*	*12.7*	*29.3*	100	*5.5*	*33.0*	*61.5*
Manner-Suomi	2,321	6.3	27.7	65.6	12.7	29.3	100	5.5	33.1	61.4
Åland	12	9.1	11.6	78.5	2.6	15.0	119	8.8	13.1	78.0
France	**22,755**	*4.3*	*26.3*	*69.4*	*11.4*	*44.5*	99	*2.4*	*27.4*	*70.1*
Île de France	4,903	0.5	19.7	79.8	10.4	45.4	153	0.2	23.0	76.8
Bassin Parisien	3,936	5.9	30.4	63.7	11.8	44.3	88	4.1	33.2	62.6
Nord-Pas-de-Calais	1,342	1.7	29.4	68.9	15.9	48.3	82	1.4	30.1	68.5
Est	2,075	3.0	35.3	61.7	8.7	38.2	91	2.4	34.0	63.6
Ouest	3,117	7.6	29.2	63.2	10.2	42.2	83	4.8	27.7	67.5
Sud-Ouest	2,365	7.7	23.1	69.2	11.4	46.2	85	4.7	24.6	70.7
Centre-Est	2,722	4.4	29.5	66.1	9.8	42.4	93	2.0	33.2	64.7
Méditerranée	2,295	4.0	19.3	76.7	15.7	46.0	83	2.8	20.2	77.0
Germany	**36,089**	*2.9*	*33.8*	*63.3*	*9.8*	*51.7*	108	*1.1*	*34.9*	*64.0*
Baden-Württemberg	4,807	2.5	41.4	56.1	6.0	53.5	123	1.1	41.3	57.7
Bayern	5,787	4.1	35.6	60.3	5.7	48.3	126	1.0	35.3	63.7
Berlin	1,459	0.8	23.4	75.8	13.7	49.0	109	0.2	32.2	67.6
Brandenburg	1,145	5.4	32.2	62.3	17.6	43.2	75	2.2	39.3	58.6
Bremen	274	1.3	26.4	72.3	11.9	63.1	145	0.2	31.2	68.5
Hamburg	788	0.5	22.4	77.1	8.4	59.2	197	0.3	20.9	78.8
Hessen	2,684	1.7	31.7	66.7	7.1	57.9	140	0.5	26.8	72.7
Mecklenburg-Vorpommern	757	6.8	26.9	66.3	19.6	40.2	66	2.9	29.8	67.3
Niedersachsen	3,370	4.0	32.5	63.5	8.8	61.9	99	2.9	34.1	63.0
Nordrhein-Westfalen	7,511	1.8	34.3	63.9	8.7	64.3	108	0.7	36.6	62.7
Rheinland-Pfalz	1,759	2.5	36.3	61.3	6.8	55.7	96	1.4	38.4	60.2
Saarland	436	0.9	34.8	64.3	9.6	67.6	98	0.3	36.0	63.7
Sachsen	1,927	3.0	34.8	62.3	17.9	40.9	68	1.2	36.9	61.9
Sachsen-Anhalt	1,074	4.2	31.9	63.9	21.5	42.8	64	2.1	37.0	60.9
Schleswig-Holstein	1,223	3.2	24.8	71.9	7.3	54.6	102	2.2	29.4	68.4
Thüringen	1,090	3.8	34.0	62.1	18.5	38.7	65	1.9	36.1	62.0
Greece	**3,967**	*17.8*	*23.0*	*59.2*	*10.8*	*54.5*	66	*14.9*	*25.0*	*60.0*
Voreia Ellada	1,265	26.0	23.8	50.2	10.3	53.1	63	22.1	28.8	49.2
Kentriki Ellada	805	32.5	20.7	46.8	10.4	65.2	57	24.6	27.2	48.2
Attiki	1,493	1.0	25.3	73.7	12.2	52.2	75	2.2	24.9	72.9
Nisia Aigaiou, Kriti	404	24.6	16.6	58.7	7.5	44.6	68	23.5	14.1	62.3
Ireland	**1,593**	*8.5*	*28.3*	*62.5*	*7.9*	*56.0*	102	*7.2*	*40.4*	*52.4*

2.3 *(continued)*

	Persons in employ-ment[1], 1999[2] (thousands)	Employment[1], 1999[2] percentage in			Unem-ployment rate[2] (percent-ages) 1998	Long-term unemployed[2] as a percentage of the unem-ployed, 1998[3]	Gross domestic product per head (PPS)[4] EUR 15=100 1997	Estimates [5,6] of the percentage of GDP in 1996 derived from		
		Agriculture	Industry	Services				Agriculture	Industry	Services
Italy	20,618	5.4	32.4	62.2	12.3	58.5	102	3.5	28.9	67.6
Nord Ovest	2,363	3.8	35.2	61.0	9.3	50.4	118	2.7	31.9	65.4
Lombardia	3,830	2.1	41.1	56.8	5.7	40.9	131	1.8	37.4	60.8
Nord Est	2,756	5.3	39.0	55.7	5.0	35.6	124	3.8	32.4	63.8
Emilia-Romagna	1,723	6.7	36.4	56.9	5.7	29.3	131	4.1	32.7	63.1
Centro	2,260	3.8	35.8	60.4	7.9	47.3	107	3.0	30.8	66.1
Lazio	1,868	2.9	19.0	78.1	12.3	63.0	113	1.7	17.7	80.6
Abruzzo-Molise	541	7.7	33.2	59.1	11.2	54.6	87	5.3	28.5	66.2
Campania	1,561	7.5	24.5	68.0	24.9	76.5	65	3.8	20.1	76.1
Sud	1,883	12.0	24.9	63.1	22.7	61.7	67	7.4	20.2	72.4
Sicilia	1,318	9.1	19.3	71.6	25.6	65.0	65	6.2	19.1	74.7
Sardegna	515	8.1	22.7	69.2	21.5	57.5	72	5.4	22.5	72.1
Luxembourg	176	1.9	21.9	75.8	2.8	31.2	174	1.3	23.9	74.8
Netherlands	7,605	..	..	..	4.0	47.0	113	3.2	26.8	70.0
Noord-Nederland	751	..	..	..	5.8	50.2	107	3.9	38.6	57.5
Oost-Nederland	1,578	..	..	..	3.7	46.2	98	3.9	27.0	69.1
West-Nederland	3,579	..	..	..	3.9	48.6	121	2.7	20.1	77.2
Zuid-Nederland	1,698	..	..	..	3.6	41.6	111	3.7	32.5	63.8
Portugal	4,830	12.6	35.3	52.1	4.7	43.5	73	4.1	33.9	62.0
Continente	4,619	12.5	35.6	52.0	4.8	43.4	74	4.0	34.5	61.4
Açores	96	18.4	26.0	55.7	4.0	40.5	51	11.8	19.7	68.4
Madeira	115	15.0	31.2	53.8	3.7	51.2	56	3.8	18.2	78.0
Spain	13,773	7.4	30.6	62.0	19.1	49.2	80	4.5	29.0	66.5
Noroeste	1,423	15.6	30.3	54.1	18.1	57.6	68	7.1	30.4	62.5
Noreste	1,502	5.4	37.2	57.5	14.5	50.9	93	4.0	37.3	58.8
Madrid	1,933	1.0	25.8	73.2	17.0	58.7	101	0.2	24.6	75.1
Centro	1,715	12.0	30.3	57.7	20.2	46.1	69	9.9	30.7	59.4
Este	4,124	4.0	35.4	60.6	15.2	49.2	91	2.1	32.0	65.9
Sur	2,493	11.7	25.3	63.0	28.1	44.3	60	8.6	22.8	68.6
Canarias	584	6.6	19.8	73.7	19.8	44.5	76	3.5	16.2	80.3
Sweden	4,054	3.0	25.0	72.0	8.9	37.8	102	2.2	30.0	66.9
United Kingdom[7]	27,107	1.6	26.0	72.3	6.2	32.6	102	1.8	29.8	68.4
North East	1,035	1.1	28.4	70.2	9.1	34.7	84	0.7	37.3	62.0
North West	3,018	1.2	28.2	70.5	6.4	36.3	93	1.5	35.7	62.8
Yorkshire and the Humber	2,258	1.2	28.5	70.3	7.1	32.1	91	2.2	36.2	61.6
East Midlands	1,992	2.0	31.6	66.3	4.7	27.4	98	2.6	38.6	58.8
West Midlands	2,438	1.4	33.0	65.4	6.2	33.1	94	2.1	39.0	58.9
East	2,616	1.6	26.4	72.0	4.9	32.4	99	2.6	28.6	68.8
London	3,284	0.3	16.0	83.6	8.1	34.3	146	0.1	17.1	82.8
South East	3,946	1.3	22.9	75.7	4.1	31.3	109	1.0	24.9	74.1
South West	2,354	2.4	25.6	72.0	4.5	29.1	100	3.7	30.3	66.1
Wales	1,216	2.7	29.5	67.6	7.0	36.1	84	2.3	38.7	59.0
Scotland	2,273	2.0	25.3	72.5	7.3	28.5	97	2.8	34.0	63.2
Northern Ireland	676	5.0	26.5	68.5	8.8	37.0	82	4.7	30.7	64.5

1 See Notes and Definitions.
2 1998 for EUR15 and Greece.
3 1997 for EUR15 and Ireland.
4 Purchasing Power Standard; see Notes and Definitions.
5 Estimates for GDP by sector are based on the Gross Value Added (GVA) figures for each area.
6 Estimates for Government Office regions are provided by the Office for National Statistics.
7 Government Office Regions for the United Kingdom equal NUTS-1 regions for the European Union. See Notes and Definitions.

Source: Eurostat; Office for National Statistics

2.4 Agricultural statistics, 1997

	Agricultural land as a percentage of total land area[1]	Arable land as a percentage of agricultural land[2]	Average yield[3]		Livestock per 1,000 ha of utilised agricultural land[4]			Economic value of farms(SGM)[5,6] EUR 15=100
			Wheat 100kg/ha	Barley 100kg/ha	All cattle	All sheep and lambs	All pigs	
EUR 15	42.0	56.2	54	46	624	733	864	100
Austria	40.8	40.8	52	48	642	112	1,075	70
Ostösterreich	49.5	74.0	50	47	475	59	1,009	88
Südösterreich	32.4	25.9	52	43	677	138	1,456	51
Westösterreich	41.2	22.4	60	55	759	140	904	68
Belgium	45.3	61.7	79	74	2,233	82	5,175	280
Bruxelles-Brussels	4.3	85.7	90	..	571	143	0	..
Vlaams Gewest	46.6	67.4	81	69	2,580	113	10,946	282
Région Wallonne	44.6	56.8	79	76	1,945	56	365	276
Denmark	64.1	92.8	72	54	733	37	4,158	340
Finland	6.4	98.8	37	34	523	48	672	102
Manner-Suomi	..	..	..	..	..	..	..	..
Åland	..	..	..	..	..	..	..	..
France	54.6	60.4	66	60	672	343	516	206
Île de France	49.2	95.5	78	69	..	..	..	512
Bassin Parisien	64.9	70.3	73	66	..	..	..	282
Nord-Pas-de-Calais	71.2	76.0	83	76	1,508	..	876	292
Est	47.0	52.1	65	61	..	..	..	183
Ouest	69.6	75.9	61	53	981	..	1,800	212
Sud-Ouest	49.1	54.2	45	39	632	847	..	147
Centre-Est	46.6	34.8	52	45	771	..	228	133
Méditerranée	33.6	23.2	22	29	..	..	..	190
Germany	48.1	68.3	73	59	909	134	1,401	184
Baden-Württemberg	41.5	57.0	66	55	937	194	1,512	102
Bayern	48.0	63.2	66	53	1,252	114	1,043	120
Berlin	2.0	72.7	..	..	455	227	545	224
Brandenburg	44.1	77.6	53	47	531	89	532	511
Bremen	24.5	19.1	85	70	1,362	32	213	..
Hamburg	18.5	44.0	79	67	617	113	234	..
Hessen	37.1	64.3	73	59	772	205	1,122	117
Mecklenburg-Vorpommern	56.0	78.7	74	69	472	51	434	871
Niedersachsen	57.5	65.7	83	60	1,106	84	2,566	253
Nordrhein-Westfalen	46.2	70.2	86	65	1,098	148	3,702	204
Rheinland-Pfalz	36.5	55.4	67	54	681	185	554	148
Saarland	28.9	52.8	67	53	866	230	331	137
Sachsen	46.4	79.2	66	59	694	128	625	521
Sachsen-Anhalt	52.1	85.9	70	64	375	108	608	856
Schleswig-Holstein	67.8	55.7	90	80	1,328	211	1,229	292
Thüringen	48.7	77.9	67	56	572	290	799	590
Greece	29.9	57.5	22	23	151	2,414	238	40
Voreia Ellada	33.5	82.4	21	25	235	1,788	186	48
Kentriki Ellada	26.3	41.2	26	25	75	2,805	350	37
Attiki	25.8	15.1	13	32	122	1,871	176	36
Nisia Aigaiou, Kriti	29.9	18.6	14	14	57	3,736	134	29
Ireland	63.1	23.4	77	57	1,577	1,271	387	107

2.4 (continued)

	Agricultural land as a percentage of total land area[1]	Arable land as a percentage of agricultural land[2]	Average yield[3]		Livestock per 1,000 ha of utilised agricultural land[4]			Economic value of farms(SGM)[5,6] EUR 15=100
			Wheat 100kg/ha	Barley 100kg/ha	All cattle	All sheep and lambs	All pigs	
Italy	*58.3*	*51.3*	35	37	503	662	533	49
Nord Ovest	*44.9*	*44.9*	46	47	872	113	503	68
Lombardia	*50.1*	*66.8*	58	52	1,716	140	2,462	117
Nord Est	*44.1*	*46.5*	56	51	946	63	518	55
Emilia-Romagna	*62.8*	*72.1*	57	52	788	136	1,628	91
Centro	*51.9*	*64.8*	41	39	238	682	515	56
Lazio	*59.1*	*53.9*	33	31	325	1,260	198	37
Abruzzo-Molise	*59.1*	*52.3*	30	35	213	954	195	32
Campania	*63.8*	*49.4*	32	32	441	468	252	41
Sud	*69.9*	*47.9*	26	28	170	544	121	35
Sicilia	*77.5*	*52.0*	22	21	214	671	60	33
Sardegna	*69.7*	*21.1*	20	19	198	2,358	169	44
Luxembourg	*49.5*	*47.7*	59	55	1,616	51	580	198
Netherlands	*47.4*	*46.8*	93	57	2,392	896	7,388	518
Noord-Nederland	*48.3*	*44.1*	89	58	1,867	1,085	1,027	554
Oost-Nederland	*50.9*	*39.1*	93	56	3,297	620	8,784	412
West-Nederland	*40.5*	*50.1*	98	60	1,468	1,344	1,681	625
Zuid-Nederland	*52.2*	*58.1*	86	50	2,991	461	21,749	514
Portugal	*42.9*	*57.2*	12	9	326	866	600	36
Continente	*43.0*	*58.6*	12	9	281	891	605	37
Açores	*51.8*	*11.6*	..	..	1,722	33	339	32
Madeira	*9.7*	*46.1*	10	..	789	1,316	2,237	14
Spain	*50.7*	*54.5*	30	30	231	936	728	56
Noroeste	*30.5*	*38.2*	20	20	1,302	332	741	26
Noreste	*51.5*	*56.5*	28	34	162	1,257	892	71
Madrid	*47.7*	*51.8*	19	21	203	485	119	51
Centro	*57.9*	*62.3*	30	30	165	1,020	349	60
Este	*37.3*	*38.9*	40	34	355	963	2,896	50
Sur	*55.1*	*46.2*	30	17	108	716	618	69
Canarias	*15.9*	*52.7*	15	15	169	377	661	67
Sweden	*7.0*	*88.7*	60	43	564	140	745	152
United Kingdom[7]	*70.2*	*35.5*	81	61	660	1,627	443	343
North East	..	..	..	..	..	..	..	..
North West	..	..	..	..	..	..	..	..
Yorkshire and the Humber	*65.5*	*57.9*	85	70	627	1,671	1,918	413
East Midlands	*71.1*	*72.9*	82	65	554	948	507	547
West Midlands	*67.6*	*53.3*	83	66	980	2,173	425	293
East	..	..	..	..	..	..	..	..
London	..	..	..	..	..	..	..	..
South East	..	..	..	..	..	..	..	..
South West	*70.0*	*42.7*	79	58	1,203	1,800	493	260
Wales	*68.6*	*14.6*	75	56	890	4,812	66	173
Scotland	*70.7*	*16.3*	83	60	365	1,129	102	241
Northern Ireland	*74.8*	*22.9*	76	54	1,477	1,420	524	130

1 1988 for Italy. 1992 for EUR15. 1993 for Germany. 1994 for the Netherlands. 1996 for Spain and the UK.
2 1988 for Italy. 1992 for EUR15. 1994 for the Netherlands. 1996 for Germany, Spain and the UK.
3 1994 for EUR15 (wheat) and Italy. 1995 for the Netherlands. 1996 for EUR15 (barley), Greece, Spain and the UK.
4 1988 for Italy. 1992 for EUR15. 1994 for France (sheep) and the Netherlands. 1996 for Belgium, Germany, Spain and the UK.
5 The economic value of farms is measured in Standard Gross Margins (SGMs). Data relates to 1995. See Notes and Definitions.
6 Vlaams Gewest includes Brussels. Berlin includes Bremen and Hamburg.
7 Government Office Regions for the United Kingdom equal NUTS-1 regions for the European Union. See Notes and Definitions.

Source: Eurostat

2.5 Gross domestic product per head[1]: by NUTS level 1 areas, 1997

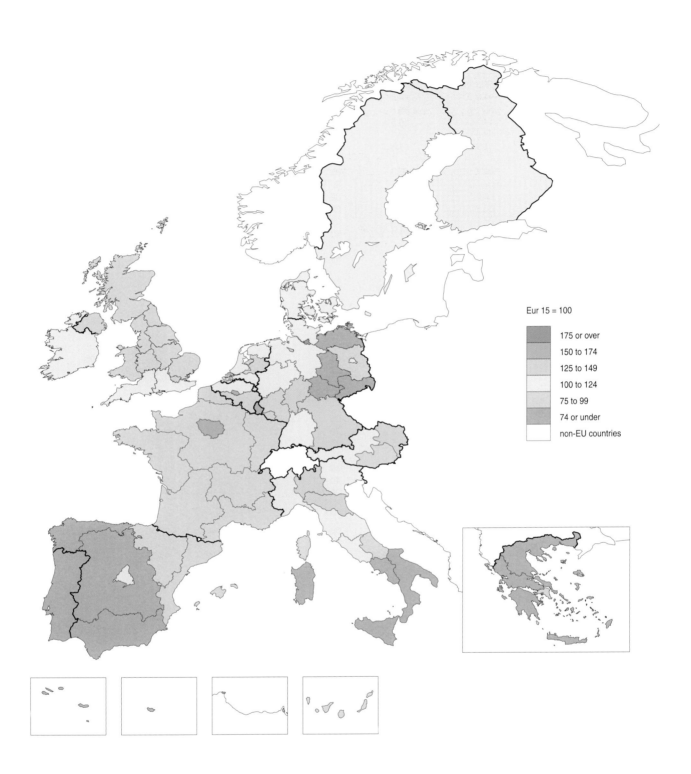

Eur 15 = 100

	175 or over
	150 to 174
	125 to 149
	100 to 124
	75 to 99
	74 or under
	non-EU countries

1 Purchasing Power Standard; see Notes and Definitions.

Source: Eurostat

3 Population and Households

Population

The population of the United Kingdom rose by just over 1.4 million people between 1991 and 1998 to 59.2 million. Population growth over this period was higher amongst males than females, at around 3 per cent and nearly 2 per cent respectively.

(Table 3.1)

Age

Around 37 per cent of the population in Northern Ireland are people under 25 year old. A higher proportion than in any other region in the United Kingdom.

(Table 3.2)

Population density

Portsmouth UA in the South East has the highest population density of any Unitary Authority or equivalent in the United Kingdom; the Highlands and Eilean Siar council areas in Scotland have the lowest.

(Map 3.3)

The region of United Kingdom with the highest proportion of pensioners was the South West, at just over one in five of the population.

(Map 3.5)

Social characteristics

London and the South East had the highest proportions of people of working age in professional occupations in Spring 1999. London also had the lowest percentage in unskilled occupations.

(Table 3.6)

One in four people in London are from an ethnic minority, compared with one in 50 people in the South West, Wales and Scotland.

(Table 3.7)

Population change

Milton Keynes UA had the greatest increase in population of any Unitary Authority, county or equivalent between 1981 and 1998. The population rose by over 60 per cent. Bracknell Forest UA, which had the second largest increase, grew by 30 per cent over the same period.

(Map 3.8)

Milton Keynes UA in the South East has the highest projected population growth between 1998 and 2011, at 16 per cent. The population of the New Council of Inverclyde in Scotland is projected to decline the most over this period with a fall of just over 14 per cent.

(Map 3.9)

London and Northern Ireland have higher crude live birth rates and lower death rates, than any other region of the United Kingdom.

(Table 3.10)

Births

The number of births between mid 1997 and mid 1998 was higher than the number of deaths over the same period in all regions other than the North East, South West and Scotland.

(Table 3.11)

Conceptions

The North East and Wales had the highest conception rates for women aged under 18. Over 64 per cent of these pregnancies led to a maternity compared with around 46 per cent in London.

(Table 3.12)

Births

In 1998, in the South East the birth rate was highest among women aged 30 to 34. In all other regions, the birth rate was higher for women aged between 25 and 29 than for any other age group.

(Table 3.13)

Deaths
Death rates are highest in the North East, North West and Scotland for those aged between 35 and 44 years. These regions also have the highest death rates across the older age bands.
(Table 3.14)

Migration
More international migrants came to London in 1998 than to any other region. London also had the highest net outflow of people to the rest of the United Kingdom.
(Table 3.15)

When moving region, people from the North East, North West or Yorkshire and the Humber tended to move to other regions in the North of England. The next most popular destinations were London and the South East.
(Table 3.16)

Marriage and cohabitation
The number of marriages in the United Kingdom decreased by a quarter between 1976 and 1998, with the greatest decline being in the North East.
(Table 3.17)

Unmarried people aged between 16 and 59 in the South East, East Midlands and Wales were slightly more likely to cohabit than those anywhere else in the United Kingdom.
(Chart 3.18)

Households
The South East and South West are projected to have 23 per cent more households in 2121 than in 1998. This compares with an increase of 7 per cent projected for the North East over the same period.
(Table 3.19)

Scotland and London have the highest proportions of one person households in the United Kingdom, and the lowest proportions of married couples with dependent children.
(Table 3.20)

3.1 Resident population[1]: by gender

Thousands and percentages

	Population (thousands)				Total population growth (percentages)		
	1971	1981	1991	1998	1971-1981	1981-1991	1991-1998
Males							
United Kingdom	27,167.3	27,409.2	28,247.0	29,128.4	0.9	3.0	3.1
North East	1,304.0	1,283.1	1,267.5	1,268.4	-1.6	-1.2	0.1
North West	3,422.4	3,357.6	3,348.7	3,384.1	-1.9	-0.3	1.1
Yorkshire and the Humber	2,384.9	2,395.0	2,441.7	2,489.5	0.4	2.0	2.0
East Midlands	1,797.8	1,894.8	1,989.6	2,063.0	5.4	5.0	3.7
West Midlands	2,542.4	2,555.6	2,596.3	2,638.9	0.5	1.6	1.6
East	2,194.6	2,385.5	2,536.9	2,654.2	8.7	6.3	4.6
London	3,611.4	3,277.6	3,352.0	3,547.9	-9.2	2.3	5.8
South East	3,321.1	3,528.6	3,759.8	3,934.3	6.2	6.6	4.6
South West	1,989.9	2,117.2	2,295.6	2,397.7	6.4	8.4	4.4
England	22,568.5	22,795.0	23,588.1	24,378.0	1.0	3.5	3.3
Wales	1,328.5	1,365.1	1,407.0	1,438.6	2.8	3.1	2.2
Scotland	2,515.7	2,494.9	2,469.5	2,484.4	-0.8	-1.0	0.6
Northern Ireland	754.6	754.2	783.2	827.3	-0.1	3.8	5.6
Females							
United Kingdom	28,760.7	28,943.0	29,566.0	30,108.1	0.6	2.1	1.8
North East	1,374.5	1,353.1	1,335.0	1,321.2	-1.6	-1.3	-1.0
North West	3,685.4	3,582.7	3,536.7	3,506.7	-2.8	-1.3	-0.8
Yorkshire and the Humber	2,517.4	2,523.5	2,541.1	2,553.4	0.2	0.7	0.5
East Midlands	1,854.1	1,958.0	2,045.8	2,106.3	5.6	4.5	3.0
West Midlands	2,603.6	2,631.1	2,669.1	2,693.6	1.1	1.4	0.9
East	2,259.7	2,468.5	2,613.0	2,722.8	9.2	5.9	4.2
London	3,918.0	3,528.0	3,538.0	3,639.3	-10.0	0.3	2.9
South East	3,508.6	3,716.8	3,919.1	4,069.5	5.9	5.4	3.8
South West	2,121.9	2,264.1	2,422.1	2,503.6	6.7	7.0	3.4
England	23,843.2	24,025.8	24,619.9	25,116.5	0.8	2.5	2.0
Wales	1,411.8	1,448.4	1,484.5	1,494.7	2.6	2.5	0.7
Scotland	2,719.9	2,685.3	2,637.5	2,635.6	-1.3	-1.8	-0.1
Northern Ireland	785.8	783.5	824.1	861.3	-0.3	5.2	4.5
All persons							
United Kingdom	55,928.0	56,352.2	57,813.8	59,236.5	0.8	2.5	2.5
North East	2,678.5	2,636.2	2,602.5	2,589.6	-1.6	-1.3	-0.5
North West	7,107.8	6,940.3	6,885.4	6,890.8	-2.4	-0.8	0.1
Yorkshire and the Humber	4,902.3	4,918.4	4,982.8	5,042.9	0.3	1.3	1.2
East Midlands	3,651.9	3,852.8	4,035.4	4,169.3	5.5	4.7	3.3
West Midlands	5,146.0	5,186.6	5,265.5	5,332.5	0.8	1.5	1.3
East	4,454.3	4,854.1	5,149.8	5,377.0	9.0	6.1	4.4
London	7,529.4	6,805.6	6,889.9	7,187.2	-9.6	1.2	4.3
South East	6,829.7	7,245.4	7,678.9	8,003.8	6.1	6.0	4.2
South West	4,111.8	4,381.4	4,717.8	4,901.3	6.6	7.7	3.9
England	46,411.7	46,820.8	48,208.1	49,494.6	0.9	3.0	2.7
Wales	2,740.3	2,813.5	2,891.5	2,933.3	2.7	2.8	1.4
Scotland	5,235.6	5,180.2	5,107.0	5,120.0	-1.1	-1.4	0.3
Northern Ireland	1,540.4	1,537.7	1,607.3	1,688.6	-0.2	4.5	5.1

1 See Notes and Definitions.

Source: Office for National Statistics; General Register Office for Scotland; Northern Ireland Statistics and Research Agency

3.2 Resident population[1]: by age and gender, 1998

Thousands and percentages

	0-4	5-15	16-19	20-24	25-44	45-59	60-64	65-79	80 and over	All ages
Males (thousands)										
United Kingdom	1,882.1	4,328.1	1,508.9	1,803.2	9,015.2	5,387.3	1,379.7	3,100.1	723.8	29,128.4
North East	77.7	191.8	69.5	77.9	380.1	237.1	63.7	143.5	27.0	1,268.4
North West	215.9	523.0	176.8	202.0	1,029.8	635.2	164.7	358.7	78.0	3,384.1
Yorkshire and the Humber	158.7	373.6	130.1	157.1	763.4	460.1	119.3	266.8	60.3	2,489.5
East Midlands	128.4	304.2	105.4	123.9	622.9	397.4	99.4	229.6	51.8	2,063.0
West Midlands	171.9	400.8	137.5	156.6	794.7	499.5	129.2	287.0	61.7	2,638.9
East	170.8	383.7	130.9	151.9	819.4	505.4	127.8	293.1	71.3	2,654.2
London	259.8	502.9	177.1	267.8	1,245.3	574.8	142.0	302.4	75.9	3,547.9
South East	250.6	573.1	200.3	220.6	1,223.5	750.1	185.3	420.6	110.2	3,934.3
South West	143.0	342.3	120.3	136.2	702.5	459.7	119.2	295.0	79.4	2,397.7
England	1,576.7	3,595.5	1,248.0	1,494.1	7,581.6	4,519.4	1,150.5	2,596.7	615.6	24,378.0
Wales	89.1	217.9	76.2	85.4	411.4	275.2	73.5	171.2	38.7	1,438.6
Scotland	154.2	365.0	132.8	163.2	776.6	457.4	121.7	260.0	53.5	2,484.4
Northern Ireland	62.2	149.8	51.8	60.6	245.6	135.2	34.0	72.2	16.0	827.3
Females (thousands)										
United Kingdom	1,788.6	4,111.2	1,430.6	1,717.2	8,721.7	5,433.0	1,438.3	3,854.3	1,613.4	30,108.1
North East	74.0	182.2	65.8	72.5	375.3	238.7	67.5	179.9	65.3	1,321.2
North West	205.5	497.3	169.1	190.8	996.0	634.0	171.7	455.3	187.0	3,506.7
Yorkshire and the Humber	151.4	355.2	121.9	144.2	724.9	458.7	126.0	332.1	139.0	2,553.4
East Midlands	121.6	287.0	99.5	117.0	603.8	392.4	101.2	273.7	110.0	2,106.3
West Midlands	163.5	379.9	130.5	147.1	760.4	494.1	131.3	349.7	137.1	2,693.6
East	162.5	365.7	124.6	144.0	791.4	506.7	130.3	351.5	146.3	2,722.8
London	245.5	480.1	170.1	266.6	1,194.7	592.6	145.6	371.5	172.7	3,639.3
South East	237.9	542.2	187.8	216.8	1,175.7	755.8	190.8	523.1	239.5	4,069.5
South West	135.8	323.6	112.4	128.0	678.1	467.4	125.1	365.1	168.1	2,503.6
England	1,497.7	3,413.1	1,181.7	1,427.0	7,300.4	4,540.4	1,189.4	3,201.8	1,365.0	25,116.5
Wales	84.7	207.4	72.2	77.7	399.1	278.3	76.5	211.9	87.0	1,494.7
Scotland	146.7	348.4	127.2	156.5	775.9	475.0	135.4	345.0	125.6	2,635.6
Northern Ireland	59.4	142.4	49.4	56.0	246.3	139.4	37.0	95.6	35.8	861.3
All persons (percentages)										
United Kingdom	6.2	14.2	5.0	5.9	29.9	18.3	4.8	11.7	3.9	100.0
North East	5.9	14.4	5.2	5.8	29.2	18.4	5.1	12.5	3.6	100.0
North West	6.1	14.8	5.0	5.7	29.4	18.4	4.9	11.8	3.8	100.0
Yorkshire and the Humber	6.2	14.5	5.0	6.0	29.5	18.2	4.9	11.9	4.0	100.0
East Midlands	6.0	14.2	4.9	5.8	29.4	18.9	4.8	12.1	3.9	100.0
West Midlands	6.4	14.6	5.0	5.7	29.2	18.6	4.9	11.9	3.7	100.0
East	6.2	13.9	4.8	5.5	30.0	18.8	4.8	12.0	4.0	100.0
London	7.0	13.7	4.8	7.4	33.9	16.2	4.0	9.4	3.5	100.0
South East	6.1	13.9	4.8	5.5	30.0	18.8	4.7	11.8	4.4	100.0
South West	5.7	13.6	4.7	5.4	28.2	18.9	5.0	13.5	5.1	100.0
England	6.2	14.2	4.9	5.9	30.1	18.3	4.7	11.7	4.0	100.0
Wales	5.9	14.5	5.1	5.6	27.6	18.9	5.1	13.1	4.3	100.0
Scotland	5.9	13.9	5.1	6.2	30.3	18.2	5.0	11.8	3.5	100.0
Northern Ireland	7.2	17.3	6.0	6.9	29.1	16.3	4.2	9.9	3.1	100.0

1 See Notes and Definitions.

Source: Office for National Statistics; General Register Office for Scotland; Northern Ireland Statistics and Research Agency

3.3 Population density[1], 1998

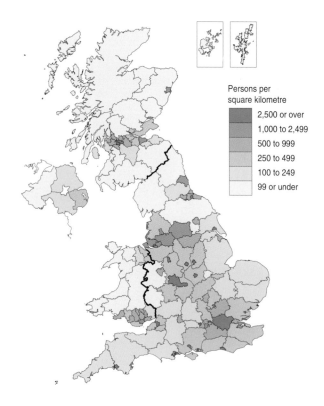

Persons per
square kilometre

- 2,500 or over
- 1,000 to 2,499
- 500 to 999
- 250 to 499
- 100 to 249
- 99 or under

1 See Notes and Definitions.

Source: Office for National Statistics; General Register Office for Scotland; Northern Ireland Statistics and Research Agency

3.4 Population under 16[1], 1998

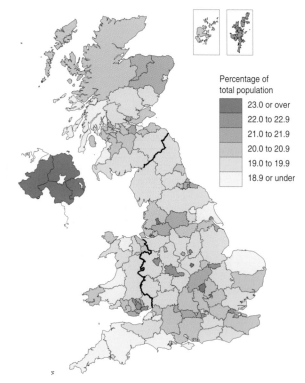

Percentage of
total population

- 23.0 or over
- 22.0 to 22.9
- 21.0 to 21.9
- 20.0 to 20.9
- 19.0 to 19.9
- 18.9 or under

1 See Notes and Definitions.

*Source: Office for National Statistics; General Register Office for
Scotland; Northern Ireland Statistics and Research Agency*

3.5 Population of retirement age[1], 1998

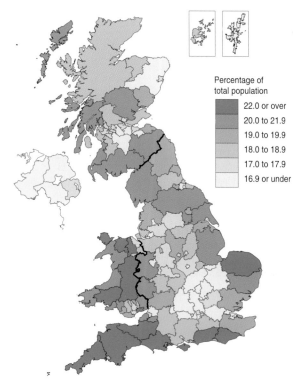

Percentage of
total population

- 22.0 or over
- 20.0 to 21.9
- 19.0 to 19.9
- 18.0 to 18.9
- 17.0 to 17.9
- 16.9 or under

1 Males aged 65 or over, females aged 60 or over. See Notes and Definitions.

*Source: Office for National Statistics; General Register Office for
Scotland; Northern Ireland Statistics and Research Agency*

3.6 Social class[1] of working-age[2] population, Spring 1999

Percentages and thousands

| | Social class | | | | | | | Total working-age population (=100%) (thousands) |
	Professional occupations (I)	Managerial and technical (II)	Skilled occupations non-manual (IIIN)	Skilled occupations manual (IIIM)	Partly skilled occupations (IV)	Unskilled occupations (V)	Other[3]	
United Kingdom	5.3	25.4	20.4	17.3	14.4	4.5	12.6	36,177
North East	3.0	21.2	19.0	18.2	16.2	5.9	16.5	1,572
North West	4.7	23.2	20.9	17.7	15.0	4.6	14.0	4,175
Yorkshire and the Humber	4.3	22.2	20.8	19.1	16.2	5.3	12.2	3,067
East Midlands	5.0	23.4	19.4	20.1	16.3	4.8	10.9	2,555
West Midlands	4.1	22.0	20.0	19.6	16.8	4.7	12.7	3,234
East	5.6	28.4	21.8	16.5	13.5	4.2	9.9	3,284
London	7.2	29.9	21.1	13.5	10.6	3.2	14.5	4,596
South East	7.2	30.8	21.5	14.6	12.7	3.8	9.4	4,877
South West	5.4	26.2	20.6	18.1	15.0	4.8	9.9	2,902
England	5.5	26.0	20.8	17.1	14.3	4.4	12.1	30,263
Wales	4.1	22.4	17.5	19.3	16.4	5.3	15.0	1,743
Scotland	5.1	23.4	19.4	17.9	14.7	5.6	13.9	3,160
Northern Ireland	2.8	21.3	19.1	18.3	14.7	3.6	20.1	1,011

1 Based on occupation. See Notes and Definitions.
2 Men aged 16-64 and women aged 16-59.
3 Includes members of the armed forces, those who did not state their social class, and those whose previous occupation was more than eight years ago, or who have never had a job.

Source: Labour Force Survey, Office for National Statistics and Department of Economic Development, Northern Ireland

3.7 Resident population[1]: by ethnic group[2], 1999/2000[3]

Percentages and thousands

| | Ethnic minority population | | | | | | | Ethnic minority population as a percent-age of total population |
| | Percentage in each group | | | | Total (=100%) (thousands) | White population (thousands) | Total population (thousands) | |
	Black	Indian	Pakistani/ Bangladeshi	Mixed/ other				
Great Britain	31	24	24	20	3,805	53,074	56,892	7
North East	..	..	48	..	43	2,509	2,552	2
North West	20	21	42	18	258	6,529	6,790	4
Yorkshire and the Humber	17	14	54	16	290	4,694	4,985	6
East Midlands	23	53	13	12	216	3,922	4,138	5
West Midlands	21	33	38	8	518	4,759	5,278	10
East	24	35	20	21	198	5,142	5,340	4
London	41	22	15	22	1,812	5,305	7,122	25
South East	23	25	19	33	263	7,651	7,914	3
South West	36	..	..	36	75	4,757	4,833	2
England	31	25	24	20	3,672	45,268	48,952	8
Wales	27	..	..	51	53	2,846	2,900	2
Scotland	14	..	44	30	80	4,960	5,040	2

1 Population in private households, students in halls of residence and those in NHS accommodation. See Notes and Definitions.
2 For some ethnic origins in some regions, sample sizes are too small to provide a reliable estimate.
3 Four quarter average Spring 1999 to Winter 1999/00.

Source: Labour Force Survey, Office for National Statistics

3.8 Population change, mid 1981-1998[1]

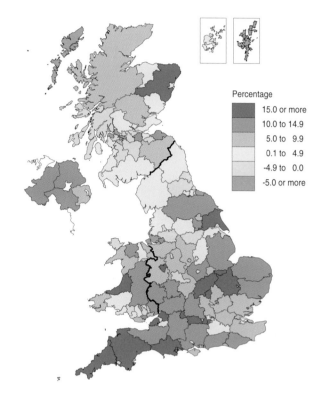

Percentage

- 15.0 or more
- 10.0 to 14.9
- 5.0 to 9.9
- 0.1 to 4.9
- -4.9 to 0.0
- -5.0 or more

1 See Notes and Definitions.

Source: Office for National Statistics; General Register Office for Scotland; Northern Ireland Statistics and Research Agency

3.9 Projected population change[1], 1998-2011

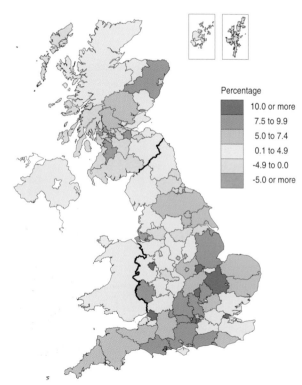

Percentage

- 10.0 or more
- 7.5 to 9.9
- 5.0 to 7.4
- 0.1 to 4.9
- -4.9 to 0.0
- -5.0 or more

1 1996-based sub-national projections for England and Scotland; 1998-based national projections for Wales and Northern Ireland. See Notes and Definitions.

Source: Office for National Statistics; National Assembly for Wales; General Register Office for Scotland; Northern Ireland Statistics and Research Agency

3.10 Live births, deaths and natural change in population

Thousands and rates

	Thousands				Rates per 1,000 population			
	1981	1986	1991	1998	1981	1986	1991	1998
Live births[1]								
United Kingdom	730.8	755.0	792.5	717.1	13.0	13.3	13.7	12.1
North East	34.2	34.7	34.9	28.7	13.0	13.3	13.4	11.1
North West	90.4	93.4	97.5	81.2	13.0	13.6	14.2	11.8
Yorkshire and the Humber	62.6	65.3	68.6	59.7	12.7	13.3	13.8	11.8
East Midlands	49.2	50.3	54.0	48.3	12.8	12.8	13.4	11.6
West Midlands	67.5	70.4	74.2	65.0	13.0	13.5	14.1	12.2
East	62.6	64.4	68.4	64.4	12.9	12.8	13.3	12.0
London	92.4	97.7	105.8	105.3	13.6	14.4	15.4	14.7
South East	89.0	92.9	99.8	95.6	12.3	12.4	13.0	11.9
South West	50.4	54.5	57.6	53.8	11.5	12.0	12.2	11.0
England	598.2	623.6	660.8	602.1	12.8	13.2	13.7	12.2
Wales	35.8	37.0	38.1	33.4	12.7	13.1	13.2	11.4
Scotland	69.1	65.8	67.0	57.3	13.3	12.8	13.1	11.2
Northern Ireland	27.3	28.2	26.3	23.9	17.8	18.0	16.3	14.0
Deaths[2]								
United Kingdom	658.0	660.7	646.2	627.6	11.7	11.6	11.2	10.6
North East	32.1	32.0	31.8	30.6	12.2	12.3	12.2	11.8
North West	86.6	85.5	82.7	78.8	12.5	12.5	12.0	11.4
Yorkshire and the Humber	59.1	58.9	57.3	55.5	12.0	12.0	11.5	11.0
East Midlands	42.8	43.5	43.9	44.1	11.1	11.1	10.9	10.6
West Midlands	56.4	57.7	57.0	55.6	10.9	11.1	10.8	10.4
East	50.7	52.4	53.3	54.0	10.4	10.5	10.3	10.0
London	77.6	73.9	68.9	62.3	11.4	10.9	10.0	8.7
South East	81.3	84.2	83.0	81.8	11.2	11.2	10.8	10.2
South West	54.4	56.4	56.2	55.3	12.4	12.4	11.9	11.3
England	541.0	544.5	534.0	518.1	11.6	11.5	11.1	10.5
Wales	35.0	34.7	34.1	33.9	12.4	12.3	11.8	11.6
Scotland	63.8	63.5	61.0	59.2	12.3	12.4	12.0	11.6
Northern Ireland	16.3	16.1	15.1	15.0	10.6	10.3	9.4	8.9
Natural change								
United Kingdom	72.8	94.2	146.3	89.5	1.3	1.7	2.5	1.5
North East	2.1	2.7	3.1	-1.8	0.8	1.0	1.2	-0.7
North West	3.8	7.9	14.8	2.3	0.5	1.2	2.2	0.4
Yorkshire and the Humber	3.5	6.4	11.3	4.2	0.7	1.3	2.3	0.8
East Midlands	6.4	6.8	10.1	4.2	1.7	1.7	2.5	1.0
West Midlands	11.1	12.7	17.2	9.4	2.1	2.4	3.3	1.8
East	11.9	12.0	15.1	10.4	2.5	2.4	3.0	2.0
London	14.8	23.8	36.9	43.0	2.2	3.5	5.4	6.0
South East	7.7	8.8	16.8	13.8	1.1	1.2	2.2	1.7
South West	-4.0	-1.9	1.4	-1.5	-0.9	-0.4	0.3	-0.3
England	57.2	79.1	126.8	84.0	1.2	1.7	2.6	1.7
Wales	0.8	2.3	4.0	-0.5	0.3	0.8	1.4	-0.2
Scotland	5.3	2.3	6.0	-1.8	1.0	0.5	1.1	-0.4
Northern Ireland	10.9	11.9	10.9	8.7	6.5	7.2	6.6	5.2

1 Based on the usual area of residence of the mother. See Notes and Definitions for details of the inclusion or exclusion of births to non-resident mothers in the individual countries and regions of England. The United Kingdom figures have been calculated on all births registered in the United Kingdom, ie: including births to mothers usually resident outside the United Kingdom. Data relate to year of occurrence in England and Wales, and year of registration in Scotland and Northern Ireland.

2 Based on the usual area of residence of the deceased. See Notes and Definitions for details of the inclusion or exclusion of deaths of non-resident persons in the individual countries and regions of England. The figures for the United Kingdom have been calculated on all deaths registered in the United Kingdom in 1998, ie: including deaths of persons usually resident outside the United Kingdom.

Source: Office for National Statistics; General Register Office for Scotland; Northern Ireland Statistics and Research Agency

3.11 Components of population change, mid-1997 to mid-1998[1]

Thousands

	Resident population mid-1997	Births	Deaths	Net natural change	Net migration and other changes	Total change	Resident population mid-1998
United Kingdom	59,014.0	718.3	618.0	100.3	122.3	222.6	59,236.5
North East	2,594.4	28.9	29.9	-1.0	-3.7	-4.8	2,589.6
North West	6,884.6	81.4	78.0	3.4	2.7	6.2	6,890.8
Yorkshire and the Humber	5,037.0	60.1	54.9	5.1	0.8	5.9	5,042.9
East Midlands	4,156.3	48.1	43.2	4.9	8.1	13.0	4,169.3
West Midlands	5,320.8	65.7	55.0	10.7	1.1	11.7	5,332.5
East	5,334.2	64.2	52.4	11.9	30.9	42.8	5,377.0
London	7,122.2	105.3	61.5	43.8	21.4	65.1	7,187.2
South East	7,958.8	95.2	81.1	14.1	31.0	45.0	8,003.8
South West	4,876.0	53.8	54.9	-1.1	26.4	25.3	4,901.3
England	49,284.2	602.6	510.9	91.8	118.6	210.3	49,494.6
Wales	2,926.9	33.9	33.6	0.3	6.1	6.4	2,933.3
Scotland	5,122.5	58.0	58.5	-0.5	-2.0	-2.5	5,120.0
Northern Ireland	1,680.3	23.8	15.1	8.7	-0.4	8.3	1,688.6

1 See Notes and Definitions.

Source: Office for National Statistics; General Register Office for Scotland; Northern Ireland Statistics and Research Agency

3.12 Conceptions[1] to women aged under 18[2]: by outcome

Percentages, thousands and rates

	1992				1998			
	Percentage of conceptions				Percentage of conceptions			
	Leading to maternities	Leading to abortions	Total number (thousands)	Rate per 1,000 popoulation[3]	Leading to maternities	Leading to abortions	Total number (thousands)	Rate per 1,000 population[3]
England and Wales	60.9	39.1	37,552	43.6	58.0	42.0	44,061	47.0
North East	68.3	31.7	2,423	54.1	64.8	35.2	2,727	56.3
North West	65.5	34.5	6,066	50.9	61.6	38.4	6,464	50.1
Yorkshire and the Humber	65.6	34.4	4,316	51.2	63.5	36.5	4,807	53.0
East Midlands	62.8	37.2	3,098	44.5	60.9	39.1	3,628	37.7
West Midlands	60.9	39.1	4,525	49.5	58.9	41.1	5,073	51.4
East	54.0	46.0	3,046	34.0	56.4	43.6	3,582	37.7
London	53.7	46.3	4,621	45.0	45.8	54.2	6,023	51.0
South East	55.7	44.3	4,338	33.0	55.4	44.6	5,374	37.7
South West	57.8	42.2	2,732	34.5	55.2	44.8	3,354	39.3
England	60.5	39.5	35,165	43.3	57.5	42.5	41,032	46.5
Wales	66.4	33.6	2,387	48.0	64.1	35.9	3,029	54.8

1 Conception statistics are derived from numbers of registered births and registered abortions. They do not include spontaneous miscarriages and illegal abortions.
2 Based on place of usual residence. Information about usual residence of women undergoing abortions is known to be not wholly accurate. Some women living outside London and other big citiews may have given a temporary address in the city as their usual place of residence.
3 The rates for girls aged under 18 are based on the population of girls aged 15 to 17.

Source: Office for National Statistics

3.13 Age specific birth rates[1]

Rates

	Under 20	20-24	25-29	30-34	35-39	40 and over	All ages	TFR[3]
	Live births per 1,000 women in age groups[2]							
1981								
United Kingdom	28	107	130	70	22	5	62	1.81
North East	34	114	128	60	18	4	62	1.79
North West	35	114	130	65	21	5	63	1.85
Yorkshire and the Humber	31	117	128	59	18	6	62	1.80
East Midlands	30	113	127	63	19	4	61	1.79
West Midlands	32	108	133	69	20	7	62	1.84
East	22	110	138	70	20	4	61	1.82
London	29	83	114	80	31	6	62	1.71
South East	20	97	138	73	23	4	59	1.77
South West	24	103	131	63	18	3	57	1.71
England	28	104	129	69	22	5	61	1.78
Wales	30	121	127	67	21	6	63	1.86
Scotland	31	112	131	66	21	4	63	1.84
Northern Ireland	27	135	172	117	52	13	86	2.59
1991								
United Kingdom	33	89	120	87	32	5	64	1.82
North East	44	102	119	72	23	4	63	1.82
North West	42	101	124	84	29	5	67	1.93
Yorkshire and the Humber	41	99	122	78	26	4	64	1.85
East Midlands	34	95	126	81	26	4	63	1.83
West Midlands	39	102	126	84	31	5	67	1.93
East	24	86	129	91	31	5	62	1.83
London	29	69	97	96	47	10	64	1.74
South East	23	78	122	95	35	5	61	1.80
South West	25	84	125	86	30	5	60	1.77
England	33	89	119	87	32	5	64	1.81
Wales	39	103	127	77	27	5	64	1.88
Scotland	33	82	117	78	27	4	60	1.69
Northern Ireland	29	97	146	105	46	10	75	2.16
1998								
United Kingdom	31	74	102	90	40	8	59	1.71
North East	42	84	99	73	28	5	54	1.66
North West	35	85	101	83	34	7	58	1.73
Yorkshire and the Humber	37	86	110	79	31	6	58	1.75
East Midlands	32	76	104	85	34	7	57	1.69
West Midlands	35	89	107	85	37	7	61	1.80
East	25	72	98	96	42	8	59	1.70
London	26	61	102	99	53	13	63	1.77
South East	24	65	96	100	46	9	59	1.70
South West	25	70	104	91	39	8	57	1.68
England	30	75	102	90	40	8	59	1.72
Wales	39	90	109	82	33	6	59	1.79
Scotland	30	63	94	83	34	6	53	1.55
Northern Ireland	28	70	118	108	47	9	65	1.89

1 Based on the usual area of residence of the mother. See Notes and Definitions for details of the inclusion or exclusion of births to non-resident mothers in the individual countries and regions of England. The United Kingdom figures have been calculated on all births registered in the United Kingdom, ie: including births to mothers usually resident outside the United Kingdom. Data relate to year of occurrence in England and Wales, and year of registration in Scotland and Northern Ireland.

2 The rates for women aged under 20, 40 and over and all ages are based upon the population of women aged 15-19, 40-44 and 15-44 respectively. See Notes and Definitions.

3 The Total Fertility Rate (TFR) is the average number of children which would be born to a women if the current pattern of fertility persisted throughout her child-bearing years. Previously known as Total Period Fertility . See Notes and Definitions.

Source: Office for National Statistics; General Register Office for Scotland; Northern Ireland Statistics and Research Agency

3.14 Age specific death rates: by gender, 1998[1]

Rates and Standardised Mortality Ratios

	Deaths per 1,000 population for specific age groups											SMR[2] (UK = 100)
	Under 1[3]	1-4	5-15	16-24	25-34	35-44	45-54	55-64	65-74	75-84	85 and over	
Males												
United Kingdom	6.3	0.3	0.2	0.8	1.0	1.6	4.1	11.6	33.0	81.7	187.3	100
North East	6.5	0.4	0.2	0.8	0.9	1.8	4.6	14.2	40.0	91.7	202.5	115
North West	6.7	0.3	0.2	0.8	1.2	1.8	4.6	13.1	36.6	87.9	193.6	109
Yorkshire and the Humber	7.2	0.4	0.2	0.9	1.1	1.5	4.2	11.8	34.3	86.3	188.3	104
East Midlands	5.9	0.4	0.2	0.8	0.9	1.4	3.7	11.1	32.6	81.0	196.9	99
West Midlands	7.1	0.3	0.2	0.8	0.9	1.5	4.0	11.7	33.4	84.8	184.7	101
East	6.2	0.2	0.2	0.7	0.8	1.3	3.1	9.3	27.6	78.7	188.4	91
London	6.4	0.3	0.2	0.6	1.1	1.7	4.5	11.9	32.3	78.3	171.4	97
South East	5.0	0.3	0.1	0.8	0.8	1.3	3.3	9.5	28.5	74.8	186.3	90
South West	5.6	0.2	0.1	0.7	0.8	1.5	3.5	9.6	28.1	73.0	181.8	89
England	6.2	0.3	0.2	0.8	1.0	1.5	3.9	11.2	32.1	80.8	186.5	98
Wales	6.1	0.2	0.2	1.0	1.1	1.5	4.0	11.7	34.6	80.8	181.8	100
Scotland	6.2	0.3	0.2	1.1	1.3	2.2	5.5	15.1	39.4	89.5	198.8	117
Northern Ireland	6.1	0.3	0.3	0.8	1.1	1.4	4.5	12.2	34.6	87.5	190.4	105
Females												
United Kingdom	5.0	0.3	0.1	0.3	0.4	1.0	2.7	7.0	20.2	54.2	152.6	100
North East	3.5	0.3	0.1	0.4	0.5	1.2	3.0	8.4	24.0	62.9	163.7	114
North West	5.6	0.3	0.1	0.3	0.6	1.1	3.0	8.2	23.2	59.9	159.1	109
Yorkshire and the Humber	6.5	0.3	0.1	0.3	0.4	1.1	2.5	7.1	21.3	57.4	152.7	103
East Midlands	5.1	0.2	0.1	0.3	0.4	1.0	2.5	6.6	20.1	54.2	154.3	100
West Midlands	5.8	0.2	0.1	0.3	0.5	1.0	2.7	6.7	19.9	54.1	154.5	100
East	3.8	0.2	0.1	0.3	0.4	0.9	2.6	5.9	17.5	50.8	152.3	94
London	5.4	0.3	0.1	0.2	0.5	1.0	2.6	6.8	20.0	51.6	133.3	93
South East	3.8	0.2	0.1	0.3	0.3	0.8	2.4	5.9	16.8	49.9	151.2	93
South West	4.1	0.2	0.1	0.3	0.4	0.9	2.4	5.5	16.4	47.0	146.1	89
England	4.9	0.2	0.1	0.3	0.4	1.0	2.6	6.7	19.6	53.5	150.7	98
Wales	5.2	0.2	0.1	0.3	0.4	1.0	2.7	7.5	21.3	54.6	149.8	101
Scotland	4.9	0.2	0.2	0.4	0.5	1.3	3.3	8.7	24.3	60.3	174.7	116
Northern Ireland	5.1	0.3	0.2	0.4	0.4	1.1	2.6	6.8	19.8	54.6	156.1	101
All persons												
United Kingdom	5.7	0.3	0.2	0.6	0.7	1.3	3.4	9.3	26.1	64.8	161.8	100
North East	5.0	0.3	0.2	0.6	0.7	1.5	3.8	11.2	31.4	73.9	173.2	114
North West	6.1	0.3	0.1	0.6	0.9	1.5	3.8	10.6	29.4	70.4	167.7	109
Yorkshire and the Humber	6.9	0.3	0.2	0.6	0.7	1.3	3.3	9.4	27.3	68.5	161.9	103
East Midlands	5.5	0.3	0.1	0.6	0.7	1.2	3.1	8.8	26.0	65.0	165.8	100
West Midlands	6.4	0.2	0.1	0.5	0.7	1.3	3.4	9.2	26.2	65.9	162.4	101
East	5.0	0.2	0.1	0.5	0.6	1.1	2.9	7.6	22.3	61.9	162.5	93
London	5.9	0.3	0.1	0.4	0.8	1.3	3.5	9.3	25.7	61.9	143.4	95
South East	4.4	0.2	0.1	0.5	0.6	1.1	2.8	7.7	22.2	59.6	160.7	91
South West	4.8	0.2	0.1	0.5	0.6	1.2	2.9	7.5	21.8	57.2	155.9	89
England	5.6	0.3	0.1	0.5	0.7	1.3	3.3	8.9	25.4	64.1	160.2	98
Wales	5.7	0.2	0.1	0.7	0.8	1.3	3.4	9.6	27.5	64.7	158.2	101
Scotland	5.6	0.3	0.2	0.8	0.9	1.7	4.4	11.8	31.0	71.0	180.9	116
Northern Ireland	5.6	0.3	0.2	0.6	0.7	1.3	3.6	9.4	26.4	67.0	165.0	101

1 Based on the usual area of residence of the deceased. See Notes and Definitions for details of the inclusion or exclusion of deaths of non-resident persons in the individual countries and regions of England. The UK figures have been calculated on all deaths registered in the UK in 1998, ie: including deaths of persons usually resident outside the UK.
2 Standardised Mortality Ratio (SMR) is the ratio of observed deaths to those expected by applying a standard death rate to the regional population. See Notes and Definitions.
3 Deaths of infants under 1 year of age per 1,000 live births.

Source: Office for National Statistics; General Register Office for Scotland; Northern Ireland Statistics and Research Agency

3.15 Migration

Thousands

	Inflow					Outflow				
	1981	1986	1991	1997	1998	1981	1986	1991	1997	1998
Inter-regional migration[1]										
North East	31	36	40	39	39	39	46	41	45	44
North West	102	112	114	107	104	122	138	123	117	116
Yorkshire and the Humber	68	79	85	93	93	73	91	85	100	98
East Midlands	77	102	90	108	108	72	85	81	97	97
West Midlands	67	87	83	93	93	79	95	88	104	101
East	121	145	122	145	143	104	128	113	125	124
London	155	183	149	167	171	187	232	202	222	218
South East	202	243	198	230	226	166	204	185	206	207
South West	108	149	121	144	139	88	103	99	112	111
England	94	116	96	111	111	93	101	112	115	111
Wales	45	55	52	59	56	42	50	47	54	54
Scotland	47	44	56	55	53	48	58	47	53	54
Northern Ireland	7	9	13	10	12	10	15	9	13	12
International migration[2,3]										
United Kingdom	153	250	267	285	332	233	213	239	225	199
North East	4	9	7	12	6	14	7	4	11	4
North West	15	26	14	21	25	24	17	19	9	15
Yorkshire and the Humber	9	13	20	14	14	14	14	14	12	10
East Midlands	5	9	12	15	13	10	5	7	12	7
West Midlands	11	11	14	16	16	13	8	18	10	9
East	9	21	26	19	27	18	20	22	18	17
London	49	78	79	101	126	55	51	67	68	66
South East	26	38	45	39	58	35	43	37	37	29
South West	10	18	18	21	19	13	16	19	16	15
England	138	223	233	257	304	196	182	207	192	172
Wales	3	8	8	11	7	11	6	6	4	4
Scotland	10	16	22	15	20	21	21	23	25	19
Northern Ireland	2	2	3	2	1	4	4	2	3	4

1 Based on patients re-registering with NHS doctors in other parts of the United Kingdom. See Notes and Definitions.
2 Subject to relatively large sampling errors where estimates are based on small numbers of contacts. See Notes and Definitions.
3 Figures for all years exclude migration to and from the Irish Republic. Data for the South East prior to 1988 include migration via the UK mainland between the Channel Islands and the Isle of Man and the rest of the world. Adjustment of the figures shown are required for 'visitor switchers' and migration to and from the Irish Republic. See Notes and Definitions.

Source: National Health Service Central Register and International Passenger Survey, Office for National Statistics; General Register Office for Scotland; Northern Ireland Statistics and Research Agency

3.16 Inter-regional movements[1], 1998

Thousands

	United Kingdom	North East	North West	York-shire and the Humber	East Mid-lands	West Mid-lands	East	London	South East	South West	Wales	Scot-land	Nor-thern Ireland
						Region of origin							
Region of destination													
United Kingdom	.	44	116	98	97	101	124	218	207	111	54	54	12
North East	39	.	6	8	3	2	3	4	4	2	1	4	1
North West	104	6	.	18	10	13	8	10	12	8	9	8	2
Yorkshire and the Humber	93	10	18	.	16	8	8	8	11	6	3	5	1
East Midlands	108	4	11	18	.	15	16	10	18	8	3	4	1
West Midlands	93	3	13	8	13	.	8	10	14	13	8	3	1
East	143	3	8	8	14	8	.	55	28	10	3	5	1
London	171	5	14	11	10	11	32	.	57	17	6	8	2
South East	226	5	14	11	15	14	29	85	.	34	8	8	1
South West	139	3	10	7	9	16	13	20	45	.	10	4	1
Wales	56	1	11	3	3	9	3	5	9	9	.	2	-
Scotland	53	4	8	5	4	3	5	7	8	4	2	.	2
Northern Ireland	12	-	1	1	1	1	1	2	1	1	-	2	.

1 Based on patients re-registering with NHS doctors in other parts of the United Kingdom. See Notes and Definitions.

Source: Office for National Statistics; General Register Office for Scotland; Northern Ireland Statistics and Research Agency

3.17 Marriages[1,2]

Thousands

	1976	1986	1998
United Kingdom	406.0	393.9	304.8
North East	20.1	17.6	11.7
North West	50.3	46.3	32.2
Yorkshire and the Humber	36.3	35.2	24.1
East Midlands	26.7	27.4	20.9
West Midlands	36.6	35.2	26.0
East	32.2	34.7	27.8
London	58.4	47.5	39.5
South East	48.5	52.0	43.6
South West	30.1	32.5	27.4
England	339.0	328.4	253.1
Wales	19.5	19.5	14.2
Scotland	37.5	35.8	29.7
Northern Ireland	9.9	10.2	7.8

1 Marriages solemnised outside the United Kingdom are not included.
2 Region of occurrence of marriage.

Source: Office for National Statistics; General Register Office for Scotland; Northern Ireland Statistics and Research Agency

3.18 Cohabitation amongst non-married people aged 16-59, 1996-1999[1]

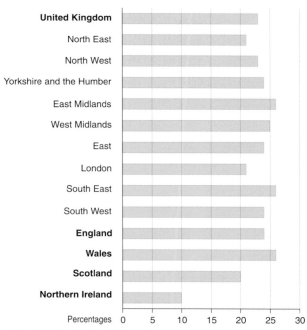

1 Combined data from the 1996-97 and 1998-99 surveys.

Source: General Household Survey, Office for National Statistics; Continuous Household Survey, Northern Ireland Statistics and Research Agency

3.19 Household numbers and projections[1]

Millions

	Household numbers			Household projections				
	1981	1991	1998	2001[1]	2006[1]	2011[1]	2016[1]	2021[1]
Great Britain	20.17	22.39	23.90	24.41	25.26	..	..	..
North East	0.98	1.05	1.09	1.10	1.12	1.14	1.15	1.17
North West	2.55	2.72	2.84	2.87	2.93	3.00	3.06	3.11
Yorkshire and the Humber	1.83	1.99	2.10	2.14	2.20	2.26	2.32	2.37
East Midlands	1.41	1.60	1.72	1.76	1.83	1.90	1.97	2.03
West Midlands	1.86	2.04	2.16	2.19	2.24	2.30	2.35	2.40
East	1.76	2.03	2.22	2.28	2.39	2.49	2.60	2.70
London	2.63	2.84	3.06	3.13	3.25	3.38	3.52	3.64
South East	2.64	3.03	3.30	3.40	3.57	3.74	3.91	4.06
South West	1.64	1.90	2.05	2.11	2.21	2.32	2.42	2.52
England	17.31	19.21	20.54	20.99	21.73	22.52	23.31	24.00
Wales	1.02	1.13	1.19	1.20	1.24	1.27	1.31	1.34
Scotland	1.85	2.05	2.17	2.22	2.29	..	..	..

1 1996-based projections. See Notes and Definitions.

Source: Department of the Environment, Transport and the Regions; National Assembly for Wales; Scottish Executive

3.20 Households: by type, Spring 1999

Percentages and thousands

	Types of households (percentages)								Total households (=100%) (thousands)
			Married couple			Lone parent			
	One person	Two or more un-related adults	With dependent children	With non-dependent children only	With no children	With dependent children	With non-dependent children only	Two or more families[1]	
United Kingdom	28.2	3.2	23.7	6.7	27.8	6.6	3.0	0.9	24,309
North East	28.7	2.4	24.3	7.0	25.3	8.1	3.3	..	1,063
North West	28.4	2.6	23.6	7.1	26.2	8.3	3.2	0.6	2,863
Yorkshire and the Humber	28.2	2.8	23.6	6.5	28.9	6.4	3.1	0.6	2,102
East Midlands	25.5	2.5	25.4	7.5	30.4	5.2	2.9	0.6	1,713
West Midlands	26.9	2.7	24.3	6.8	28.6	6.4	3.2	1.2	2,164
East	26.9	2.2	25.1	6.8	30.9	5.0	2.4	0.6	2,200
London	31.1	5.6	21.3	5.4	22.3	9.1	3.4	1.8	2,890
South East	27.0	3.0	25.0	6.8	30.3	4.9	2.3	0.8	3,238
South West	27.6	3.2	23.0	5.4	32.4	5.3	2.5	0.7	2,060
England	27.9	3.1	23.8	6.6	28.3	6.5	2.9	0.9	20,294
Wales	27.6	2.4	23.6	6.9	27.8	7.5	3.2	0.9	1,205
Scotland	32.2	4.0	20.8	6.9	25.5	6.7	3.4	0.5	2,210
Northern Ireland	25.8	2.7	30.2	8.9	19.3	7.4	4.8	..	600

1 For some regions, sample sizes are too small to provide a reliable estimate.

Source: Labour Force Survey, Office for National Statistics; Department of Economic Development, Northern Ireland

4 Education and Training

Teachers

In 1999/00 there were around 27 thousand secondary school teachers in the North West and South East, more than any other region in England.

(Table 4.1)

Class sizes

Around 40 per cent of primary school key stage 2 classes in the North West, East Midlands and the South West contained more than 30 pupils in 1999/00, compared with almost 12 per cent in Northern Ireland.

(Table 4.2)

Under fives

In 1999/00, 86 per cent of children under five years old in the North East region were enrolled in education in schools compared with 47 per cent in the South East and 49 per cent in the South West, the highest and lowest proportions in the United Kingdom.

(Table 4.3)

Examination results

In 1998/99 around two-fifths of pupils in the North East and Yorkshire and the Humber achieved 5 or more GCSE grades A* to C. In Scotland almost three-fifths of pupils achieved the equivalent at SCE Standard Grade.

(Table 4.4)

The proportion of pupils in their last year of compulsory education in 1998/99 who achieved no graded GCSE's or SCE Standard Grades was lowest in Northern Ireland at 3.5 per cent.

(Table 4.4)

In 1998/99 over a third of pupils in the South East and Northern Ireland achieved 2 or more A levels compared with around a fifth in the North East, the highest and lowest proportions in the United Kingdom.

(Table 4.4)

Key stages

68 per cent of pupils in Summer 1999 in the South East reached or exceeded the expected standards in Key Stage 3 English, compared with 58 per cent in the North East.

(Table 4.6)

Post-compulsory education

Sixteen year olds in Wales and Scotland were more likely than those in England to be in full-time education or on a government-supported training scheme in 1997/98.

(Table 4.7)

Further education

Around a fifth of students from the North East were studying courses in England leading to Level 1 NVQ/GNVQ qualifications, compared with around 13 per cent of students from the South East and 15 per cent of students from London studying at the same level.

(Table 4.8)

Expenditure

In 1998-99 over £27 billion was spent by Local Education Authorities in the United Kingdom.

(Table 4.9)

In 1998-99 £1,880 of Local Education Authority expenditure in the United Kingdom was spent per nursery and primary school pupil, £210 more than 1994-95.

(Table 4.10)

Higher education

Higher education students whose home is in the East of England are the least likely to study within their own region, while those in Scotland are the most likely.

(Table 4.11)

Graduates First degree graduates who studied in Northern Ireland were more likely to continue in education or training than graduates from any other part of the United Kingdom in 1999.

(Table 4.12)

Qualifications More than one in five people of working age in London are qualified to degree level or equivalent, compared with around one in seven in the United Kingdom overall.

(Table 4.13)

National Learning Targets In Spring 2000, 35 per cent of adults living in London had an NVQ level 4 qualification (degree level), exceeding the National Learning Target by 7 percentage points.

(Table 4.14)

Training Among female employees of working age, those in Yorkshire and the Humber were the most likely to have received job-related training in Spring 2000.

(Table 4.15)

Two in five employers in the East Midlands and the South West regions in 1999 provided off-the-job training in the last 12 months, higher than any other region in England.

(Table 4.16)

Employers in the Yorkshire and Humber region were the most likely to have been involved in training related to national vocational qualifications while employers in London and the South East were the least likely to have been involved in this element of training in the last 12 months.

(Table 4.16)

In 1998-99, 56 per cent of those who had left Work-Based Learning for Adults in Northern Ireland were in employment six months later while 34 per cent in the North East were in employment, the highest and lowest proportions.

(Table 4.17)

In the East of England, South East and South West regions around three quarters of leavers from Work-Based Training for Young People were in employment six months later, higher than in any other region of England, Wales and Northern Ireland.

(Table 4.17)

4.1 Pupils and teachers: by type of school[1], 1999/00

Thousands and numbers

	Public sector schools				Non-maintained schools	All special schools	All schools
	Nursery schools	Primary schools[2]	Secondary schools	Pupil Referral Units			
Pupils[3] (thousands)							
United Kingdom	76.5	5,176.7	3,854.2	..	603.8	113.2	9,833.0
North East	2.4	225.4	181.3	0.6	15.9	6.1	431.7
North West	5.1	644.9	456.0	0.8	56.1	15.4	1,178.4
Yorkshire and the Humber	2.2	456.2	340.4	1.0	32.5	8.4	840.8
East Midlands	1.7	366.0	283.9	0.4	34.3	5.9	692.3
West Midlands	3.8	488.1	364.1	1.1	44.1	12.7	913.9
East	2.4	445.2	365.1	0.6	60.0	9.0	882.3
London	5.4	612.0	399.1	2.0	121.2	12.5	1,152.3
South East	2.6	643.9	483.9	1.5	139.3	17.3	1,288.6
South West	1.4	396.0	308.0	0.6	59.7	8.1	773.8
England	27.2	4,277.8	3,181.8	8.5	563.3	95.4	8,153.9
Wales	1.8	280.6	204.2	0.3	9.6	3.8	500.2
Scotland	42.9	437.0	313.2	.	29.6	9.3	832.1
Northern Ireland	4.6	181.3	155.0	.	1.2	4.7	346.8
Teachers[3] (thousands)							
United Kingdom	3.2	227.8	232.4	..	61.2	18.2	544.8
North East	0.1	9.7	10.4	0.1	1.4	0.9	22.6
North West	0.3	27.6	27.1	0.2	5.1	2.4	62.7
Yorkshire and the Humber	0.1	19.2	19.5	0.2	3.0	1.3	43.3
East Midlands	0.1	15.2	16.3	0.1	3.5	1.0	36.1
West Midlands	0.2	20.8	21.3	0.2	4.5	1.8	48.8
East	0.1	19.3	21.2	0.2	6.3	1.3	48.4
London	0.3	27.4	24.3	0.5	11.4	2.1	66.1
South East	0.2	27.7	27.9	0.3	15.6	2.6	74.3
South West	0.1	16.7	17.5	0.1	6.5	1.2	42.1
England	1.5	183.7	185.4	2.0	57.2	14.7	444.6
Wales	0.1	12.6	12.4	..	1.0	0.6	26.6
Scotland	1.4	22.5	24.1	.	2.9	2.2	53.1
Northern Ireland	0.2	9.0	10.5	.	0.1	0.8	20.6
Pupils per teacher[3] (numbers)							
United Kingdom	24.2	22.7	16.6	..	9.9	6.2	18.1
North East	20.0	23.1	17.3	5.7	11.7	7.0	19.1
North West	18.7	23.4	16.8	4.6	11.0	6.3	18.8
Yorkshire and the Humber	18.0	23.7	17.5	4.8	10.9	6.4	19.4
East Midlands	17.1	24.0	17.5	4.3	9.8	6.2	19.1
West Midlands	21.9	23.5	17.1	4.3	9.9	6.9	18.7
East	17.8	23.1	17.2	2.7	9.6	6.8	18.2
London	16.3	22.3	16.5	4.2	10.6	5.8	17.4
South East	16.3	23.2	17.3	4.4	8.9	6.6	17.3
South West	18.1	23.7	17.7	4.3	9.2	6.6	18.4
England	18.1	23.3	17.2	4.3	9.9	6.5	18.3
Wales[4]	18.5	22.3	16.5	..	9.8	6.8	18.8
Scotland	31.0	19.4	13.0	.	10.3	4.2	15.7
Northern Ireland	25.7	20.2	14.7	.	8.8	6.2	16.8

1 See Notes and Definitions. Data for Wales and Scotland refer to 1998/99.
2 For Northern Ireland, figures include pupils and teachers in the preparatory departments of grammar schools.
3 Full-time equivalents.
4 Data for all schools excludes Pupil Referral Units as information on teachers is not collected for Wales.

Source: Department for Education and Employment; National Assembly for Wales; Scottish Executive; Northern Ireland Department of Education

4.2 Class sizes for all classes[1], 1999/00[2]

Numbers and percentages

| | Primary schools | | | | | | Secondary schools | |
| | Key Stage 1[3] | | Key Stage 2[3] | | All Primary schools | | | |
	Average number in class	Percentage of classes of more than 30 pupils	Average number in class	Percentage of classes of more than 30 pupils	Average number in class	Percentage of classes of more than 30 pupils	Average number in class	Percentage of classes of more than 30 pupils
Great Britain	25.7	10.0	27.9	30.8	26.9	21.5	21.8	..
North East	25.2	7.5	27.4	27.0	26.3	18.3	22.2	7.4
North West	25.6	10.3	28.7	39.2	27.3	26.0	22.1	8.9
Yorkshire and the Humber	25.8	9.5	28.5	35.6	27.4	24.5	22.4	8.6
East Midlands	25.5	9.1	28.8	38.7	27.5	25.9	22.3	7.7
West Midlands	25.8	10.0	28.1	31.1	27.1	21.9	22.1	9.0
East	25.8	10.1	28.2	32.5	26.9	21.4	21.9	7.1
London	27.0	9.4	27.9	19.5	27.5	15.2	22.2	6.2
South East	26.3	11.4	28.5	34.8	27.4	24.0	22.0	7.1
South West	25.9	10.2	28.9	40.6	27.4	25.9	22.3	10.5
England	26.0	9.9	28.4	33.2	27.3	22.6	22.2	8.0
Wales	24.0	6.0	26.7	25.0	25.4	16.7	20.6	..
Scotland[3]	24.8	13.1	25.0	15.1	24.9	14.3	19.2	..
Northern Ireland[4]	23.3	4.3	24.8	11.7	23.8	7.2	..	..

1 Maintained schools only. Primary figures for Scotland include composite classes covering more than one year group. In Northern Ireland a class a defined as a group of pupils normally under therol of one teacher. See Notes and Definitions.
2 Primary data for Scotland relate to 1998/99. Secondary data for Wales and Scotland relate to 1997/98.
3 In Scotland primary P1-P3 is interpreted to be key stage 1 and P4-P7, key stage 2. See Notes and Definitions.
4 Pupils in composite classes which overlap Key Stage 1 and Key Stage 2 are included in the 'All Primary Schools' total, but are excluded from all other categories.

Source: Department for Education and Employment; National Assembly for Wales; Scottish Executive; Northern Ireland Department of Education

4.3 Children under five in schools

Thousands and percentages

| | 1990/91 | | | | 1999/00[1] | | | |
| | | Participation rates[2] (percentages) | | | | Participation rates[2] (percentages) | | |
	Children under 5 in school (thousands)	Maintained nursery & primary schools	Independent and special schools	All schools	Children under 5 in school (thousands)	Maintained nursery & primary schools[3]	Independent and special schools	All schools
United Kingdom	776.8	47	4	51	936.0	60	4	64
North East	53.4	76	3	78	52.3	84	2	86
North West	119.9	61	3	64	118.2	67	3	71
Yorkshire and the Humber	81.4	60	2	62	90.7	71	3	73
East Midlands	55.4	49	4	53	63.2	59	3	63
West Midlands	83.0	55	4	59	92.9	66	4	70
East	51.7	34	4	38	71.9	49	5	54
London	101.8	50	5	56	136.2	62	7	69
South East	63.9	26	7	33	92.5	39	8	47
South West	44.0	34	5	38	54.8	44	5	49
England	654.5	48	4	52	772.8	58	5	63
Wales	52.7	68	1	69	55.6	77	1	78
Scotland[4]	45.2	33	1	34	79.8	64	2	66
Northern Ireland	24.5	46	1	46	27.8	57	1	57

1 Data for Wales and Scotland refer to 1998/99. Excludes private and voluntary education providers in England.
2 Pupils under five in schools as a percentage of the three and four year old population.
3 Pupils in funded places in Voluntary and Private Pre-School Education centres funded under the Pre-School Education Expansion Programme are included for Northern Ireland.
4 Figures for 1990/91 exclude pupils aged four in primary schools.

Source: Department for Education and Employment; National Assembly for Wales; Scottish Executive; Department of Education, Northern Ireland

4.4 Examination achievements[1]: by gender, 1998/99

	Pupils in their last year of compulsory education[1]					Pupils/students in education[4] achieving 2 or more A levels/ 3 or more SCE Highers (percentages)	Average A/AS level point scores
	Percentage achieving GCSE or SCE Standard Grade[2]				Total (=100%) (thousands)		
	5 or more grades A*-C	1-4 grades A*-C	Grades D-G only[3]	No graded GCSEs/SCEs			
Males							
United Kingdom[5]	43.8	25.2	24.1	6.9	359.6	26.5	17.9
North East	35.5	24.6	31.2	8.6	16.2	19.4	17.2
North West	41.3	24.9	26.2	7.5	44.1	25.2	19.0
Yorkshire and the Humber	36.8	23.4	30.9	8.9	31.2	22.9	18.7
East Midlands	42.1	23.9	27.5	6.5	25.6	25.9	17.8
West Midlands	39.9	25.2	27.8	7.1	34.0	24.7	17.8
East	46.8	24.7	22.7	5.8	32.0	30.2	17.7
London	41.3	28.2	23.4	7.1	37.4	26.1	16.8
South East	48.6	23.3	21.2	6.9	47.7	32.8	18.1
South West	47.3	24.4	22.8	5.5	28.9	29.2	18.3
England	42.8	24.8	25.4	7.0	297.1	26.9	18.0
Wales	42.1	24.2	24.0	9.8	18.0	22.2	15.9
Scotland	52.1	29.6	13.8	4.4	31.5	24.0	.
Northern Ireland	48.9	25.6	20.5	5.0	13.0	27.9	..
Females							
United Kingdom[5]	54.6	24.3	16.3	4.8	344.0	33.2	18.3
North East	46.1	24.8	22.9	6.2	15.8	25.7	17.5
North West	51.0	24.5	19.0	5.5	42.4	30.5	19.4
Yorkshire and the Humber	47.3	24.2	22.2	6.3	29.7	28.1	19.0
East Midlands	52.4	24.2	18.7	4.7	24.1	31.6	17.7
West Midlands	50.6	25.6	18.9	4.9	32.1	30.9	17.9
East	57.8	23.7	14.3	4.1	30.6	36.2	18.4
London	52.2	27.9	14.8	5.1	37.0	33.1	16.9
South East	59.4	22.6	13.7	4.3	44.8	39.3	18.8
South West	58.7	22.5	14.7	4.0	27.3	36.5	18.9
England	53.4	24.5	17.2	5.0	283.9	33.1	18.3
Wales	53.1	23.5	17.1	6.2	17.5	30.1	17.0
Scotland	63.6	24.0	8.8	3.5	30.2	32.3	.
Northern Ireland	63.4	22.2	12.6	1.9	12.5	40.5	..
All pupils/students							
United Kingdom[5]	49.1	24.8	20.3	5.9	703.6	29.7	18.1
North East	40.8	24.7	27.1	7.4	32.0	22.5	17.4
North West	46.0	24.7	22.7	6.5	86.5	27.8	19.2
Yorkshire and the Humber	41.9	23.8	26.7	7.6	60.9	25.4	18.9
East Midlands	47.1	24.1	23.2	5.6	49.7	28.6	17.7
West Midlands	45.1	25.4	23.5	6.0	66.1	27.7	17.8
East	52.2	24.2	18.6	5.0	62.5	33.2	18.1
London	46.7	28.1	19.1	6.1	74.4	29.5	16.8
South East	53.8	23.0	17.6	5.6	92.6	36.0	18.5
South West	52.8	23.5	18.9	4.8	56.2	32.7	18.6
England	47.9	24.6	21.4	6.0	581.0	30.0	18.2
Wales	47.5	23.9	20.6	8.0	35.5	26.1	16.5
Scotland	57.8	26.9	11.4	4.0	61.7	28.1	.
Northern Ireland	56.0	23.9	16.6	3.5	25.4	34.1	..

1 See Notes and Definitions.
2 England figures include GNVQ equivalents.
3 No grades above D and at least one in the D-G range. Figures for Wales, England and the English regions include pupils with one GCSE short course only.
4 Pupils in schools and students in further education institutions aged 17-19 at the end of the academic year in England, Wales and Northern Ireland as a percentage of the 18 year old population. Pupils in Scotland generally sit Highers one year earlier and the figures tend to relate to the results of pupils in Year S5/S6 as a percentage of the 17 year old population.
5 England and Wales only for 'Average A/AS level point scores'.

Source: Department for Education and Employment; National Assembly for Wales; Scottish Executive; Northern Ireland Department of Education

4.5 Pupils[1] achieving GCSE grades A*-C[2]: by selected subjects and gender, 1998/99[3]

Percentages

	English	Mathematics	Science Any science[4]	Science Single award[5]	Science Double award	Any modern language[6]	French	Geography	History	Craft Design Technology	All core subjects[7]
Males											
United Kingdom	44.7	43.8	44.3	1.3	35.7	31.4	21.6	23.4	17.8	29.5	24.3
North East	38.4	36.6	37.1	0.6	28.8	22.8	15.9	19.2	14.8	26.9	17.9
North West	43.8	42.6	42.8	1.0	34.1	30.0	20.9	21.4	17.1	29.2	23.3
Yorkshire and the Humber	38.8	38.3	39.2	0.9	32.6	25.9	17.1	21.1	15.5	28.3	19.8
East Midlands	42.8	43.5	44.7	1.0	38.8	30.8	21.0	23.4	16.2	30.1	23.4
West Midlands	42.3	41.3	41.3	0.9	33.3	29.2	19.4	22.4	17.1	28.8	22.0
East	47.9	47.8	48.6	1.1	42.0	33.5	22.6	26.3	19.8	32.6	26.5
London	44.5	42.5	40.9	1.9	31.1	32.5	20.9	20.0	17.6	23.9	23.6
South East	50.0	49.1	49.6	2.0	37.4	37.7	27.2	26.8	20.8	30.1	29.9
South West	48.3	47.9	50.1	1.1	41.3	33.7	24.1	28.4	18.6	35.4	27.2
England	44.7	43.8	44.3	1.3	35.7	31.4	21.6	23.4	17.8	29.5	24.3
Wales	43.6	41.5	45.2	2.0	33.0	19.8	15.3	26.3	18.5	21.9	32.4
Scotland	63.1	49.2	58.0	7.2	.	41.1	27.3	26.1	17.7	30.1	32.5
Northern Ireland	49.4	45.3	41.2	9.2	31.7	35.1	25.9	27.8	18.7	2.8	23.4
Females											
United Kingdom	62.9	46.7	48.7	2.6	36.4	48.0	33.8	21.7	22.0	34.1	35.2
North East	54.9	40.0	39.9	1.5	32.0	38.4	27.5	17.3	19.0	38.3	27.7
North West	59.5	44.6	45.3	1.4	37.8	45.7	32.7	18.3	19.6	35.5	33.0
Yorkshire and the Humber	55.8	40.8	41.7	1.3	35.8	42.3	28.6	19.4	19.3	39.2	29.5
East Midlands	59.7	44.5	46.2	1.9	41.3	47.1	34.0	19.7	20.2	41.5	32.8
West Midlands	58.4	42.7	44.5	1.1	38.9	44.3	30.2	21.2	20.5	38.2	30.9
East	65.3	50.0	51.5	2.2	45.1	50.2	35.0	23.5	24.6	44.9	37.2
London	60.2	45.2	45.8	3.1	38.1	47.2	30.4	19.6	21.2	34.3	32.6
South East	66.4	51.2	53.0	2.2	43.8	53.4	38.2	25.1	23.7	37.2	40.1
South West	66.0	49.8	52.4	1.7	44.4	51.9	37.7	26.6	22.9	46.8	38.6
England	61.2	45.9	47.2	1.9	40.1	47.4	33.1	21.4	21.4	39.1	34.1
Wales	60.9	44.3	48.8	3.3	37.8	33.9	26.9	22.4	23.8	17.4	38.7
Scotland	78.4	52.6	63.5	5.1	.	60.2	42.2	21.2	25.2	10.9	44.2
Northern Ireland	68.5	52.5	47.3	11.8	39.2	53.1	40.5	28.0	25.0	0.6	33.7
All pupils											
United Kingdom	52.7	44.9	45.7	1.6	37.8	39.2	27.2	22.5	19.6	34.2	29.1
North East	46.5	38.3	38.5	1.0	30.4	30.5	21.6	18.3	16.9	32.5	22.7
North West	51.5	43.6	44.0	1.2	35.9	37.7	26.7	19.9	18.3	32.3	28.0
Yorkshire and the Humber	47.1	39.5	40.4	1.1	34.2	33.9	22.7	20.3	17.3	33.6	24.5
East Midlands	51.0	44.0	45.4	1.4	40.0	38.7	27.3	21.6	18.2	35.6	28.0
West Midlands	50.1	42.0	42.8	1.0	36.0	36.5	24.6	21.8	18.7	33.3	26.3
East	56.4	48.9	50.0	1.7	43.5	41.7	28.7	24.9	22.1	38.6	31.7
London	52.3	43.8	43.4	2.5	34.6	39.8	25.6	19.8	19.4	29.1	28.1
South East	57.9	50.1	51.3	2.1	40.5	45.3	32.5	26.0	22.2	33.5	34.8
South West	56.9	48.8	51.2	1.4	42.8	42.5	30.7	27.5	20.7	40.9	32.8
England	52.7	44.9	45.7	1.6	37.8	39.2	27.2	22.5	19.6	34.2	29.1
Wales	52.1	42.9	47.0	2.6	35.4	26.8	21.0	24.4	21.1	19.7	35.5
Scotland	70.6	50.9	60.7	6.2	.	50.4	34.6	23.7	21.4	20.7	38.2
Northern Ireland	58.7	48.8	44.2	10.5	35.4	43.9	33.0	27.9	21.8	1.8	28.4

1 Pupils in their last year of compulsory education.
2 SCE Standard Grade awards at levels 1-3 in Scotland.
3 See Notes and Definitions.
4 Includes double award, single award and individual science subjects. In Scotland, 'Any science' includes Biology, Chemistry, Physics or General Science Standard Grade. See Notes and Definitions.
5 General Science in Scotland.
6 Including French.
7 Figures for England, Scotland and Northern Ireland refer to English, mathematics, a science and a modern language. Figures for Wales are based on achievements in mathematics, a science and either English or Welsh (as a first language). In 1998/99, 7.7 per cent of pupils achieved GCSE grade A* to C in Welsh as a first language and a further 14.7 per cent in Welsh as a second language.

Source: Department for Education and Employment; National Assembly for Wales; Scottish Executive; Northern Ireland Department of Education

4.6 Pupils reaching or exceeding expected standards[1]: by Key Stage Teacher Assessment, Summer 1999

Percentages

	Key Stage 1[2]			Key Stage 2[3]			Key Stage 3[4]		
	English	Mathematics	Science	English	Mathematics	Science	English	Mathematics	Science
North East	82	86	87	65	67	74	58	59	54
North West	82	86	87	69	70	76	63	63	58
Yorkshire and the Humber	82	85	86	64	66	72	60	61	57
East Midlands	83	86	88	66	68	74	64	65	62
West Midlands	81	85	85	66	67	74	62	61	57
East	83	87	88	68	69	74	67	69	66
London	79	84	84	65	67	73	60	58	54
South East	83	87	88	70	71	78	68	69	66
South West	83	87	87	69	70	77	67	68	66
England[5]	82	86	87	68	69	75	64	64	60

1 See Notes and Definitions.
2 Percentage of pupils achieving level 2 or above at key stage 1
3 Percentage of pupils achieving level 4 or above at key stage 2.
4 Percentage of pupils achieving level 5 or above at key stage 3.
5 Includes non-LEA maintained schools. These are not included in the regions figures.

Source: Department for Education and Employment

4.7 16 and 17 year olds participating in post-compulsory education[1] and government-supported training, 1997/98

Percentages[2]

	16 year olds					17 year olds				
	At school[1]	In further education[1,3] Full-time	In further education[1,3] Part-time	Government-supported training (GST)	All in full-time education and GST[4]	At school[1]	In further education[1,3] Full-time	In further education[1,3] Part-time	Government-supported training (GST)	All in full-time education and GST[4]
Region of study										
United Kingdom	37.9	32.5	8.0	..	..	28.7	28.6	9.4	..	..
North East	25.4	34.9	7.4	17.9	84.0	19.5	29.7	9.4	17.8	74.7
North West	24.3	40.8	8.5	13.3	83.5	19.4	33.9	10.2	14.7	74.6
Yorkshire and the Humber	30.9	33.6	9.1	12.8	83.5	23.4	27.6	11.0	14.7	73.8
East Midlands	36.7	29.7	7.9	11.2	82.6	29.1	25.9	9.3	13.2	73.9
West Midlands	30.7	35.4	8.3	11.0	82.1	24.5	30.7	10.3	12.4	74.7
East	40.2	33.1	5.7	7.0	84.0	32.0	28.4	7.4	9.2	74.7
London	39.7	35.0	4.4	5.4	83.4	28.9	32.2	6.0	6.2	72.2
South East	39.7	35.2	4.7	5.8	84.0	30.6	29.7	6.1	7.7	72.4
South West	38.4	34.4	6.1	9.0	84.7	30.9	29.7	7.5	10.5	75.4
England	34.4	35.1	6.7	9.7	83.5	26.8	30.1	8.4	11.2	73.9
Wales	36.7	31.9	12.7	16.6	88.9	28.2	27.2	9.7	15.1	72.1
Scotland[5]	68.1	10.9	12.6	9.1	88.1	38.8	10.9	15.2	15.8	65.5
Northern Ireland[6]	45.0	27.5	17.8	..	..	36.6	28.2	13.2	..	..

1 See Notes and Definitions.
2 As a percentage of the estimated 16 and 17 year old population respectively.
3 Including sixth form colleges and a small element of further education in higher education institutions in England.
4 Figures for England and Wales exclude overlap between full-time education and government-supported training.
5 The estimates of 16 year olds at school exclude those pupils who leave school in the winter term at the minimum statutory school-leaving age.
6 Participation in part-time FE should not be aggregated with full-time FE or schools activity due to the unquantifiable overlap of these activities.

Source: Department for Education and Employment; National Assembly for Wales; Scottish Executive; Northern Ireland Department of Education

4.8 Home students in further education[1] in England: by level of course of study[2], 1999/00

Percentages and thousands

	Courses leading to NVQ/GNVQ or equivalent academic qualifications (percentages)						Total FE students[1] studying in England (=100%) (thousands)
	Level 1	Level 2	Level 3	Other recognised qualifications	All courses leading to recognised qualifications	Other courses	
Region of domicile							
United Kingdom	17.0	26.9	35.0	11.3	90.1	9.9	2,029.7
North East	20.6	29.6	34.9	9.7	94.8	5.2	114.4
North West	17.8	27.2	34.4	11.1	90.4	9.6	321.4
Yorkshire and the Humber	19.4	27.7	30.6	12.9	90.5	9.5	219.4
East Midlands	18.1	28.4	33.3	13.1	92.9	7.1	157.5
West Midlands	16.9	25.1	33.7	15.1	90.7	9.3	234.8
East	16.8	27.7	38.3	8.7	91.4	8.6	174.6
London	15.1	26.3	35.2	11.6	88.1	11.9	227.4
South East	13.3	26.9	39.6	8.9	88.6	11.4	274.6
South West	18.5	26.0	36.4	8.4	89.2	10.8	193.2
England	17.1	27.0	35.2	11.1	90.4	9.6	1,916.9
Other[3]	15.2	25.2	30.7	14.5	85.7	14.3	112.8

1 Further education institutions only. See Notes and Definitions.
2 Highest level of qualification aimed for by students.
3 Includes those from Wales, Scotland, Northern Ireland, Channel Islands, Isle of Man and home students whose region of domicile was unknown or unclassified.

Source: Department for Education and Employment

4.9 Local education authority expenditure[1], 1998-99[2]

	Percentage of total LEA expenditure							Total expend-iture (=100%) (£ million)	Current expenditure per pupil[6] (£ per pupil)	
	Pre-primary & primary schools[3]	Secon-dary schools[3]	Special schools[3]	Continuing education	Adminis-tration and inspection[4]	Other educa-tional services[4,5]	Capital expend-iture		Nursery/ primary pupils	Second-ary pupils
United Kingdom	39.4	36.4	6.5	10.8	1.4	1.5	4.1	27,504.5	1,880	2,530
North East	38.9	38.0	6.5	11.2	0.4	-	5.0	1,223.0	1,790	2,320
North West	40.6	36.7	6.8	11.9	0.6	-	3.4	3,292.3	1,790	2,430
Yorkshire and the Humber	41.0	37.7	4.9	12.5	0.4	-	3.5	2,288.7	1,870	2,310
East Midlands	39.8	39.2	5.3	11.3	0.5	-	3.8	1,838.0	1,810	2,420
West Midlands	41.1	37.0	6.7	11.0	0.6	-	3.5	2,535.8	1,910	2,440
East	38.7	40.2	6.2	10.8	0.9	-	3.2	2,343.9	1,850	2,470
London	41.1	31.5	8.0	13.5	1.1	-	4.8	3,734.1	2,220	2,910
South East	39.2	36.2	7.3	12.0	0.7	-	4.6	3,359.8	1,820	2,380
South West	38.9	37.8	6.1	11.9	0.7	-	4.6	2,048.4	1,770	2,380
England	40.1	36.6	6.6	11.9	0.7	-	4.0	22,664.0	1,880	2,450
Wales	37.6	35.1	4.8	10.0	3.5	4.6	4.5	1,420.7	1,890	2,610
Scotland	35.2	37.2	7.0	1.2	6.0	8.4	4.9	2,504.1	1,890	2,960
Northern Ireland	35.6	29.7	4.9	9.4	2.6	13.8	4.0	915.7	1,900	2,780

1 See Notes and Definitions.
2 Data for Scotland and Northern Ireland (other than expenditure per pupil figures for Northern Ireland which are 1996/97) refer to 1997/98.
3 Includes LEA expenditure on grant-maintained schools.
4 The bulk of expenditure on central services under these headings has been recharged to columns 1-4, (which are 1996/97) along with transport of pupils for Wales.
5 Includes school catering services in Wales, Scotland and Northern Ireland.
6 These figures must be interpreted carefully in the light of different educational structures between regions. Excludes LEA expenditure on grant-maintained schools. Rounded to the nearest £10.

Source: Department for Education and Employment; National Assembly for Wales; Scottish Executive; Northern Ireland Department of Education

4.10 Local education authority expenditure per school pupil[1]: by type of school

£ per pupil[2] (cash terms)

	Nursery and Primary		Secondary	
	1994-95	1998-99[3]	1994-95	1998-99[3]
United Kingdom	1,670	1,880	2,340	2,530
North East	1,620	1,790	2,130	2,320
North West	1,570	1,790	2,260	2,430
Yorkshire and the Humber	1,620	1,870	2,170	2,310
East Midlands	1,630	1,810	2,320	2,420
West Midlands	1,650	1,910	2,240	2,440
East	1,670	1,850	2,250	2,470
London	1,940	2,220	2,630	2,910
South East	1,640	1,820	2,240	2,380
South West	1,550	1,770	2,180	2,380
England	1,660	1,880	2,270	2,450
Wales	1,640	1,890	2,260	2,610
Scotland	1,850	1,890	2,890	2,960
Northern Ireland	1,550	1,900	2,330	2,780

1 These figures must be interpreted carefully in the light of different educational structures between the regions. See Notes and Definitions.
2 Figures are rounded to nearest £10.
3 Data for Scotland refer to 1997-98.

Source: Department for Education and Employment; National Assembly for Wales; Scottish Executive; Northern Ireland Department of Education

4.11 Home domiciled higher education students[1]: by region of study and domicile, 1999/00[2]

Percentages and thousands

	Region of study												All students (=100%) (thousands)
	North East	North West	York-shire and the Humber	East Mid-lands	West Mid-lands	East	London	South East	South West	Wales	Scot-land	Nor-thern Ireland	
Region of domicile													
United Kingdom[3]	4.4	11.6	9.9	6.7	8.3	5.7	14.9	10.6	6.7	5.3	13.1	2.9	1,656.8
North East	69.6	5.8	10.3	3.0	1.9	1.2	2.0	1.5	0.9	0.5	3.2	0.1	60.9
North West	3.5	65.9	11.1	3.5	5.0	1.2	2.3	2.0	1.4	1.9	2.0	-	171.0
Yorkshire and the Humber	5.9	7.8	67.3	5.7	3.3	1.5	2.4	1.9	1.3	1.0	1.8	-	119.9
East Midlands	2.6	6.5	16.6	47.5	9.0	3.6	4.4	4.3	2.6	1.6	1.1	-	95.8
West Midlands	1.2	6.9	5.9	7.6	59.3	1.9	3.6	4.1	4.3	4.1	0.9	-	131.5
East	1.8	3.8	6.3	9.2	5.0	41.5	15.4	9.8	4.3	1.7	1.3	-	125.7
London	0.8	2.5	2.4	2.6	2.8	5.1	69.0	9.6	3.1	0.9	1.1	-	219.9
South East	1.5	3.2	3.8	5.3	4.5	4.2	16.4	48.8	8.1	2.8	1.3	-	207.5
South West	1.1	3.2	3.2	3.8	5.4	2.5	6.8	13.4	52.4	6.9	1.2	-	115.7
England[3]	5.3	13.9	12.1	8.2	10.1	7.0	18.5	13.0	8.1	2.4	1.5	-	1,314.1
Wales	0.6	6.8	2.7	2.6	4.1	1.2	3.1	3.8	5.6	68.6	0.8	-	80.2
Scotland	0.7	0.9	0.6	0.4	0.3	0.4	0.7	0.6	0.3	0.2	94.7	-	201.7
Northern Ireland	1.4	3.7	1.4	1.0	0.9	0.9	1.7	1.2	0.5	0.8	9.6	76.8	60.7

1 Including higher education students in further education institutions for England, Wales, Scotland and Northern Ireland. Excluding Open University students. These data are not comparable with those shown in previous editions. See Notes and Definitions.
2 Data for higher education students in further education institutions relates to 1998/99.
3 Including students from the Channel Islands and Isle of Man. Higher education institution figures also include students whose region of domicile was unknown or unclassified, but further education figures do not.

Source: Department for Education and Employment; Higher Education Statistics Agency; National Assembly for Wales; Scottish Executive; Northern Ireland Department of Education

4.12 Destination of 1999 full-time first degree graduates[1]

Percentages and thousands

	UK employment		Overseas employment[2]	Total Employment	Continuing education or training	Believed unemployed	Other destinations[3]	All first degree graduates[4] (=100%) (thousands)
	Permanent	Temporary						
Region of study								
United Kingdom	43.6	17.8	2.3	63.6	20.5	5.3	10.7	235.1
North East	40.7	21.8	2.4	64.9	20.5	5.8	8.8	11.9
North West	45.3	17.2	2.2	64.7	20.6	5.0	9.6	26.0
Yorkshire and the Humber	44.9	18.1	2.9	65.8	17.2	4.8	12.2	24.9
East Midlands	47.0	19.6	1.7	68.3	16.7	5.1	10.0	18.2
West Midlands	43.0	19.2	1.6	63.8	20.1	6.2	9.9	19.9
East	41.5	13.7	2.7	57.9	26.7	4.6	10.8	12.9
London	45.1	15.5	1.3	62.0	20.3	6.7	11.0	34.5
South East	43.3	18.6	2.2	64.0	20.2	4.7	11.1	28.2
South West	46.6	18.4	2.7	67.6	16.7	5.4	10.4	16.2
England	44.4	17.8	2.1	64.4	19.7	5.4	10.6	192.7
Wales	42.6	16.8	2.7	62.0	23.8	4.9	9.3	13.6
Scotland	39.1	19.2	2.8	61.0	22.4	4.9	11.7	22.9
Northern Ireland	36.2	12.9	4.5	53.6	29.4	5.0	12.0	5.8

1 Graduating from higher education institutions in 1999. As a percentage of known destinations.
2 Home students only.
3 Includes overseas graduates leaving the United Kingdom and graduates not available for employment.
4 Includes known and unknown destinations.

Source: Department for Education and Employment; Higher Education Statistics Agency

4.13 Population of working age[1]: by highest qualification[2], Spring 2000

Percentages and thousands

	Degree or equivalent	Higher education qualifications[3]	GCE A Level or equivalent[4]	GCSE grades A*-C or equivalent	Other qualifications	No qualifications	Total[5] (=100%) (thousands)
United Kingdom	14.8	8.4	24.0	22.3	14.1	16.4	36,312
North East	9.1	8.2	24.6	23.5	14.8	19.8	1,573
North West	12.8	8.6	25.4	24.1	11.9	17.2	4,180
Yorkshire and the Humber	12.2	8.3	25.2	22.6	14.4	17.2	3,078
East Midlands	13.1	7.4	24.1	23.9	14.0	17.5	2,569
West Midlands	12.9	7.9	21.9	22.9	14.8	19.6	3,239
East	14.5	7.4	23.6	25.2	14.0	15.4	3,301
London	22.9	6.7	18.8	17.5	19.4	14.7	4,619
South East	17.8	8.7	24.8	24.0	13.4	11.3	4,907
South West	14.6	9.4	24.8	24.9	13.8	12.6	2,921
England	15.3	8.0	23.5	22.9	14.6	15.7	30,386
Wales	12.1	9.4	20.9	24.2	13.7	19.7	1,750
Scotland	12.9	11.8	30.4	15.8	11.0	18.1	3,160
Northern Ireland	12.5	6.4	24.4	21.9	8.8	26.1	1,017

1 Males aged 16-64 and females aged 16-59.
2 See Notes and Definitions.
3 Below degree level.
4 Includes recognised trade apprenticeship.
5 Population in private households, students in halls of residence and those in NHS accommodation. Includes those who did not state their qualifications, but percentages are based on figures excluding them.

Source: Department for Education and Employment, from the Labour Force Survey

4.14 Progress towards achieving the National Learning Targets for England for 2002, Spring 2000[1,2]

Percentages

	85 per cent of 19 year olds with an NVQ level 2 qualification or equivalent			60 per cent of 21 year olds with and NVQ level 3 qualification or equivalent			50 per cent of adults[3] with an NVQ level 3 qualification or equivalent			28 per cent of adults[3] with an NVQ level 4 qualification		
	Males	Females	All	Males	Females	All	Males	Females	All	Males	Females	All
Region of residence												
North East	66	81	74	46	44	45	47	35	41	22	21	21
North West	77	75	76	54	49	52	52	40	47	25	26	25
Yorkshire and the Humber	69	75	72	52	48	50	48	40	45	23	25	24
East Midlands	68	76	71	52	48	50	48	37	43	24	23	24
West Midlands	69	75	72	56	51	54	48	38	43	25	24	25
East	68	81	74	53	52	52	49	39	44	25	25	25
London	73	73	73	62	53	57	54	49	52	35	35	35
South East	79	82	81	62	58	60	54	43	50	31	28	30
South West	72	81	76	59	56	58	51	40	46	27	26	27
England	72	77	75	56	52	54	51	41	47	27	27	27

1 See Notes and Definitions for details of the targets.
2 The questions on qualifications in the Labour Force Survey were changed substantially in Spring 1996. Figures are therefore not directly comparable with those for earlier years.
3 Males aged 18-64 and females aged 18-59, who are in employment or actively seeking employment.

Source: Department for Education and Employment from Labour Force Survey

4.15 Employees of working age[1] receiving job-related training[2]: by gender, Spring 2000

Percentages[3]

	Males				Females			
	On-the-job training only	Off-the-job training only	Both on and off-the-job training	Any job-related training	On-the-job training only	Off-the-job training only	Both on and off-the-job training	Any job-related training
United Kingdom	4.4	7.4	2.9	14.7	4.9	9.3	3.5	17.8
North East	3.7	7.4	2.9	14.0	4.6	8.6	4.1	17.3
North West	4.5	6.9	3.8	15.3	5.5	9.3	3.8	18.6
Yorkshire and the Humber	4.6	6.8	2.4	13.8	5.3	10.4	3.6	19.2
East Midlands	5.4	6.9	2.1	14.5	5.0	8.5	3.5	17.0
West Midlands	4.3	7.0	3.0	14.4	4.7	9.4	3.8	17.9
East	4.1	7.9	2.8	14.8	4.3	9.0	3.4	16.7
London	4.3	8.8	3.2	16.3	4.1	10.3	3.2	17.6
South East	4.0	6.8	2.7	13.5	5.5	9.7	3.5	18.8
South West	4.7	9.6	3.1	17.4	4.6	9.7	3.7	18.1
England	4.6	8.5	3.2	14.9	4.9	9.5	3.6	18.0
Wales	3.7	8.3	2.6	14.5	6.0	8.1	3.4	17.5
Scotland	4.7	6.4	3.2	14.3	5.1	8.4	3.2	16.7
Northern Ireland	2.9	4.2	..	8.5	4.4	6.9	..	12.9

1 Males aged 16-64 and females aged 16-59.
2 Job-related education or training received in the four weeks before interview. In some cases sample sizes are too small to provide reliable estimates.
3 As a percentage of all employees of working age.

Source: Department for Education and Employment from Labour Force Survey

4.16 Employers' provision of training[1,2], 1999

Percentages

	Proportion of employers:				Involved in the last 12 months with:			
	Reporting an increase in skill needs	With a training plan	With a training budget	Providing off-the-job training in the last 12 months	NVQs/ SVQs	Other Government supported training for young people	Modern Apprenticeship	National Traineeships
North East	52	31	24	34	26	1	11	5
North West	60	35	27	34	25	3	4	1
Yorkshire and the Humber	57	36	23	37	28	4	6	3
East Midlands	65	34	30	42	24	4	6	2
West Midlands	58	29	25	31	22	1	5	1
East	60	33	28	34	24	1	4	2
London	63	29	19	28	12	1	2	3
South East	70	31	24	32	15	-	3	1
South West	64	39	32	41	25	2	6	2
England	62	32	25	34	21	2	4	2

1 Care should be taken when making comparisons with earlier years due to the difference in coverage and focus between this survey and its predecessor, *Skill Needs in Britain*. See Notes and Definitions.

2 The LTW Survey was carried out in November and December 1999 and asked if any training provision had been made 12 months prior to the interview.

Source: Learning and Training at Work 1999, IFF Research Limited for the Department of Education and Employment

4.17 Work-based training[1], 1998-99

Percentages and thousands

	Work-based Learning for Adults						Work-based Training for Young People[2]					
	Status six months after leaving[3] (percentages)						Status six months after leaving[3] (percentages)					
	In employment	In further education or training	Unemployed	Other	Gained full qualification[4] (percentages)	All leavers[5] (thousands)	In employment	In further education or training	Unemployed	Other	Gained full qualification[4] (percentages)	All leavers[5] (thousands)
England and Wales	40	5	47	8	41	111.4	69	12	13	7	49	224.1
North East	34	5	53	7	42	8.8	61	14	18	7	49	17.0
North West	40	4	49	7	43	19.1	67	13	14	7	48	42.2
Yorkshire and the Humber	37	5	52	7	37	12.3	65	13	15	7	45	25.4
East Midlands	40	4	48	8	40	7.4	71	11	11	7	51	20.7
West Midlands	38	6	49	7	39	11.9	70	12	11	7	50	24.7
East	43	4	44	9	49	6.8	77	8	8	7	54	17.6
London	42	5	46	7	38	22.2	62	13	17	8	47	17.9
South East	45	5	42	8	39	9.4	75	9	8	8	49	26.3
South West	45	5	41	9	39	8.8	74	11	8	8	51	18.6
England	40	5	47	8	40	106.7	69	11	12	7	49	210.3
Wales	40	7	46	7	54	4.7	64	12	17	7	47	13.8
Northern Ireland[1]	56	11	28	6	90	2.6	66	8	17	8	91	6.9

1 Schemes in Northern Ireland differ from those in England and Wales; see Notes and Definitions.

2 Work Based Training for Young People data for England and Wales excludes National Traineeships.

3 Status on completion of courses in Northern Ireland.

4 In Northern Ireland, full qualifications gained by completers expressed as a percentage of completers.

5 All those who left the programme during 1998-99 except in Northern Ireland where the figure covers completers of courses only and does not include early leavers.

Source: Department for Education and Employment; Department of Higher and Further Education and Training and Employment; Northern Ireland

5 Labour Market

Employment rates

In 1999, the South East had the highest employment rate at 80 per cent compared with 65 per cent in the North East.

(Table 5.1)

Men had a higher employment rate than women in 1999 in the United Kingdom, at 78 per cent and 68 per cent respectively.

(Table 5.1)

Labour force

In Spring 1999, 4 per cent of the labour force (those in work or available for work) were females aged 60 or over or males aged 65 and over in the South West compared with around 2 per cent in the North East, North West and Yorkshire and the Humber.

(Table 5.2)

Economic activity

Around 83 per cent of people of working age in the South East and South West were economically active in Spring 1999 compared with 72 per cent in the North East and Northern Ireland, the highest and lowest proportions.

(Table 5.23)

Employment status

In Spring 1999, 2 in 3 people in the South East were economically active compared with 4 in 7 in North East and Wales, the lowest proportion of any other region.

(Table 5.3)

The proportion of employees who were working part-time in Spring 1999 was highest in the South West.

(Table 5.3)

Jobs

The highest number of self-employment jobs for both males and females in the United Kingdom in 1999 were in the South East.

(Table 5.5)

The proportion of men and women who said that the reason why they were working part-time was because they could not find a full-time job was highest in the North East and Wales at around 31 per cent, double the proportion in the South East and the East of England.

(Table 5.7)

In Spring 1999, females in employment in the South West were the most likely to have a second job.

(Table 5.9)

Hours of work

In Spring 1999, in the United Kingdom, those working as managers and administrators worked on average more hours in a week than any other occupational group.

(Table 5.10)

Earnings

In April 1999, average weekly earnings for full-time employees varied from £584 in London to £377 in Northern Ireland for men, and from £423 in London to £287 in the East Midlands for women.

(Table 5.11)

Unemployment

Between Spring 1999 and Winter 1999/2000, the ILO unemployment rates for males and females in Scotland who have been unemployed for two years or over were more than twice the national rates.

(Table 5.15)

Unemployment	In Spring 1999, the North East had the highest ILO unemployment rate at 10.1 per cent, compared with 3.6 per cent in the South East, the lowest rate in the United Kingdom.

(Table 5.16)

Claimant count	In 1999, the seasonally adjusted claimant count rate in the South East was 2.4 per cent, lower than in any other region.

(Table 5.17)

In March 2000, claimant count rates in North Ayrshire, West Dunbartonshire and Hartlepool were amongst the highest in the United Kingdom.

(Map 5.13)

Vacancies	In 1999, there were around 38 thousand vacancies remaining unfilled in the South East, compared with 16 thousand vacancies in the North East and 17 thousand in Wales.

(Table 5.18)

Redundancies	In Spring 1999, redundancy rates in the West Midlands, Wales and Scotland were around three fifths higher than in London.

(Chart 5.19)

Jobseekers's Allowance	In March 2000, the highest proportion of those claiming Jobseeker's Allowance in the United Kingdom were aged between 20-29 years old; among this age group, the proportion was highest in the North West.

(Table 5.20)

New Deal	In 1999, the highest number of new starts on the New Deal for young people aged 18 to 24 in Great Britain were in London and North West.

(Table 5.21)

In 1999, males in the East of England and the South West were the most likely to enter unsubsidised employment after leaving the New Deal for 18 to 24 year olds.

(Table 5.22)

In Spring 1999 in the North East, 22.5 per cent of households with at least one member of working age had no one in employment, double the proportion in the South East.

(Table 5.24)

Sickness	Sickness absence from work in Spring 1999 was highest in Wales and the West Midlands and lowest in Northern Ireland.

(Chart 5.25)

Labour disputes	In 1999, the number of working days lost due to labour disputes in Scotland was 22 days for every thousand employees, compared with only 4 days in Wales.

(Table 5.26)

Trade Unions	In Autumn 1999, Trade Union membership among manual employees ranged from 38 percent in the North East to 22 per cent in the South East, South West and the East of England. For non-manual employees it ranged from 40 per in the North East to 23 per cent in the South East.

(Table 5.27)

Glossary of terms

**Employees
(Labour Force Survey)**

A household-based measure of persons aged 16 or over who regard themselves as paid employees. People with two or more jobs are counted only once.

**Employee jobs
(New Earnings Survey)**

A measure of employees in employment, obtained from surveys of employers, of jobs held by civilians who are paid by an employer who runs a PAYE tax scheme. Those people with two or more jobs are represented for each of those jobs in the same survey.

(Employer Survey)

A measure, obtained from surveys of employees, of jobs held by civilians. People with two or more jobs are counted in each job.

Self-employed

A household-based measure (from the Labour Force Survey (LFS)) of persons aged 16 or over who regard themselves as self-employed in their main job.

**People on government-supported training and employment programmes
(Labour Force Survey)**

A household-based measure of persons aged 16 or over participating in Work-based learning for Adults and Young People, Work Trial and Project Work as well as other similar programmes organised by a Training Enterprise Council (England and Wales), Local Enterprise Company (Scotland) or the Training and Employment Agency (Northern Ireland). Because of the nature of many of these programmes, the LFS has difficulty in identifying scheme participants.

**Labour force in employment
(Labour Force Survey)**

A household-based measure of employees, self-employed persons, participants in government-supported training and employment programmes, and persons doing unpaid work for a family business.

Workforce jobs

A measure of employee jobs (obtained from employer surveys), self-employment jobs (obtained from the Labour Force Survey), all HM Forces, and government-supported trainees (obtained from the Employment Service).

ILO unemployed

An International Labour Organisation (ILO) recommended measure, used in household surveys such as the LFS, which counts as unemployed those aged 16 or over who are without a job, are available to start work in the next two weeks and who have been seeking a job in the last four weeks, or were waiting to start a job already obtained in the next two weeks.

Claimant count

A count derived from administrative sources, of those people who are claiming unemployment-related benefits at Employment Service local offices (formerly Unemployment Benefit Offices).

Economically active/labour force

The **labour force in employment** *plus* the **ILO unemployed**.

ILO unemployment rate

The percentage of the **economically active** who are **ILO unemployed**.

Claimant count rate

The number of people claiming unemployment-related benefit as a percentage of **workforce jobs** plus the **claimant count**.

Economically inactive

Persons who are neither part of the labour force in employment nor ILO unemployed. For example, all people under 16, those retired or looking after a home, or those permanently unable to work.

Population of working age

Males aged 16 to 64 years and females aged 16 to 59 years.

Economic activity rate

The percentage of the population in a given age group which is in the **labour force**.

Some of these items are covered in more detail in the Notes and Definitions.

5.1 Labour force and employment rates[1]

Thousands and percentages

	Labour force (thousands)					Employment rates[2] (percentages)				
	1995	1996	1997	1998	1999	1995	1996	1997	1998	1999
Males										
United Kingdom	15,713	15,776	15,818	15,813	15,937	76.1	76.4	77.5	78.1	78.4
North East	651	665	657	653	634	67.5	70.4	70.3	71.6	68.3
North West	1,766	1,781	1,773	1,737	1,788	72.5	73.3	74.4	73.3	75.4
Yorkshire and the Humber	1,361	1,337	1,328	1,337	1,349	76.5	75.3	74.6	76.0	76.7
East Midlands	1,132	1,147	1,142	1,146	1,147	78.4	79.2	79.5	81.4	80.4
West Midlands	1,441	1,453	1,452	1,459	1,447	76.3	76.3	78.9	79.7	78.4
East	1,494	1,484	1,481	1,510	1,513	81.7	82.5	81.5	84.0	84.0
London	1,920	1,938	1,973	1,949	2,005	72.7	72.6	75.9	75.0	76.9
South East	2,190	2,212	2,226	2,251	2,267	82.2	82.4	83.5	84.6	85.9
South West	1,298	1,279	1,308	1,307	1,323	79.7	79.3	82.0	82.7	82.6
England	13,253	13,295	13,340	13,350	13,474	76.8	77.1	78.3	79.0	79.4
Wales	710	726	727	705	725	70.7	72.0	72.2	71.2	72.1
Scotland	1,356	1,351	1,344	1,341	1,329	74.8	73.8	74.2	74.9	74.0
Northern Ireland	395	404	407	417	409	67.8	70.0	72.0	73.4	71.8
Females										
United Kingdom	11,960	12,098	12,208	12,284	12,422	65.6	66.5	67.2	67.6	68.3
North East	501	502	510	489	502	61.6	61.2	63.0	61.3	61.2
North West	1,342	1,372	1,349	1,355	1,371	63.2	65.1	64.9	64.8	66.0
Yorkshire and the Humber	1,020	1,032	1,015	1,030	1,030	65.7	67.1	65.9	67.4	67.7
East Midlands	861	878	895	892	901	68.1	68.8	70.7	70.5	71.0
West Midlands	1,063	1,070	1,064	1,087	1,110	65.0	65.1	65.8	67.3	68.3
East	1,116	1,114	1,129	1,162	1,164	69.1	68.8	69.7	70.8	71.5
London	1,464	1,502	1,528	1,518	1,547	61.8	63.1	64.4	63.9	65.4
South East	1,667	1,715	1,734	1,751	1,770	69.9	71.8	71.9	72.7	73.0
South West	996	996	1,024	1,040	1,058	69.1	70.0	71.8	72.1	73.2
England	10,031	10,180	10,251	10,324	10,454	66.0	67.0	67.7	68.1	68.9
Wales	559	549	563	553	559	63.2	62.8	63.8	63.5	64.3
Scotland	1,081	1,070	1,085	1,098	1,091	66.1	65.7	66.1	67.7	67.4
Northern Ireland	290	298	310	309	318	56.9	58.6	60.7	60.3	61.1
All persons										
United Kingdom	27,673	27,873	28,026	28,097	28,359	71.1	71.6	72.6	73.1	73.6
North East	1,152	1,167	1,167	1,142	1,137	64.7	66.0	66.8	66.7	64.9
North West	3,108	3,153	3,123	3,092	3,160	68.1	69.4	69.9	69.3	70.9
Yorkshire and the Humber	2,381	2,368	2,343	2,367	2,379	71.4	71.4	70.5	71.9	72.5
East Midlands	1,993	2,025	2,037	2,039	2,047	73.5	74.3	75.3	76.2	75.9
West Midlands	2,504	2,523	2,516	2,546	2,558	70.9	71.0	72.7	73.8	73.6
East	2,609	2,598	2,610	2,672	2,677	75.7	76.0	75.9	77.7	78.1
London	3,384	3,440	3,502	3,467	3,552	67.4	68.0	70.3	69.7	71.4
South East	3,857	3,927	3,960	4,002	4,037	76.3	77.3	77.9	78.9	79.7
South West	2,294	2,275	2,332	2,347	2,381	74.7	74.9	77.1	77.7	78.1
England	23,283	23,475	23,591	23,674	23,928	71.7	72.3	73.3	73.8	74.4
Wales	1,268	1,275	1,290	1,259	1,285	67.2	67.7	68.2	67.5	68.4
Scotland	2,436	2,421	2,428	2,439	2,420	70.6	69.9	70.3	71.4	70.8
Northern Ireland	685	702	717	726	726	62.5	64.4	66.5	67.0	66.6

1 At Spring of each year. Based on the population of working age in private households, students halls of residence and NHS accommodation. See Notes and Definitions.
2 Total in employment as a percentage of all persons of working age in each region.

Source: Labour Force Survey, Office for National Statistics

5.2 Labour force[1]: by age, Spring 1999

Percentages and thousands

	Percentage aged					All ages (=100%) (thousands)
	16-24	25-34	35-44	Females 45-59, Males 45-64	Females 60 or over, Males 65 or over	
United Kingdom	15.1	26.0	25.1	31.0	2.9	29,194
North East	14.8	26.0	26.9	30.1	2.2	1,162
North West	15.2	26.0	25.5	31.1	2.2	3,231
Yorkshire and the Humber	15.8	26.0	25.7	30.5	2.1	2,429
East Midlands	15.1	25.0	25.3	32.1	2.4	2,098
West Midlands	14.8	25.8	24.4	32.2	2.8	2,631
East	14.4	25.8	24.1	32.3	3.3	2,767
London	15.8	28.8	25.9	26.7	2.9	3,657
South East	14.0	25.3	24.3	32.7	3.6	4,190
South West	14.6	24.0	23.9	33.4	4.0	2,481
England	14.9	26.0	25.0	31.2	2.9	24,645
Wales	15.6	24.7	25.3	31.6	2.8	1,321
Scotland	15.6	25.9	25.9	30.2	2.4	2,480
Northern Ireland	17.1	27.6	24.8	27.8	2.8	747

1 See Notes and Definitions.

Source: Labour Force Survey, Office for National Statistics

5.3 Employment status and rates, Spring 1999

Percentages and thousands

	In employment				ILO unem- ployed	Total econom- ically active	Econom- ically inactive	All aged 16 or over[2] (=100%) (thousands)	Employment rates[3]		
	Employees		Self- employed	Total[1]					Males	Females	All persons
	Full- time	Part- time									
United Kingdom	38.7	13.0	6.9	59.1	3.8	62.9	37.1	46,431	78.4	68.3	73.6
North East	33.5	12.7	4.5	51.4	5.7	57.1	42.9	2,033	68.3	61.2	64.9
North West	37.8	12.3	6.0	56.6	3.7	60.3	39.7	5,360	75.4	66.0	70.9
Yorkshire and the Humber	37.3	14.0	5.6	57.5	4.0	61.5	38.5	3,948	76.7	67.7	72.5
East Midlands	39.3	13.9	6.9	60.5	3.3	63.8	36.2	3,291	80.4	71.0	75.9
West Midlands	39.0	13.2	6.2	58.9	4.3	63.2	36.8	4,164	78.4	68.3	73.6
East	40.7	13.2	8.2	62.5	2.7	65.2	34.8	4,245	84.0	71.5	78.1
London	40.7	10.7	8.2	60.1	4.9	65.0	35.0	5,624	76.9	65.4	71.4
South East	41.6	14.1	8.0	64.2	2.4	66.6	33.4	6,292	85.9	73.0	79.7
South West	36.7	15.0	8.5	60.9	3.0	63.9	36.1	3,881	82.6	73.2	78.1
England	39.0	13.1	7.1	59.8	3.7	63.5	36.5	38,838	79.4	68.9	74.4
Wales	34.9	11.5	6.1	53.4	4.0	57.4	42.6	2,302	72.1	64.3	68.4
Scotland	37.9	13.0	5.3	56.9	4.6	61.5	38.5	4,035	74.0	67.4	70.8
Northern Ireland	37.2	10.0	6.6	55.2	4.3	59.5	40.5	1,256	71.8	61.1	66.6

1 Includes those on government-supported employment and training schemes and unpaid family workers.
2 Based on the population of working age in private households, student halls of residence and NHS accommodation.
3 Total in employment as a percentage of all persons of working age in each region.

Source: Labour Force Survey, Office for National Statistics

5.4 Employee jobs: by industry[1] and gender, September 1998

Percentages and thousands

	Agriculture, hunting, forestry & fishing	Mining, quarrying, (inc oil & gas extraction)	Manu- facturing	Electricity, gas, water	Construction	Distribution, hotels & catering, repairs
Males						
Great Britain	1.8	0.5	24.4	0.8	7.4	20.9
North East	1.1	0.6	30.9	1.1	9.9	17.0
North West	1.3	0.2	28.8	0.8	7.7	21.3
Yorkshire and the Humber	1.7	0.7	30.3	0.9	7.9	19.6
East Midlands	2.4	0.8	33.6	0.8	7.7	19.5
West Midlands	1.6	0.3	35.9	1.0	6.6	19.1
East	2.8	0.3	23.6	0.9	8.0	22.3
London	0.1	0.2	9.9	0.3	5.2	21.8
South East	1.9	0.2	19.5	0.8	7.0	23.2
South West	3.0	0.5	23.9	1.1	7.7	22.0
England	1.7	0.4	24.4	0.8	7.2	21.1
Wales	2.9	0.7	31.3	0.6	8.1	18.4
Scotland	3.1	2.2	21.5	1.2	9.9	19.6
Females						
Great Britain	0.8	0.1	10.2	0.3	1.6	24.9
North East	0.3	0.1	11.6	0.4	1.6	25.2
North West	0.5	-	11.3	0.3	1.5	26.4
Yorkshire and the Humber	0.7	0.1	11.9	0.3	1.9	25.2
East Midlands	1.2	0.1	16.6	0.3	1.7	24.2
West Midlands	0.8	0.1	14.3	0.4	1.4	24.3
East	1.6	0.1	9.9	0.4	1.8	26.0
London	0.1	0.1	5.7	0.1	1.1	23.1
South East	1.3	-	7.9	0.3	1.7	24.2
South West	1.1	0.1	9.1	0.4	1.6	26.6
England	0.8	0.1	10.2	0.3	1.6	24.9
Wales	0.8	0.1	11.7	0.2	1.9	24.6
Scotland	0.9	0.4	9.5	0.5	1.6	25.2
All persons						
Great Britain	1.3	0.3	17.5	0.6	4.6	22.8
North East	0.7	0.3	21.2	0.7	5.7	21.1
North West	0.9	0.1	20.2	0.6	4.7	23.8
Yorkshire and the Humber	1.2	0.4	21.4	0.6	5.0	22.3
East Midlands	1.8	0.5	25.5	0.5	4.8	21.8
West Midlands	1.2	0.2	25.8	0.8	4.2	21.5
East	2.2	0.2	17.0	0.6	5.0	24.1
London	0.1	0.1	7.8	0.2	3.2	22.5
South East	1.6	0.1	13.8	0.6	4.4	23.7
South West	2.1	0.3	16.6	0.7	4.7	24.3
England	1.3	0.2	17.5	0.6	4.4	22.9
Wales	1.9	0.4	21.8	0.4	5.1	21.5
Scotland	2.0	1.3	15.6	0.9	5.8	22.4

5.4 *(continued)*

Percentages and thousands

	Transport, storage & communi- cation	Financial & business services	Public admin- istration & defence	Education, social work & health services	Other	Whole economy (=100%) (thousands)
Males						
Great Britain	8.1	17.9	5.9	8.2	4.0	11,976
North East	6.9	11.3	7.1	10.0	4.0	458
North West	8.0	14.3	5.6	8.2	3.9	1,327
Yorkshire and the Humber	7.8	13.4	5.3	8.7	3.6	1,012
East Midlands	7.3	12.6	4.5	7.8	3.1	878
West Midlands	6.8	13.4	4.8	7.2	3.3	1,158
East	8.9	17.5	4.4	7.6	3.6	1,075
London	10.6	32.3	6.4	7.4	5.8	1,853
South East	8.0	21.6	5.7	8.1	3.9	1,703
South West	7.0	15.2	7.4	8.7	3.5	973
England	8.2	18.6	5.7	8.0	4.0	10,436
Wales	6.1	10.6	7.4	9.5	4.3	518
Scotland	7.6	14.1	7.1	9.4	4.2	1,022
Females						
Great Britain	3.4	19.4	5.9	28.4	5.0	11,358
North East	2.9	12.5	8.1	32.1	5.1	462
North West	3.4	16.3	6.3	29.4	4.4	1,294
Yorkshire and the Humber	3.0	15.7	6.4	30.1	4.7	945
East Midlands	3.0	14.8	5.0	28.8	4.4	790
West Midlands	2.9	17.0	5.1	29.1	4.6	1,022
East	4.0	19.3	4.5	27.7	4.8	1,005
London	4.6	31.7	5.7	21.3	6.5	1,760
South East	4.0	22.1	4.9	28.4	5.1	1,648
South West	2.8	16.6	6.5	30.8	4.6	938
England	3.6	20.1	5.7	27.9	5.0	9,861
Wales	2.1	12.3	7.3	33.9	5.0	489
Scotland	2.8	16.0	7.2	30.9	4.9	1,008
All persons						
Great Britain	5.8	18.6	5.9	18.0	4.5	23,351
North East	4.9	11.9	7.6	21.1	4.6	920
North West	5.7	15.3	6.0	18.7	4.1	2,622
Yorkshire and the Humber	5.5	14.5	5.8	19.0	4.1	1,958
East Midlands	5.2	13.7	4.7	17.7	3.7	1,669
West Midlands	5.0	15.1	4.9	17.5	3.9	2,181
East	6.5	18.4	4.4	17.3	4.2	2,081
London	7.7	32.0	6.1	14.2	6.2	3,616
South East	6.1	21.9	5.3	18.1	4.5	3,353
South West	4.9	15.9	7.0	19.5	4.1	1,912
England	6.0	19.3	5.7	17.7	4.5	20,312
Wales	4.1	11.5	7.4	21.3	4.7	1,007
Scotland	5.2	15.1	7.2	20.1	4.6	2,031

1 Based on SIC 1992. See Notes and Definitions.

Source: Annual Employment Survey, Office for National Statistics

5.5 Employee jobs and self-employment jobs[1]: by gender

Thousands

| | Employee jobs | | | | | | Self-employment jobs[2] | | | | | |
| | Males | | | Females | | | Males | | | Females | | |
	1997	1998	1999	1997	1998	1999	1997	1998	1999	1997	1998	1999
United Kingdom	11,898	12,244	12,380	11,490	11,728	11,835	2,642	2,524	2,484	996	979	946
North East	453	458	447	450	464	464	80	73	81	25	22	24
North West	1,309	1,328	1,331	1,285	1,299	1,314	275	251	268	85	85	83
Yorkshire and the Humber	997	1,014	1,032	946	952	953	196	191	188	77	77	61
East Midlands	847	869	851	803	778	799	174	168	169	63	63	68
West Midlands	1,126	1,155	1,141	1,005	1,030	1,054	208	211	195	69	68	73
East	1,053	1,071	1,103	999	1,012	1,023	260	263	261	101	104	103
London	1,777	1,864	1,903	1,699	1,776	1,779	341	361	345	141	147	132
South East	1,616	1,682	1,722	1,577	1,654	1,653	440	392	377	184	181	174
South West	947	960	994	931	951	968	279	256	249	118	110	106
England	10,124	10,402	10,524	9,696	9,915	10,006	2,254	2,166	2,132	863	856	823
Wales	498	518	537	484	492	493	124	111	118	45	42	38
Scotland	978	1,022	1,013	1,008	1,015	1,025	193	175	167	69	67	69
Northern Ireland	299	303	306	302	307	311	72	72	68	18	14	16

1 At September each year. See Notes and Definitions.
2 With or without employees.

Source: Short-term Employment and Labour Force Surveys, Office for National Statistics; Quarterly Employment and Labour Force Surveys.

5.6 Self-employment[1]: by broad industry group[2], Spring 1999

Percentages and thousands

| | | Industry | | | | Total self-employed[4] (=100%) (thousands) |
	Agriculture & fishing	Manufacturing	Construction	All industry[3]	Services	
United Kingdom	6.4	7.3	20.8	28.2	65.4	3,202
North East	..	..	17.9	24.3	70.0	92
North West	6.4	8.5	19.6	28.2	65.4	321
Yorkshire and the Humber	5.9	9.3	20.0	29.7	64.4	223
East Midlands	7.6	7.9	19.3	27.6	64.9	227
West Midlands	5.5	8.9	22.5	31.4	63.1	256
East	4.3	8.1	24.9	33.2	62.5	348
London	..	5.8	17.6	23.4	75.6	462
South East	3.3	6.7	21.6	28.3	68.3	504
South West	7.9	7.4	24.1	31.6	60.5	331
England	4.8	7.5	21.0	28.7	66.5	2,765
Wales	16.8	7.2	21.9	29.3	53.9	141
Scotland	10.1	..	17.4	22.1	67.8	214
Northern Ireland	31.3	..	17.7	25.0	43.7	83

1 Main job only.
2 Based on SIC 1992. In some cases, sample sizes are too small to provide a reliable estimate.
3 Includes SIC groups C and E: Quarrying, Energy and Water.
4 Total includes those who did not state their industry and those whose workplace is outside the United Kingdom, but percentages are based on figures which exclude them.

Source: Labour Force Survey, Office for National Statistics

5.7 Reasons given for working part-time[1], Spring 1999

Percentages and thousands

	Males				Females			
	Did not want a full-time job	Could not find a full-time job	Student or at school	All part-time workers[2,3] (=100%) (thousands)	Did not want a full-time job	Could not find a full-time job	Student or at school	All part-time workers[2,3] (=100%) (thousands)
United Kingdom	41.7	21.1	34.0	1,316	80.1	7.6	10.7	5,394
North East	35.2	31.0	30.8	54	79.9	10.7	8.7	224
North West	38.9	21.6	37.7	125	78.7	8.2	11.6	592
Yorkshire and the Humber	37.3	26.1	32.0	111	79.4	8.6	10.2	488
East Midlands	43.2	20.9	32.8	95	83.5	7.1	7.9	418
West Midlands	39.9	21.3	34.7	109	81.2	6.2	10.9	495
East	49.4	14.3	34.1	125	83.3	5.6	9.3	516
London	39.7	25.5	32.5	190	74.3	8.0	15.8	510
South East	47.6	13.2	36.0	201	82.8	5.1	10.0	815
South West	45.2	18.2	31.4	128	81.2	7.4	9.7	527
England	42.6	20.4	33.8	1,139	80.6	7.1	10.6	4,584
Wales	38.1	30.2	27.2	47	78.3	10.1	10.6	244
Scotland	35.1	22.6	39.8	107	76.4	10.0	11.9	458
Northern Ireland[4]	..	..	..	24	76.7	12.5	8.9	108

1 Based on respondents' own definition of part-time.
2 Employees and the self-employed only.
3 Includes people who said they worked part-time because they were ill or disabled. Hence percentages shown do not add to 100 per cent.
4 Some sample sizes are too small to provide reliable estimates.

Source: Labour Force Survey, Office for National Statistics

5.8 Part-time[1] working: by gender, Spring 1999

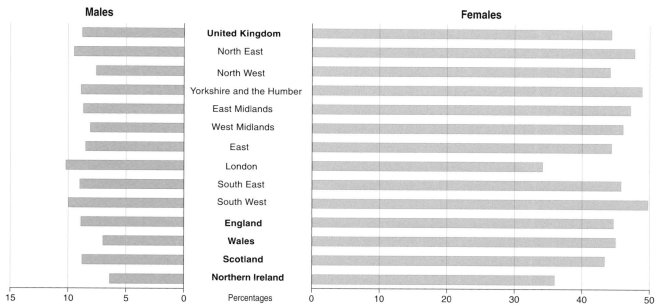

1 Part-time workers as a percentage of all in employment. Based on respondents' own definition of part-time.

Source: Labour Force Survey, Office for National Statistics

5.9 People in employment with a second job: by gender, Spring 1999

Thousands and percentages

	People with a second job (thousands)			As a percentage of all in employment		
	Males	Females	All persons	Males	Females	All persons
United Kingdom	552	749	1,301	3.6	6.1	4.7
North East	19	29	48	3.4	6.1	4.6
North West	50	78	128	3.0	5.7	4.2
Yorkshire and the Humber	46	55	100	3.6	5.4	4.4
East Midlands	43	50	93	3.9	5.6	4.7
West Midlands	44	77	121	3.2	7.1	4.9
East	49	74	122	3.3	6.3	4.6
London	59	70	129	3.1	4.6	3.8
South East	90	125	215	4.0	7.0	5.3
South West	60	88	148	4.6	8.2	6.3
England	460	645	1,105	3.6	6.2	4.8
Wales	29	33	62	4.3	6.0	5.0
Scotland	48	59	107	3.9	5.5	4.7
Northern Ireland	16	12	28	4.2	3.8	4.0

Source: Labour Force Survey, Office for National Statistics

5.10 Average usual weekly hours[1] of work of full-time employees: by occupational group, Spring 1999

Hours

	Managers & administrators	Professional, associate professional & technical	Clerical & secretarial	Craft & related	Personal & protective services	Sales	Plant & machine operative	Other[2]	All occupations[3]
United Kingdom	46.3	44.0	39.6	44.2	42.4	42.2	45.1	43.6	43.7
North East	45.0	44.1	39.2	44.0	41.9	41.1	44.1	41.1	43.1
North West	45.8	43.9	39.3	43.7	41.7	41.3	45.3	41.8	43.2
Yorkshire and the Humber	46.2	44.0	39.4	44.8	41.8	41.1	44.7	43.9	43.6
East Midlands	47.1	44.0	39.9	44.2	42.4	42.1	46.4	43.4	44.0
West Midlands	46.1	43.8	39.7	43.6	42.6	42.9	43.6	44.0	43.4
East	46.6	44.2	40.1	44.5	43.2	42.5	46.1	44.4	44.1
London	46.3	44.7	39.5	44.2	42.8	42.6	44.8	44.8	43.8
South East	46.8	44.5	39.6	45.1	43.3	42.9	46.6	45.0	44.4
South West	46.7	44.4	39.5	44.7	42.0	43.4	44.9	43.5	43.9
England	46.4	44.3	39.6	44.3	42.5	42.3	45.1	43.8	43.8
Wales	45.2	43.5	39.0	43.9	42.9	42.0	45.3	40.9	43.3
Scotland	46.4	42.8	39.8	44.2	42.3	41.7	44.8	44.0	43.3
Northern Ireland	44.0	41.9	39.4	43.3	39.6	42.0	44.3	40.7	42.0

1 Includes paid and unpaid overtime and excludes meal breaks. The analysis also excludes those who did not state the number of hours they worked.
2 See Notes and Definitions.
3 Includes those whose workplace is outside the United Kingdom, and those who did not specify their occupation.

Source: Labour Force Survey, Office for National Statistics

5.11 Average weekly earnings[1]: by industry[2] and gender, April 1999

£ per week

	Whole economy			Agriculture, forestry, fishing & hunting		Manufacturing	
	Males	Females	All persons	Males	Females	Males	Females
United Kingdom	440.7	325.6	398.7	298.0	232.0	422.5	290.3
North East	384.6	289.8	349.6	269.4	..	394.0	270.2
North West	415.1	299.4	372.6	..	..	425.2	277.9
Yorkshire and the Humber	395.8	297.9	361.0	..	..	395.6	261.0
East Midlands	398.3	286.7	361.7	332.4	..	395.8	250.6
West Midlands	414.6	301.0	375.6	..	..	403.0	262.0
East	436.0	323.9	396.6	306.5	..	444.6	306.9
London	584.4	422.8	520.0	..	..	550.4	423.7
South East	471.2	341.0	423.2	316.5	..	478.6	343.3
South West	402.9	297.8	364.9	259.9	..	407.0	293.2
England	448.1	330.6	405.4	302.4	224.3	427.9	296.2
Wales	384.0	298.3	353.6	..	..	397.7	271.9
Scotland	406.0	297.7	364.9	300.6	..	402.3	262.0
Northern Ireland	376.8	295.1	344.9	212.9	..	342.3	231.5

	Mining, quarrying & electricity, gas, water		Construction		Distribution, hotels & catering, repairs		Transport, storage & communication	
	Males	Females	Males	Females	Males	Females	Males	Females
United Kingdom	517.8	364.9	398.2	303.7	381.0	261.1	422.4	342.7
North East	..	..	366.4	..	326.3	220.2	360.3	..
North West	519.9	..	378.8	..	362.6	243.6	399.8	292.5
Yorkshire and the Humber	476.5	..	365.4	344.4	342.0	238.1	389.3	308.0
East Midlands	464.5	..	..	..	372.1	240.2	357.3	281.1
West Midlands	513.1	..	385.5	..	369.9	238.3	368.0	..
East	476.3	321.9	408.5	..	393.2	275.2	433.3	329.2
London	..	..	500.3	..	435.7	312.2	537.5	441.2
South East	558.3	..	442.7	..	441.3	300.6	429.2	354.2
South West	516.2	..	357.5	..	331.8	231.1	372.4	298.5
England	524.6	362.8	404.3	308.1	388.6	266.4	429.6	349.9
Wales	473.8	..	346.6	..	308.9	223.2	355.3	275.5
Scotland	530.2	..	389.2	..	342.3	234.4	379.2	299.7
Northern Ireland	376.4	..	321.8	..	317.5	208.4	381.8	295.6

	Financial & business services		Public administration & defence		Education, social work & health services		Other	
	Males	Females	Males	Females	Males	Females	Males	Females
United Kingdom	551.6	363.2	439.3	328.0	443.6	347.3	420.3	327.1
North East	403.8	293.4	397.5	306.8	410.3	320.6	..	..
North West	463.8	303.2	415.7	322.9	429.5	329.5	..	283.9
Yorkshire and the Humber	445.4	305.6	422.4	322.4	427.4	335.2	340.8	243.3
East Midlands	437.6	304.4	421.9	308.8	423.3	325.4	..	..
West Midlands	483.0	323.7	453.0	305.2	459.0	339.8	..	265.6
East	497.4	330.9	418.6	333.4	447.8	358.6	374.4	..
London	747.3	486.1	513.0	394.8	501.8	412.6	564.9	448.9
South East	551.6	366.2	445.6	336.3	447.3	349.8	387.2	289.5
South West	481.5	303.4	423.9	291.8	434.1	332.2	343.3	..
England	563.6	370.8	444.5	333.0	447.1	349.3	427.1	332.4
Wales	408.4	288.8	408.3	316.8	434.9	336.1	..	..
Scotland	467.3	317.7	411.9	311.7	413.9	326.8	..	285.2
Northern Ireland	405.6	268.2	450.1	302.3	448.0	354.7	340.2	301.9

1 Average gross weekly earnings; data relate to full-time employees on adult rates whose pay for the survey pay-period was not affected by absence. See Notes and Definitions.
2 Classification is based on SIC 1992.

Source: New Earnings Survey, Office for National Statistics and Department of Enterprise, Trade and Investment, Northern Ireland.

5.12 Average weekly earnings and hours: by gender, April 1999[1]

| | Average gross weekly earnings | | | | | | | | Percentage of employees who received overtime pay | Average weekly hours | |
| | Total (£) | of which | | | Percentage earning under | | | | | Total including overtime (hours) | Overtime (hours) |
		Overtime pay (£)	PBR pay[2] (£)	Shift etc premium pay (£)	£200	£300	£400	£500			
All full-time male employees											
United Kingdom	440.7	26.7	17.2	6.5	8.1	31.9	55.8	72.9	32.4	41.4	2.8
North East	384.6	27.3	17.0	9.4	10.0	38.4	63.1	80.5	35.0	41.2	2.8
North West	415.1	28.0	14.5	8.5	8.9	34.9	58.6	76.3	33.3	41.4	2.9
Yorkshire and the Humber	395.8	27.9	17.5	6.6	9.2	36.7	62.2	79.8	34.1	41.7	3.1
East Midlands	398.3	30.1	14.9	5.9	8.8	35.1	62.2	79.2	36.9	42.1	3.3
West Midlands	414.6	27.3	16.2	6.6	7.8	33.7	60.0	77.3	33.6	41.5	2.8
East	436.0	29.0	16.7	5.6	7.2	30.0	55.3	72.5	34.2	41.9	3.0
London	584.4	22.0	22.6	5.0	4.7	19.5	38.9	55.9	24.5	40.2	2.0
South East	471.2	25.2	21.5	5.0	6.2	27.6	50.3	67.4	29.8	41.3	2.5
South West	402.9	24.7	12.9	6.5	9.9	35.5	60.0	76.5	33.9	41.4	2.6
England	448.1	26.4	17.7	6.3	7.6	31.0	54.9	72.1	32.0	41.3	2.7
Wales	384.0	25.5	12.8	10.0	10.7	38.4	63.5	80.5	33.3	41.6	2.7
Scotland	406.0	29.1	14.9	7.3	9.0	34.8	59.5	76.8	35.8	41.4	3.0
Northern Ireland	376.8	28.4	11.4	5.6	14.3	42.8	64.0	79.0	35.4	41.3	2.9
Full-time manual male employees											
United Kingdom	333.9	43.8	13.3	10.9	11.4	45.9	75.5	90.0	50.1	44.4	4.8
North East	325.6	41.0	19.9	13.1	13.1	49.9	77.1	90.9	49.1	43.3	4.3
North West	332.6	43.2	12.6	13.6	11.4	47.5	75.1	90.0	49.3	44.0	4.7
Yorkshire and the Humber	321.5	45.0	15.8	10.7	12.3	49.0	78.5	92.4	51.1	44.6	5.2
East Midlands	329.7	47.0	15.8	9.1	11.0	46.2	77.4	91.2	53.3	44.9	5.3
West Midlands	328.2	41.1	14.0	11.2	10.2	45.6	77.9	91.9	48.9	43.8	4.6
East	341.2	48.2	12.2	9.5	9.6	42.2	74.0	89.2	54.2	45.2	5.4
London	376.9	46.1	11.1	10.1	8.6	34.2	64.1	82.2	46.2	44.8	4.8
South East	347.1	45.6	10.6	8.6	9.1	41.8	72.2	87.5	50.0	45.0	4.9
South West	315.4	40.0	8.3	10.5	15.0	50.5	80.7	93.2	51.7	44.3	4.6
England	336.0	44.3	13.0	10.6	11.0	44.9	75.1	89.7	50.4	44.5	4.9
Wales	326.4	37.1	12.9	15.4	13.8	49.4	75.3	90.6	45.2	43.9	4.2
Scotland	328.6	43.8	16.6	11.0	11.2	48.2	76.7	91.4	50.8	44.3	4.9
Northern Ireland	296.6	39.6	13.9	8.0	19.4	61.5	85.5	93.7	49.0	43.8	4.6
Full-time non-manual male employees											
United Kingdom	523.8	13.4	20.3	3.1	5.5	21.0	40.5	59.7	18.6	39.0	1.1
North East	449.5	12.3	13.8	5.3	6.6	25.7	47.6	69.1	19.4	38.9	1.2
North West	487.8	14.7	16.2	3.9	6.6	23.7	44.1	64.2	19.2	39.1	1.2
Yorkshire and the Humber	468.1	11.3	19.1	2.7	6.2	24.8	46.4	67.5	17.5	38.9	1.0
East Midlands	467.5	13.0	14.1	2.7	6.6	24.0	46.9	67.0	20.3	39.3	1.2
West Midlands	497.0	14.2	18.4	2.2	5.5	22.3	42.9	63.3	19.1	39.3	1.2
East	508.5	14.3	20.1	2.6	5.3	20.6	41.1	59.7	18.9	39.4	1.2
London	664.6	12.7	27.1	3.1	3.2	13.8	29.2	45.8	16.1	38.4	0.9
South East	546.5	12.8	28.2	2.8	4.4	19.1	37.0	55.1	17.6	39.1	1.1
South West	477.0	11.6	16.8	3.1	5.6	22.9	42.5	62.4	18.7	38.9	1.0
England	532.1	13.1	21.2	3.0	5.2	20.5	39.7	58.8	18.1	39.0	1.1
Wales	449.7	12.2	12.8	3.8	7.2	25.8	50.1	69.1	19.7	39.0	1.0
Scotland	475.6	15.9	13.4	4.0	7.1	22.8	43.9	63.8	22.3	38.7	1.3
Northern Ireland	450.8	18.1	9.2	3.3	9.7	25.5	44.3	65.5	23.0	39.0	1.4

5.12 *(continued)*

| | Average gross weekly earnings | | | | | | | | Percentage of employees who received overtime pay | Average weekly hours | |
| | Total (£) | of which | | | Percentage earning under | | | | | Total including overtime (hours) | Overtime (hours) |
		Overtime pay (£)	PBR pay[2] (£)	Shift etc premium pay (£)	£200	£300	£400	£500			
All full-time female employees											
United Kingdom	325.6	7.5	7.9	2.9	20.4	54.7	75.3	88.2	18.0	37.5	0.8
North East	289.8	6.5	5.8	3.0	27.1	63.9	81.1	93.0	17.7	37.4	0.8
North West	299.4	7.8	7.3	3.5	23.4	60.9	80.2	91.7	18.4	37.6	0.9
Yorkshire and the Humber	297.9	7.6	6.2	2.7	26.1	62.2	79.7	91.8	18.7	37.5	0.9
East Midlands	286.7	8.0	7.9	2.8	28.0	64.5	83.5	93.4	20.5	37.8	0.9
West Midlands	301.0	6.3	9.3	2.5	23.8	62.3	80.6	90.9	17.4	37.6	0.8
East	323.9	8.4	8.3	2.8	18.8	54.7	75.8	88.8	18.8	37.7	0.9
London	422.8	8.0	10.5	2.1	7.6	30.2	56.6	75.0	15.1	37.2	0.7
South East	341.0	8.1	8.7	2.8	15.1	50.7	74.0	87.0	18.6	37.8	0.9
South West	297.8	7.2	6.4	3.3	24.7	62.2	81.6	91.9	19.6	37.6	0.9
England	330.6	7.7	8.2	2.8	19.5	53.6	74.7	87.6	18.0	37.6	0.8
Wales	298.3	5.8	5.2	3.8	25.0	61.9	79.7	91.5	16.5	37.3	0.7
Scotland	297.7	7.4	6.3	3.2	23.8	61.2	79.0	92.4	19.2	37.0	0.8
Northern Ireland	295.1	6.7	5.1	4.9	28.4	59.7	78.4	90.7	16.5	37.4	0.8
Full-time manual female employees											
United Kingdom	221.0	13.5	7.2	6.3	49.4	85.5	96.3	98.9	29.1	39.9	1.9
North East	207.4	13.6	4.6	6.2	54.9	87.8	96.2	99.1	27.3	39.8	1.9
North West	221.5	14.4	7.8	6.4	49.1	84.6	96.0	99.2	28.9	39.8	2.1
Yorkshire and the Humber	209.9	13.8	7.8	6.8	56.5	89.9	97.6	98.9	29.0	39.9	2.0
East Midlands	209.1	12.9	10.6	4.3	55.0	89.2	97.3	100.0	30.5	39.6	1.8
West Midlands	217.0	12.2	9.4	5.4	48.0	88.2	98.3	99.4	29.0	39.9	1.8
East	222.1	15.7	7.0	5.3	47.4	86.1	97.0	99.2	32.5	40.1	2.2
London	261.2	16.5	4.8	6.1	34.6	70.0	90.5	95.7	29.7	40.3	2.1
South East	235.7	15.6	4.9	7.9	39.6	82.9	95.2	98.3	30.8	40.5	2.1
South West	213.6	13.1	4.7	6.2	55.1	86.2	96.7	99.6	30.8	39.6	1.9
England	223.3	14.2	7.0	6.2	48.2	84.8	96.1	98.8	29.9	40.0	2.0
Wales	209.5	9.9	6.9	7.7	55.3	90.3	98.1	100.0	26.9	39.4	1.4
Scotland	216.2	11.2	8.4	7.5	52.1	86.6	96.6	98.8	25.4	39.3	1.5
Northern Ireland	189.8	9.0	11.7	3.4	66.5	94.1	99.5	100.0	23.6	39.3	1.4
Full-time non-manual female employees											
United Kingdom	346.1	6.4	8.0	2.2	14.7	48.7	71.2	86.1	15.9	37.0	0.6
North East	308.9	4.9	6.1	2.2	20.6	58.3	77.6	91.5	15.5	36.8	0.5
North West	315.2	6.5	7.2	2.9	18.2	56.1	77.0	90.2	16.3	37.1	0.7
Yorkshire and the Humber	317.6	6.2	5.8	1.8	19.2	56.0	75.7	90.3	16.4	36.9	0.7
East Midlands	309.1	6.5	7.1	2.3	20.2	57.3	79.5	91.5	17.6	37.2	0.7
West Midlands	321.3	4.8	9.2	1.8	18.0	56.0	76.3	88.8	14.7	37.0	0.5
East	341.1	7.1	8.5	2.4	13.9	49.4	72.2	87.1	16.5	37.2	0.7
London	439.1	7.2	11.1	1.7	4.9	26.2	53.2	72.9	13.6	36.9	0.6
South East	359.4	6.8	9.3	1.9	10.8	45.0	70.3	85.0	16.5	37.4	0.7
South West	315.3	5.9	6.7	2.6	18.3	57.2	78.4	90.2	17.3	37.2	0.7
England	350.8	6.4	8.4	2.1	14.1	47.7	70.6	85.5	15.8	37.1	0.6
Wales	319.7	4.8	4.8	2.9	17.7	55.0	75.2	89.4	14.0	36.8	0.5
Scotland	317.7	6.5	5.8	2.1	16.8	55.0	74.7	90.8	17.7	36.4	0.6
Northern Ireland	317.4	6.2	3.7	5.2	20.3	52.4	74.0	88.7	15.0	37.0	0.7

1 Data relate to full-time employees on adult rates whose pay for the survey pay-period was not affected by absence. See Notes and Definitions.
2 PBR pay is payments-by-results, bonuses, commission and all other incentive payments plus profit-related payments.

Source: New Earnings Survey, Office for National Statistics and Department of Enterprise, Trade and Investment, Northern Ireland

5.13 Claimant count rate[1]: by sub-region[2], March 2000

Percentages

- 9.0 or over
- 6.5 to 8.9
- 5.5 to 6.4
- 4.5 to 5.4
- 3.5 to 4.4
- 2.5 to 3.4
- 2.4 or under

1 The claimant count rate is the number of people claiming unemployment-related benefit as a proportion of claimant count and workforce jobs in each area.
    Not seasonally adjusted. See Notes and Definitions.
2 Travel-to-work areas for Northern Ireland. See map on page 234.

Source: Office for National Statistics

5.14 ILO unemployment rates: by age, 1999-2000[1]

Percentages and thousands

	Percentage of the economically active[2] who are unemployed and aged				All ILO unemployed of working age (thousands)
	16-24	25-34	35-49	Males 50-64, females 50-59	
United Kingdom	12.9	5.7	4.3	4.4	1,728
North East	18.9	8.4	6.9	6.7	107
North West	13.7	5.8	4.3	5.2	205
Yorkshire and the Humber	13.5	6.5	4.2	4.6	155
East Midlands	14.1	4.7	3.5	3.7	114
West Midlands	14.3	6.6	4.5	4.8	171
East	9.5	3.6	2.8	3.2	111
London	14.4	6.1	6.2	6.4	270
South East	8.4	3.7	2.7	2.8	157
South West	9.6	4.3	3.4	3.0	108
England	12.4	5.3	4.1	4.3	1,397
Wales	16.9	7.7	4.7	4.4	97
Scotland	15.4	7.2	4.9	5.8	181
Northern Ireland	10.5	7.9	6.5	..	53

1 Average of four quarters ending Winter 1999/2000. See Notes and Definitions.
2 Those of working age who are economically active.

Source: Labour Force Survey, Office for National Statistics

5.15 ILO unemployment rates[1]: by duration and gender, 1999-2000[2]

Percentages and thousands

	Males					Females				
	6 months or less	6 months and up to 12 months	1 year and up to 2 years	2 years and over	Total (= 100%) (thousands)	6 months or less	6 months and up to 12 months	1 year and up to 2 years	2 years and over	Total (= 100%) (thousands)
United Kingdom	3.3	0.9	0.7	1.0	1,065	3.3	1.0	0.9	1.4	684
North East	4.4	1.8	1.4	1.7	69	4.6	1.9	1.7	2.4	39
North West	3.3	1.0	0.7	1.2	135	3.5	1.2	0.9	1.7	71
Yorkshire and the Humber	3.6	1.0	0.6	1.1	100	3.8	1.1	0.7	1.6	55
East Midlands	3.2	0.9	0.6	0.7	67	2.9	1.1	0.7	0.9	47
West Midlands	3.6	1.1	0.8	1.0	102	3.4	1.1	1.0	1.4	71
East	2.6	0.6	0.4	0.5	67	2.5	0.7	0.6	0.6	46
London	4.0	1.1	1.1	1.3	163	4.0	1.1	1.1	1.7	112
South East	2.6	0.5	0.3	0.5	85	2.2	0.5	0.4	0.6	75
South West	2.8	0.7	0.4	0.6	62	2.6	0.7	0.5	0.8	47
England	4.0	1.3	1.0	1.0	851	4.1	1.6	1.2	1.6	564
Wales	3.9	1.1	0.9	1.4	64	4.2	1.1	1.1	2.1	34
Scotland	2.9	..	1.2	2.2	116	2.8	..	1.5	3.0	67
Northern Ireland	3.3	0.9	0.7	0.9	34	3.2	1.0	0.8	1.2	19

1 For those aged 16 or over. Not seasonally adjusted. See Notes and Definitions.
2 Average of four quarters ending Winter 1999/2000. See Notes and Definitions.

Source: Office for National Statistics

5.16 ILO unemployment rates[1]

Percentages

	Spring quarter of each year				
	1995	1996	1997	1998	1999
United Kingdom	8.6	8.2	7.1	6.1	6.0
North East	11.4	10.8	9.8	8.1	10.1
North West	9.0	8.4	6.8	6.6	6.2
Yorkshire and the Humber	8.7	8.1	8.0	7.0	6.5
East Midlands	7.4	7.4	6.3	4.9	5.2
West Midlands	8.9	9.2	6.8	6.3	6.8
East	7.5	6.1	5.8	5.0	4.1
London	11.5	11.3	9.1	8.2	7.6
South East	6.4	6.0	5.2	4.3	3.6
South West	7.8	6.3	5.2	4.5	4.7
England	8.6	8.1	6.8	6.0	5.8
Wales	8.8	8.3	8.3	6.7	7.0
Scotland	8.4	8.7	8.5	7.4	7.4
Northern Ireland	11.0	9.5	7.5	7.2	7.2

1 For those of working age. Not seasonally adjusted. See Notes and Definitions.

Source: Labour Force Survey, Office for National Statistics

5.17 Claimant count rates[1]

Percentages

	Seasonally adjusted annual averages				
	1995	1996	1997	1998	1999
United Kingdom	8.0	7.2	5.5	4.7	4.3
North East	11.2	10.2	8.4	7.5	7.3
North West	8.5	7.7	6.0	5.3	4.9
Yorkshire and the Humber	8.5	7.8	6.3	5.6	5.2
East Midlands	7.4	6.7	4.8	4.1	3.8
West Midlands	8.1	7.2	5.5	4.7	4.6
East	6.5	5.9	4.1	3.3	3.0
London	9.4	8.5	6.4	5.2	4.7
South East	5.9	5.0	3.4	2.7	2.4
South West	6.8	6.1	4.3	3.5	3.1
England	7.8	7.1	5.3	4.4	4.1
Wales	8.4	8.0	6.4	5.6	5.2
Scotland	7.9	7.6	6.4	5.7	5.4
Northern Ireland	11.2	10.7	8.1	7.3	6.5

1 See Notes and Definitions.

Source: Office for National Statistics

5.18 Vacancies[1] at jobcentres

Thousands

	1995	1996	1997	1998	1999[2]
United Kingdom	181.9	224.9	283.3	295.8	314.2
North East	6.4	8.0	10.1	11.0	16.0
North West	22.7	26.6	34.3	40.9	37.1
Yorkshire and the Humber	13.3	16.6	20.9	22.6	24.1
East Midlands	12.7	14.7	20.3	20.6	21.2
West Midlands	15.2	18.7	23.1	30.1	35.7
East	14.7	17.6	33.5	24.0	24.0
London	16.3	28.5	35.2	28.2	32.0
South East	22.7	28.0	34.4	34.8	37.5
South West	14.4	19.1	25.4	26.1	27.8
England	138.2	178.0	227.2	238.0	255.3
Wales	13.3	14.5	18.0	17.9	17.1
Scotland	23.1	25.4	31.3	31.0	32.9
Northern Ireland	7.4	7.0	6.8	8.9	..

1 Vacancies remaining unfilled, seasonally adjusted annual averages.
2 The publication of the vacancy figures for Northern Ireland has been suspended since March 1999 as a result of a difficulty caused by the introduction of a new computer system for processing vacancies to Training and Employment Agency Offices. For the purposes of the seasonally adjusted United Kingdom figures it has been assumed provisionally that the Northern Ireland figures have remained constant since February 1999.

Source: Employment Service

5.19 Redundancies[1], Spring 1999[2]

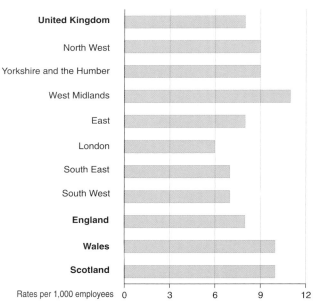

Rates per 1,000 employees

1 See Notes and Definitions. For the North East, East Midlands and Northern Ireland the sample sizes are too small to provide reliable estimates but are included in the UK total.
2 Due to a change in definition, the figures shown are not comparable with those shown in previous editions. See Notes and Definitions.

Source: Labour Force Survey, Office for National Statistics

5.20 Claimant count[1]: by age and gender, March 2000

Percentages and thousands

	Percentage aged						Total (=100%) (thousands)
	Under 20	20-29	30-39	40-49	50-59	60 or over	
Males							
United Kingdom	8.4	30.1	27.0	18.0	15.3	1.2	906.5
North East	9.4	29.5	25.0	18.6	16.3	1.2	63.4
North West	9.6	32.6	26.5	16.7	13.6	1.0	117.7
Yorkshire and the Humber	9.1	31.5	26.0	17.3	15.1	1.1	91.5
East Midlands	8.8	30.6	25.4	17.6	16.0	1.6	57.0
West Midlands	8.7	30.7	26.4	17.8	15.1	1.4	86.2
East	7.6	28.9	25.6	18.4	17.5	1.9	53.9
London	5.7	28.3	32.9	19.1	13.1	1.0	137.3
South East	7.0	27.9	26.4	19.1	17.7	1.8	66.9
South West	7.9	29.7	25.5	17.8	17.5	1.6	51.6
England	8.1	30.1	27.3	18.0	15.2	1.3	725.4
Wales	9.6	31.9	25.6	16.9	15.2	0.8	47.7
Scotland	9.8	29.9	25.6	18.0	15.5	1.2	100.3
Northern Ireland	8.0	30.3	26.5	19.1	15.1	0.9	33.1
Females							
United Kingdom	15.7	29.3	18.3	18.3	18.4	-	278.7
North East	20.9	28.5	15.8	18.6	16.2	-	15.8
North West	19.3	30.3	16.8	17.3	16.2	-	32.6
Yorkshire and the Humber	18.2	30.0	17.0	17.3	17.4	-	26.4
East Midlands	15.5	28.8	17.2	18.4	20.0	-	18.4
West Midlands	16.9	29.5	17.4	17.4	18.9	-	26.7
East	14.0	26.5	17.6	19.3	22.6	-	18.7
London	10.0	32.1	23.2	18.5	16.2	-	47.9
South East	12.7	27.1	18.8	19.8	21.7	-	21.1
South West	14.1	28.0	17.8	18.7	21.4	-	18.6
England	15.2	29.5	18.6	18.3	18.4	-	226.1
Wales	18.0	29.6	16.2	18.1	18.2	-	13.6
Scotland	18.0	26.6	17.9	19.2	18.2	-	29.3
Northern Ireland	16.2	32.4	16.7	17.2	17.5	-	9.6
All persons							
United Kingdom	10.1	29.9	24.9	18.1	16.0	1.0	1,194.3
North East	11.7	29.3	23.1	18.6	16.3	1.0	79.6
North West	11.7	32.1	24.4	16.8	14.1	0.8	151.3
Yorkshire and the Humber	11.1	31.1	24.0	17.3	15.6	0.8	118.6
East Midlands	10.4	30.1	23.4	17.8	17.0	1.2	75.9
West Midlands	10.6	30.4	24.3	17.7	16.0	1.0	113.5
East	9.3	28.3	23.6	18.7	18.9	1.4	73.1
London	6.8	29.3	30.4	18.9	13.9	0.7	187.6
South East	8.4	27.7	24.6	19.3	18.7	1.4	88.7
South West	9.5	29.2	23.5	18.0	18.5	1.2	70.6
England	9.8	29.9	25.2	18.1	16.0	1.0	958.8
Wales	11.5	31.4	23.5	17.2	15.9	0.6	61.8
Scotland	11.7	29.1	23.9	18.3	16.1	0.9	130.6
Northern Ireland	9.8	30.8	24.3	18.7	15.7	0.7	43.2

1 Not seasonally adjusted. See Notes and Definitions.

Source: Office for National Statistics

5.21 Number of starts on the New Deal 18 to 24[1]: by gender

Thousands

	Males		Females		All persons	
	1998	1999	1998	1999	1998	1999
Great Britain[2]	156.7	135.0	57.2	54.7	213.9	189.7
North East	12.3	10.4	3.6	3.8	15.9	14.2
North West	22.9	19.2	7.6	7.0	30.5	26.2
Yorkshire and the Humber	19.1	15.2	6.6	6.0	25.7	21.2
East Midlands	9.5	8.3	3.4	3.4	12.9	11.7
West Midlands	14.5	13.9	5.9	6.0	20.4	19.9
East	8.6	7.1	3.3	3.1	11.9	10.2
London	21.6	18.2	10.9	9.4	32.5	27.6
South East	10.1	8.5	3.5	3.2	13.6	11.7
South West	9.1	7.8	3.4	3.3	12.5	11.1
England	127.7	108.6	48.2	45.2	175.9	153.8
Wales	10.3	8.5	3.2	3.0	13.5	11.5
Scotland	17.2	16.0	5.3	6.0	22.5	22.0

1 See Notes and Definitions.
2 Includes clients for whom the region is recorded as unknown.

Source: Employment Service

5.22 Immediate destination on leaving New Deal 18 to 24: by gender, 1999

Percentages

	Males				Females			
	Unsubsidised job[1]	Other benefits	Other known destination[2]	Unknown destination	Unsubsidised job[1]	Other benefits	Other known destination[2]	Unknown destination
Great Britain[3]	40.8	8.8	20.6	29.8	37.0	18.2	18.9	25.9
North East	40.8	11.9	23.4	24.0	37.2	20.8	19.5	22.5
North West	41.5	10.7	20.0	27.8	38.4	21.8	17.1	22.8
Yorkshire and the Humber	40.0	9.4	20.2	30.4	34.7	19.1	19.2	27.1
East Midlands	41.4	8.2	18.5	31.9	38.8	17.4	17.1	26.7
West Midlands	39.9	8.8	20.0	31.3	34.7	19.0	20.1	26.1
East	45.5	6.8	18.0	29.7	40.4	16.3	17.6	25.6
London	35.2	5.1	20.5	39.2	34.6	13.9	19.6	31.9
South East	43.2	6.1	17.4	33.3	38.4	15.8	17.0	28.8
South West	46.6	7.4	18.5	27.5	43.1	15.2	18.1	23.6
England	40.8	8.3	19.8	31.1	37.0	17.6	18.6	26.8
Wales	43.3	10.7	21.4	24.7	39.4	19.8	18.1	22.7
Scotland	40.2	11.3	25.6	22.9	36.5	21.7	21.5	20.3

1 Those who are recorded by the Employment Service (ES) as having been placed into unsubsidised employment, plus those who are recorded as having terminated their JSA claim in order to go into a job. See Notes and Definitions.
2 Includes young people, who, on leaving New Deal continue to claim JSA.
3 Includes clients for whom the region is recorded as unknown.

Source: Employment Service

5.23 Economic activity rates[1]: by gender

Thousands and percentages

	Males			Females			All persons		
	1997	1998	1999	1997	1998	1999	1997	1998	1999
United Kingdom	84.4	83.9	84.1	71.4	71.5	72.1	78.2	78.0	78.4
North East	80.0	79.4	77.1	68.0	65.2	67.0	74.2	72.6	72.3
North West	81.3	79.4	81.5	68.3	68.5	69.2	75.1	74.2	75.7
Yorkshire and the Humber	82.5	82.8	83.2	70.4	71.4	71.3	76.8	77.4	77.6
East Midlands	85.8	85.8	85.3	74.5	74.0	74.3	80.4	80.2	80.1
West Midlands	85.5	85.7	84.7	69.9	71.3	72.7	78.1	78.9	79.1
East	87.3	88.3	87.9	73.4	74.8	74.5	80.7	81.9	81.5
London	84.4	82.3	84.0	70.2	69.1	70.0	77.5	75.9	77.3
South East	88.5	88.8	88.9	75.5	75.7	76.1	82.3	82.6	82.8
South West	87.1	86.6	87.0	75.2	75.7	76.6	81.5	81.4	82.1
England	85.0	84.6	84.9	71.8	72.0	72.6	78.7	78.6	79.1
Wales	79.9	77.2	79.1	68.6	67.2	67.7	74.5	72.5	73.7
Scotland	82.3	82.1	81.3	71.1	72.0	71.5	76.9	77.2	76.6
Northern Ireland	79.6	80.8	78.5	64.0	63.4	64.8	72.0	72.3	71.9

1 At Spring of each year. Based on the population of working age in private households, student halls of residence and NHS accommodation. See Notes and Definitions.

Source: Labour Force Survey, Office for National Statistics

5.24 Economic activity of households with at least one member of working age[1], Spring 2000

Percentages and thousands

	Percentage of households where			All households with at least one member of working age[2] (thousands)
	All members are in employment	At least one person is ILO unemployed	No one is in employment	
United Kingdom	57.8	7.7	16.6	17,767
North East	51.6	11.5	22.5	788
North West	55.2	7.2	18.4	2,084
Yorkshire and the Humber	57.6	8.5	17.5	1,553
East Midlands	60.2	7.5	13.7	1,260
West Midlands	56.3	8.6	15.9	1,544
East	62.5	5.4	12.5	1,662
London	55.8	9.1	19.1	2,170
South East	64.7	5.0	11.2	2,343
South West	62.1	6.1	13.3	1,399
England	58.9	7.4	15.7	14,804
Wales	51.8	8.5	21.9	874
Scotland	56.2	9.8	20.8	1,607
Northern Ireland	41.2	9.6	21.2	482

1 Males aged 16-64 and females aged 16-59. See Notes and Definitions.
2 Excludes those households for which no data is available.

Source: Labour Force Survey, Office for National Statistics

5.25 Employees absent due to sickness[1], Spring 1999

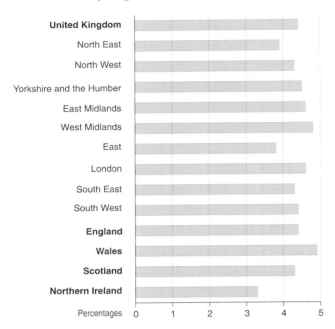

	Percentages
United Kingdom	
North East	
North West	
Yorkshire and the Humber	
East Midlands	
West Midlands	
East	
London	
South East	
South West	
England	
Wales	
Scotland	
Northern Ireland	

1 Percentage of employees absent from work due to illness or injury for at least one day in the week before interview.

Source: Labour Force Survey, Office for National Statistics

5.26 Working days lost due to labour disputes[1]

Days lost per 1,000 employees

	1996	1997	1998	1999
United Kingdom	57	10	12	10
North East	86	38	9	3
North West	56	7	9	5
Yorkshire and the Humber	46	7	1	11
East Midlands	44	3	1	1
West Midlands	56	7	7	1
East	47	5	11	2
London	85	13	13	16
South East	40	2	1	4
South West	51	-	1	2
England	57	8	6	6
Wales	62	3	2	4
Scotland	60	26	25	22
Northern Ireland	35	23	6	10

1 Regional rates are based on data for stoppages that exclude widespread disputes that cannot be allocated to a specific region. These are included in the United Kingdom strike rate only. See Notes and Definitions.

Source: Office for National Statistics

5.27 Trade union membership, Autumn 1999

Percentages[1]

	Manual			Non-manual			
	Males	Females	All manual employees	Males	Females	All non-manual employees	All employees[2]
United Kingdom	34	21	29	28	31	30	30
North East	44	28	38	40	40	40	39
North West	40	25	35	33	35	34	34
Yorkshire and the Humber	36	25	32	32	36	34	33
East Midlands	34	25	31	27	31	29	30
West Midlands	33	23	30	30	32	31	30
East	26	14	22	24	25	25	24
London	31	23	28	23	28	26	26
South East	28	12	22	21	24	23	22
South West	28	12	22	26	30	29	26
England	33	20	28	27	30	29	28
Wales	40	27	36	44	41	42	39
Scotland	35	27	32	36	38	37	35
Northern Ireland	36	30	34	38	39	39	37

1 As a percentage of all employees in each region, excluding the armed forces and those who did not say whether they belonged to a trade union.
2 Includes some people who did not state whether they were manual or non-manual.

Source: Labour Force Survey, Office for National Statistics

6 Housing

Dwelling stock

The stock of dwellings grew by more than 20 per cent in the East of England, South West and South East during the period 1981 to 1998. This rate of growth was more than double that of the North East.

(Table 6.1)

There were 433 dwelling stock per thousand of the population in the North East compared with 411 in the West Midlands in 1998.

(Table 6.1)

New homes

Private developers remained the main providers of new dwellings across the UK in 1999.

(Table 6.2)

Although private developers remained the main providers of new dwellings across the UK in 1999, there were over two and a half thousand fewer completions than in 1991.

(Table 6.2)

Council house sales

By March 1999, local authorities in the South East had sold or transferred almost three in five of their housing stock since 1979, and those in the South West over half.

(Table 6.3)

Tenure

London had the lowest proportion of owner-occupied homes compared with any other area in 1998.

(Table 6.4)

The percentage of dwellings that are rented from local authorities or new towns continued to fall between 1991 and 1998.

(Table 6.4)

Social sector renting was highest in Scotland and lowest in the South East and South West in 1998.

(Table 6.4)

Type of dwelling

Forty per cent of households in the North East lived in a semi-detached house, in comparison with 17 per cent of those in London in 1998-99.

(Table 6.5)

Over a third of households in Scotland, lived in a purpose built flat or maisonette, higher than in any other region in 1998-99.

(Table 6.5)

Household amenities

Households in Northern Ireland in 1996 were the least likely to have double-glazing whilst those in the South West the most likely.

(Table 6.6)

A greater percentage of householders in the North East, East Midlands, the East of England and the South East had central heating compared with the average percentage in England in 1996.

(Table 6.6)

Household mobility

31 per cent of households in Wales and 32 per cent of households in Northern Ireland had lived at their current address for 20 years or more in 1998-99.

(Table 6.7)

Household satisfaction	The percentage of householders living in London who were very satisfied with the area in which they lived was lower than in any other region.
	(Table 6.8)
Weekly rents	Private sector renters' average weekly rent in London in 1998-99 was more than two and a half times that in the North East.
	(Table 6.9)
Owner-occupier housing costs	In 1998-99, average weekly mortgage payments for owner-occupiers ranged from £38 in Northen Ireland to £88 in London. Average weekly costs for all owner occupiers were also lowest in Northern Ireland and highest in London.
	(Table 6.10)
Dwelling prices	The average sale price of dwellings in London rose by almost 17 per cent between the last quarter of 1998 and 1999; over the same period prices in the North East rose by 6 per cent.
	(Table 6.11)
Mortgages	The average percentage of price advanced towards home purchase for first time buyers was similar in the North East and Yorkshire and the Humber in 1999 at just under 85 per cent, and greater than anywhere else in the UK.
	(Table 6.12)
Court actions for mortgage possessions	Between 1991 and 1999, the number of actions for an order for possession of residential property by ways of summons in a county court fell across England, Wales and Northern Ireland.
	(Table 6.13)
Homelessness	In 1998-99 the breakdown of a relationship with partner was cited by 39 per cent of households accepted as homeless in the East Midlands as the reason for homelessness, compared with 10 per cent of those in Northern Ireland.
	(Table 6.14)

6.1 Stock of dwellings[1]

	Thousands							Percentage increase 1981-1998	Rate per 1,000 population 1998
	1981	1991	1994	1995	1996	1997	1998		
United Kingdom	21,596	23,714	24,271	24,457	24,644	24,835	25,009	15.8	422
North East	1,020	1,078	1,096	1,102	1,108	1,115	1,121	9.9	433
North West	2,660	2,804	2,859	2,879	2,899	2,919	2,938	10.5	426
Yorkshire and the Humber	1,901	2,031	2,074	2,089	2,103	2,118	2,131	12.1	423
East Midlands	1,484	1,646	1,693	1,710	1,725	1,739	1,754	18.2	421
West Midlands	1,941	2,090	2,134	2,150	2,165	2,179	2,194	13.0	411
East	1,859	2,111	2,175	2,197	2,219	2,240	2,259	21.5	420
London	2,682	2,927	2,978	2,996	3,010	3,025	3,040	13.3	423
South East	2,750	3,120	3,198	3,225	3,252	3,279	3,304	20.1	413
South West	1,728	1,983	2,032	2,051	2,066	2,085	2,102	21.6	429
England	18,025	19,790	20,240	20,399	20,548	20,699	20,842	15.6	421
Wales	1,099	1,191	1,221	1,231	1,241	1,250	1,257	14.4	429
Scotland	1,970	2,160	2,210	2,230	2,248	2,268	2,284	15.9	446
Northern Ireland	502	573	600	597	608	618	626	24.7	371

1 At 31 December each year. See Notes and Definitions.

Source: Department of the Environment, Transport and the Regions; National Assembly for Wales; Scottish Executive; Department for Social Development, Northern Ireland

6.2 New dwellings[1] completed: by sector

Thousands

	Private enterprise[2]		Registered social landlords		Local authorities, new towns and government departments[3]	
	1991	1999	1991	1999	1991	1999
United Kingdom	159.6	157.0	20.9	23.2	11.2	0.3
North East	5.4	6.5	1.0	0.7	0.1	-
North West	15.4	15.7	3.0	2.4	0.5	0.0
Yorkshire and the Humber	11.1	12.2	2.0	1.4	0.2	-
East Midlands	14.0	15.0	1.0	1.2	0.7	-
West Midlands	13.6	12.5	1.5	2.6	1.0	-
East	18.9	16.5	0.6	1.9	1.5	-
London	12.8	9.6	2.7	2.9	0.7	0.0
South East	23.0	20.0	2.3	3.0	2.3	-
South West	17.0	14.4	1.1	1.6	1.1	-
England	131.2	122.3	15.3	17.6	8.1	0.1
Wales	7.7	7.2	2.5	0.8	0.4	0.0
Scotland	15.5	19.6	2.3	3.7	1.7	0.1
Northern Ireland	5.2	7.9	0.8	1.1	1.0	0.2

1 Permanent dwellings only ie those with a life expectancy of 60 years or more. See Notes and Definitions.
2 Includes private landlords (persons or companies) and owner-occupiers.
3 Northern Ireland Housing Executive in Northern Ireland.

Source: Department of the Environment, Transport and the Regions; National Assembly for Wales; Scottish Executive; Department for Social Development, Northern Ireland

6.3 Sales and transfers of local authority dwellings[1]

Thousands and percentages

	April 1979 to March 1999				1998-99					Total sales and transfers April 1979 to March 1999 as a percentage of notional stock at
	Right-to-buy sales[2]	Large scale voluntary transfers[3]	Other sales and transfers	Total sales and transfers	Right-to-buy sales[2]	Large scale voluntary transfers	Other sales and transfers	Total sales and transfers	Stock at 1 April 1999	1 April 1979[4]
United Kingdom	1,760	373	372	2,506	55	74	7	138	4,073	38
North East	116	0	5	122	3	-	-	3	286	30
North West	157	29	40	227	4	15	1	19	485	32
Yorkshire and the Humber	143	8	15	165	4	-	-	5	427	28
East Midlands	126	5	16	147	4	5	-	9	280	34
West Midlands	163	46	25	234	5	13	-	19	381	38
East	152	39	42	233	5	-	-	5	290	45
London	214	36	70	320	8	9	1	18	575	36
South East	176	125	51	352	4	6	-	10	268	57
South West	120	63	19	202	3	26	-	30	184	52
England	1,367	352	282	2,001	40	74	3	118	3,178	39
Wales	105	0	7	112	2	0	-	3	197	36
Scotland	288	21	2	310	13	0	0	13	564	35
Northern Ireland[5,6]	.	.	82	82	.	.	4	4	134	43

1 Includes shared ownership deals and dwellings transferred to housing associations and private developers. Excludes New Towns. Figures for Scotland exclude sales by Scottish Homes.
2 Right-to-buy sales were introduced in Great Britain in October 1980. Figures for United Kingdom therefore relate to Great Britain.
3 Figure for United Kingdom relates to Great Britain. For Scotland includes large scale voluntary transfers and trickle transfers to housing associations. For England, includes Estate Renewal Challenge Fund transfers.
4 Calculated as sales in the period April 1979 to March 1999 expressed as a percentage of stock at 1 April 1999 plus sales in the period April 1979 to March 1999.
5 The Northern Ireland Housing Executive (NIHE) is responsible for public sector housing in Northern Ireland. Under the *Housing (NI) Order 1992* NIHE operates a voluntary house sales scheme which is comparable to the Right-to-buy schemes in Great Britain.
6 Figures relate to sales only (excluding SPED cases) and do not include transfers.

Source: Department of the Environment, Transport and the Regions; National Assembly for Wales; Scottish Executive; Department for Social Development, Northern Ireland

6.4 Tenure of dwellings[1]

Percentages

	Owner-occupied			Rented from local authority or New Town[2]			Rented from private owners or with job or business			Rented from registered social landlord		
	1991	1995	1998	1991	1995	1998	1991	1995	1998	1991	1995	1998
Great Britain	66	67	68	21	19	17	10	10	11	3	4	5
North East	58	59	63	31	29	26	7	8	7	4	4	4
North West	67	70	68	21	20	17	8	8	9	4	3	5
Yorkshire and the Humber	65	65	64	24	22	20	9	10	12	2	3	3
East Midlands	71	71	71	19	18	16	9	9	10	2	3	3
West Midlands	67	68	68	23	21	18	7	8	8	3	4	5
East	71	71	72	17	15	13	10	10	10	3	5	4
London	57	57	56	24	22	20	12	15	17	5	7	7
South East	74	74	75	12	10	8	11	11	11	3	5	6
South West	72	72	73	14	12	10	12	12	13	2	4	4
England	68	68	68	19	18	16	10	10	11	3	4	5
Wales	71	71	71	19	17	16	8	8	8	2	3	4
Scotland	52	58	61	38	31	27	7	7	7	3	4	5
Northern Ireland[3]	66	69	71	29	26	22	4	4	4	2	2	3

1 As at 31 December each year. See Notes and Definitions.
2 Including Scottish Homes, formerly the Scottish Special Housing Association and Northern Ireland Housing Executive.
3 Changes in the method of data collection mean that the 1998 figures for Northern Ireland are not comparable with either the 1998 data for Great Britain or the Northern Ireland figures before 1995. The figures are based on occupied stock.

Source: Department of the Environment, Transport and the Regions; National Assembly for Wales; Scottish Executive; Department for Social Development, Northern Ireland

6.5 Households: by type of dwellings, 1998-99

Percentages

	Detached house	Semi-detached house	Terraced house	Purpose-built flat or maisonette	Other[1]
United Kingdom	22	30	28	14	6
North East	13	40	36	9	2
North West	17	35	35	9	4
Yorkshire and the Humber	17	37	34	8	3
East Midlands	32	36	22	6	4
West Midlands	23	37	29	7	4
East	29	30	26	11	4
London	6	17	32	28	17
South East	27	30	25	11	7
South West	30	27	25	9	8
England	21	31	29	12	6
Wales	28	33	32	6	2
Scotland	21	23	16	35	4
Northern Ireland	33	24	35	6	2

1 Includes converted flats which are particularly common in London.

Source: Survey of English Housing, Department of the Environment, Transport and the Regions; General Household Survey, Office for National Statistics; Continuous Household Survey, Northern Ireland Statistics and Research Agency

6.6 Households with different types of amenity

Percentages

	1991[1]					1996[1]				
	Central heating	Double glazing	Secure win-dows and doors	Smoke detector(s)	Parking provision[2]	Central heating	Double glazing	Secure win-dows and doors	Smoke detector(s)	Parking provision[2]
North East	91.9	42.5	19.4	32.9	56.3	94.1	50.1	21.7	66.1	63.7
North West	78.0	51.2	21.9	42.0	59.8	82.9	65.4	40.9	68.0	63.3
Yorkshire and the Humber	75.8	51.4	20.6	33.6	62.7	82.5	57.4	30.2	60.9	65.6
East Midlands	88.1	56.7	19.1	39.2	66.4	92.3	62.5	30.8	68.7	72.1
West Midlands	78.5	47.2	17.9	35.0	69.4	82.2	58.4	22.5	69.6	71.6
East	91.0	56.7	17.4	44.7	74.7	92.3	65.7	22.5	73.5	75.9
London	83.8	50.0	35.3	33.6	51.2	87.5	50.7	34.8	56.9	52.9
South East	86.2	58.6	26.3	49.4	72.0	91.6	59.6	32.9	70.0	73.0
South West	83.2	49.0	22.1	44.0	69.9	88.4	68.2	25.8	68.1	77.6
England	83.4	52.1	23.1	40.1	65.3	87.8	60.1	30.3	66.8	68.9
Wales	..	50.4	..	55.8	..	89.0	67.5	..	79.7	..
Scotland	77.9	46.0	28.9	42.7	..	87.2	64.7	..	80.0	66.5
Northern Ireland	82.9	22.4	..	37.0	60.8	88.7	40.1	..	69.8	63.7

1 Data for Wales are for 1992 and 1997.
2 Includes only facilities that are an integral part of the property, ie excludes street parking. Figures for England are based on households in houses only, excluding flats.

Source: National House Condition Surveys, Department of the Environment, Transport and the Regions; National Assembly for Wales; Scottish Homes; Northern Ireland Housing Executive

6.7 Households: by length of time at current address, 1998-99

Percentages

	Less than 12 months	12 months, less than 5 years	5 years, less than 10 years	10 years, less than 20 years	20 years or more
United Kingdom	11	25	16	22	23
North East	9	26	16	20	28
North West	11	23	17	23	26
Yorkshire and the Humber	11	24	18	23	23
East Midlands	12	23	16	24	24
West Midlands	10	24	15	24	27
East	12	26	16	21	25
London	13	28	16	21	22
South East	12	28	16	23	21
South West	12	27	16	23	21
England	12	26	16	23	24
Wales	9	25	15	21	31
Scotland	9	28	20	23	20
Northern Ireland	6	21	16	24	32

Source: Survey of English Housing, Department of the Environment, Transport and the Regions; General Household Survey, Office for National Statistics; Continuous Household Survey, Northern Ireland Statistics and Research Agency

6.8 Householders' satisfaction with their accommodation and area, 1998-99[1]

Percentages

	Accomodation		Area	
	Very satisfied	Fairly satisfied	Very satisfied	Fairly satisfied
North East	64	29	53	33
North West	59	32	50	35
Yorkshire and the Humber	60	32	52	34
East Midlands	63	30	55	34
West Midlands	64	27	54	33
East	63	29	58	31
London	51	37	44	38
South East	62	31	56	33
South West	64	29	62	29
England	60	31	53	34
Scotland	58	36	59	29

1 Data for Scotland are for 1996.

Source: Survey of English Housing, Department of the Environment, Transport and the Regions; Scottish House Condition Survey, Scottish Homes

6.9 Average weekly rents[1]: by tenure, 1998-99[2]

£ per week

	Private sector average rent	Local authorities	Registered social landlords
Great Britain	90	42.50	50.00
North East	56	36.80	43.40
North West	74	40.60	43.50
Yorkshire and the Humber	59	35.10	46.10
East Midlands	68	38.10	48.20
West Midlands	71	39.80	47.70
East	86	45.60	52.30
London	142	58.00	59.30
South East	95	50.30	58.00
South West	75	43.70	50.50
England	91	43.80	51.70
Wales	62	39.10	43.60
Scotland	71	36.40	37.30
Northern Ireland	..	37.60	..

1 See Notes and Definitions.
2 Local authority rents are at April 1999; Registered social landlords are at March 1999.

Source: Department of Social Security; Department of the Environment, Transport and the Regions; National Assembly for Wales; Scottish Executive; Scottish Homes; Northern Ireland Housing Executive, Housing Corporation, Northern Ireland

6.10 Selected housing costs[1] of owner occupiers, 1998-99

£ per week

	Mortgage payments	Endowment policies	Structural insurance	Service payments	All owner occupiers[2]
Great Britain	61	19	5	6	49
North East	47	16	4	2	41
North West	48	15	4	1	39
Yorkshire and the Humber	49	16	4	2	41
East Midlands	50	17	4	3	40
West Midlands	53	16	4	5	43
East	70	22	5	10	57
London	88	27	6	14	74
South East	77	24	5	11	63
South West	61	18	4	7	45
England	63	19	5	7	51
Wales	48	15	4	2	34
Scotland	49	17	4	4	45
Northern Ireland	38	13	2	1	30

1 Those who did not make any payments within each category are excluded, this table is therefore not directly comparable with data published in previous editions of *Regional Trends* which included all owner occupiers.
See Notes and Definitions.
2 Relates to both householders with a mortgage and those who own their house outright.

Source: Family Resources Survey, Department of Social Security; Family Expenditure Survey, Northern Ireland Statistics and Research Agency

6.11 Average dwelling prices[1], 1999

£ and percentages

	Average sale price (£)				All dwellings		
	Detached houses	Semi-detached houses	Terraced houses	Flats/ maisonettes	Average price (£) 1998	Average price (£) 1999	Percentage increase 1998-1999
England and Wales	139,760	80,599	71,367	92,567	84,744	94,581	11.6
North East	101,214	55,044	42,463	44,124	56,093	59,442	6.0
North West	116,972	60,735	39,787	57,776	61,240	65,543	7.0
Yorkshire and the Humber	103,493	55,694	41,987	54,101	59,469	63,524	6.8
East Midlands	101,609	54,105	43,840	47,215	64,952	69,500	7.0
West Midlands	126,302	64,941	50,350	51,622	68,500	76,633	11.9
East	142,803	84,698	70,387	58,097	87,625	94,679	8.1
London	284,789	168,159	151,204	131,475	128,590	150,094	16.7
South East	196,487	106,101	84,186	68,910	105,342	118,385	12.4
South West	134,592	78,734	65,118	65,623	82,294	90,274	9.7
England	146,637	84,546	74,745	96,127	85,907	98,252	14.4
Wales	92,412	55,344	42,563	51,253	58,386	62,424	6.9

1 Excludes those bought at non-market prices. Averages are taken from the last quarter of each year. See Notes and Definitions.

Source: HM Land Registry

6.12 Mortgage advances, and income and age of borrowers[1], 1999

	First-time buyers				Previous owner-occupiers			
	Number of loans (thousands)	Average percentage of price advanced	Average recorded income[2] (£ per annum)	Average age of borrowers (years)	Number of loans (thousands)	Average percentage of price advanced	Average recorded income[2] (£ per annum)	Average age of borrowers (years)
United Kingdom	578	80.1	25,277	32	643	64.5	33,961	39
North East	23	84.5	20,835	32	19	69.8	28,039	39
North West	62	83.3	21,784	32	60	68.8	30,709	39
Yorkshire and the Humber	49	84.4	21,239	31	45	70.4	28,765	39
East Midlands	46	81.1	21,425	33	53	67.1	30,077	39
West Midlands	48	80.9	22,557	33	53	66.6	30,867	39
East	59	80.6	26,114	33	77	64.5	34,302	39
London	81	77.8	35,692	32	66	61.3	47,949	38
South East	77	78.9	29,861	33	125	61.0	38,457	39
South West	52	78.6	23,918	34	69	63.9	30,353	40
England	496	80.1	25,981	33	568	64.1	34,555	39
Wales	27	83.0	20,720	33	27	68.4	29,188	40
Scotland	37	79.6	21,574	34	35	70.9	30,613	40
Northern Ireland	18	79.0	20,122	31	13	63.7	27,266	36

1 See Notes and Definitions.
2 The income of borrowers is the total recorded income taken into account when the mortgage is granted.

Source: Department of the Environment, Transport and the Regions

6.13 County Court mortgage possession orders[1]

Thousands

	1991			1997			1998			1999		
	Actions entered	Sus-pended orders	Orders made	Actions entered	Sus-pended orders	Orders made	Actions entered	Sus-pended orders	Orders made	Actions entered	Sus-pended orders	Orders made
England and Wales	186.6	69.1	73.9	67.0	34.8	22.5	84.8	40.8	25.3	82.6	37.1	23.6
North East	6.0	2.9	1.9	3.0	1.6	1.0	4.3	2.3	1.2	4.6	2.5	1.1
North West	22.3	8.6	7.5	10.9	5.5	3.3	14.2	6.4	3.7	13.8	6.0	3.7
Yorkshire and the Humber	14.1	5.1	5.7	6.9	3.5	2.3	8.2	4.3	3.1	8.7	4.3	3.0
East Midlands	13.5	4.5	5.2	4.9	2.6	1.7	6.4	3.1	1.7	6.8	2.8	1.8
West Midlands	17.7	6.5	6.9	6.7	3.4	2.0	8.1	3.8	2.4	9.5	4.1	2.3
East	18.6	6.0	8.4	6.7	3.0	2.5	8.5	3.9	2.6	7.3	3.5	2.2
London	35.3	13.1	14.4	9.2	4.7	3.4	11.4	5.3	3.5	10.0	4.5	3.4
South East	32.2	13.2	13.2	9.1	5.4	3.1	11.2	5.8	3.4	9.4	4.4	2.7
South West	16.7	5.8	6.5	5.7	2.7	1.9	7.3	3.0	2.1	7.0	2.8	2.0
England	176.4	65.6	69.9	63.1	32.4	21.2	79.6	37.9	23.7	77.1	34.9	22.2
Wales	10.2	3.5	4.0	3.9	2.4	1.3	5.4	2.8	1.7	5.5	2.2	1.5
Northern Ireland[2]	3.1	..	..	1.2	..	..	1.6	0.2	0.5	1.9	0.3	0.7

1 Local authority and private. See Notes and Definitions.
2 Mortgage possession actions are heard in Chancery Division of Northern Ireland High Court.

Source: The Court Service; Northern Ireland Court Service

6.14 Households accepted as homeless: by reason[1], 1998-99

Percentages and numbers

	Reasons for homelessness						
	No longer willing or able to remain with		Break-down of relation-ship with or partner	Mortgage arrears	Rent arrears or other reason for loss of rented or tied accomm-odation	Other reasons[2]	Total (=100%) (numbers)
	Parents	Relatives or friends					
England and Wales	16	12	24	6	25	17	108,483
North East	22	10	33	7	17	12	4,450
North West	12	8	29	6	16	29	13,338
Yorkshire and the Humber	14	11	30	6	21	17	8,091
East Midlands	12	7	39	7	20	15	7,782
West Midlands	15	10	32	6	20	17	13,930
East	19	9	23	8	32	8	8,719
London	17	19	13	3	25	23	26,169
South East	20	10	18	7	35	10	12,721
South West	13	8	22	6	41	10	8,950
England	16	12	24	6	25	18	104,150
Wales	18	8	27	7	27	13	4,333
Scotland[3]	33		34	3	14	16	17,500
Northern Ireland	26		10	1	12	51	4,997

1 See Notes and Definitions.
2 A large proportion of the Northern Ireland total is classified as 'Other reasons' due to differences in the definitions used.
3 In Scotland, the basis of these figures is households assessed by the local authorities as homeless, or potentially homeless, and in priority need, as defined in section 24 of the Housing (Scotland) Act 1997. The figures relate to the calendar year 1997.

Source: Department of the Environment, Transport and the Regions; National Assembly for Wales; Scottish Executive; Department for Social Development, Northern Ireland

7 Health

Population
The South and West NHS region had the highest proportion of people aged 75 or over in 1998, whilst Northern Ireland had the lowest.

(Table 7.1)

Still births
There were 6.1 still births per 1,000 births in Scotland in 1998, compared with 4.6 in the South and West NHS region and 4.7 in Anglia and Oxford.

(Table 7.1)

Infant mortality
Infant mortality was highest in Yorkshire and the Humber in 1998 at almost seven deaths per thousand live births.

(Table 7.2)

GP consultations
Almost a quarter of under fives in the South West saw an NHS GP in 1998-99, compared with 13 per cent in the East Midlands.

(Table 7.3)

Longstanding illness
Nearly three fifths of people in Wales aged 65 or over reported a limiting long-standing illness in 1998-99, compared with just over a third in the South West.

(Table 7.3)

Cardiovascular disease
More than 11 per cent of women aged 35 or over in the West Midlands NHS region suffered from ischaemic heart disease or a stroke in 1998, compared with less than seven per cent in the North Thames NHS region.

(Table 7.4)

Nutrition
Between 37 and 40 per cent of the population's energy intake was derived from fat in every region of the United Kingdom.

(Table 7.5)

Smoking
More than half of all male smokers in the North East smoked 20 or more cigarettes per day in 1998-99, compared with less than a third of smokers in the South West and Wales.

(Table 7.6)

Alcohol consumption
61 per cent of females in Northern Ireland had not had an alcoholic drink in the last week, compared with less than two fifths in the North West, the South West and the South East in 1998-99.

(Table 7.7)

Drug use
Almost a third of 16-29 year olds in the South West used some form of illegal drug in the twelve months prior to interview in 1998, almost twice the proportion in Wales.

(Chart 7.8)

Death rates
Allowing for the age structure of the population, 47 deaths per 100,000 population were due to injuries or poisonings in Scotland in 1998, more than in any other NHS region in the UK.

(Table 7.9)

Cancer
The incidence of lung cancer in 1996 was above the UK average in the northern parts of England and in Scotland, while below average in the southern parts of England and in Northern Ireland. For breast and prostate cancers, above average incidence occurred in the southern parts of England.

(Chart 7.10)

Cervical cancer screening
Around three-quarters of women aged 25-64 in the North Thames NHS region had been screened for cervical cancer at 31 March 1999, a lower proportion than in any other English region.

(Table 7.11)

Tuberculosis

More than 30 cases of tuberculosis per 100,000 population were recorded in London in 1999, five times more than in Wales and seven times higher than in the South West.

(Table 7.12)

HIV

Of the 20,000 people in the United Kingdom who were known in 1999 to be infected with the HIV virus, more than half were residents of London, and only 95 were in Northern Ireland.

(Table 7.13)

Waiting lists

At 31 March 2000 6.6 per cent of people on NHS hospital waiting lists in the Eastern NHS region had been waiting for more than twelve months, compared with 0.8 per cent in the Northern and Yorkshire NHS region.

(Table 7.14)

There were almost two hundred thousand people on NHS hospital waiting lists in the South East at 31 March 2000, which is more than twice as many as in the West Midlands.

(Table 7.14 and Chart 7.15)

Hospital beds

In 1998-99, there were 7.1 hospital beds in Scotland per 1,000 population, compared with 3.4 in the Eastern NHS region and 3.2 in the South East.

(Table 7.16)

Prescriptions

On average 14 prescription items were dispensed per person in Wales in 1998-99; higher than in England and Scotland in 1999 and Northern Ireland in 1998-99.

(Table 7.17)

NHS community health staff

Ten per cent of the NHS hospital and community health service staff in London were direct medical and dental care staff, more than in any other region.

(Table 7.18)

GPs

At 1 October 1999, Scotland had the lowest average list size per general medical practitioner, while London had the highest.

(Table 7.19)

Dentists

Only two fifths of people in London were registered with a dentist at 1 October 1999, a lower proportion than in any other region.

(Table 7.19)

Residential care

Of the places available in residential care homes at 31 March 1999, 22 per cent of the places in Scotland were available in homes for people with physical, sensory or learning disabilities, whereas the figure for Wales was 12 per cent.

(Table 7.20)

Foster care

At 31 March 1999, almost three in four children in local authority care in the South West were living in foster homes, compared with three fifths of those in the North West.

(Table 7.21)

Child protection

In Wales, 40 children per 10,000 were on a child protection register at 31 March 1999, nearly double the proportion in the South East.

(Table 7.22)

Across the whole of the United Kingdom, the most common reason for a child being placed on a child protection register was neglect.

(Table 7.22)

7.1 Population and vital statistics: by NHS Regional Office area, 1998

	Population aged (mid year estimates) (percentages and thousands)						Vital statistics (rates)				
	0-4	5-15	16-44	45-74	75 and over	All ages	Live births[1,2]	Still births[2,3]	Deaths[4]	Perinatal mortality[5]	Infant mortality[6]
United Kingdom	6.2	14.2	40.8	31.4	7.3	59,236.5	58.7	5.4	10.6	8.2	5.7
Northern and Yorkshire	6.0	14.4	40.3	32.1	7.2	6,339.0	56.7	5.7	11.3	8.6	6.1
North West	6.1	14.9	40.2	31.6	7.2	6,604.0	58.2	5.6	11.4	8.7	6.2
Trent	6.0	14.1	40.1	32.3	7.4	5,133.8	57.2	5.4	10.9	8.7	6.0
West Midlands	6.3	14.6	39.9	32.1	7.1	5,332.5	60.7	5.6	10.5	9.3	6.5
Anglia and Oxford	6.3	14.3	41.4	31.2	6.9	5,452.3	58.4	4.7	9.5	7.2	5.1
North Thames	6.8	13.8	44.2	28.7	6.5	7,052.0	62.5	5.6	9.1	8.3	5.5
South Thames	6.4	13.6	41.6	30.6	7.9	6,908.7	60.5	5.3	10.4	8.0	5.1
South and West	5.8	13.7	38.9	32.9	8.7	6,672.3	56.9	4.6	11.0	6.8	4.7
England	6.2	14.2	40.9	31.4	7.4	49,494.6	59.0	5.3	10.5	8.2	5.6
Wales	5.9	14.5	38.3	33.3	8.0	2,933.3	58.9	5.4	11.6	8.0	5.7
Scotland	5.9	13.9	41.6	31.9	6.7	5,120.0	52.6	6.1	11.6	8.7	5.6
Northern Ireland	7.2	17.3	42.0	27.7	5.8	1,688.6	64.9	5.1	8.9	8.1	5.6

1 Per 1,000 women aged 15-44.
2 See Notes and Definitions for Population chapter.
3 Per 1,000 live and still births. A still birth relates to a baby born dead after 24 completed weeks gestation or more.
4 Per 1,000 population.
5 Still births and deaths of infants under 1 week of age per 1,000 live and still births.
6 Deaths of infants under 1 year of age per 1,000 live births.

Source: Office for National Statistics; General Register Office for Scotland; Northern Ireland Statistics and Research Agency

7.2 Still births, perinatal mortality and infant mortality[1]

Rates

	Still births[2,3]				Perinatal mortality[3,4]				Infant mortality[5]		
	1981	1993	1993	1998	1981	1993	1993	1998	1981	1993	1998
United Kingdom	6.6	4.4	5.7	5.4	12.0	7.6	9.0	8.2	11.2	6.3	5.7
North East	7.5	4.6	5.9	5.9	12.6	7.9	9.2	8.2	10.4	6.7	5.0
North West	7.0	4.5	5.8	5.6	12.7	7.7	9.0	8.7	11.3	6.5	6.1
Yorkshire and the Humber	7.8	4.6	5.9	5.6	13.5	8.0	9.4	9.2	12.1	7.3	6.9
East Midlands	6.2	3.9	5.4	5.1	11.4	7.2	8.7	8.0	11.0	6.6	5.5
West Midlands	7.0	4.4	6.0	5.6	12.9	8.4	9.9	9.3	11.7	7.1	6.4
East	5.5	3.9	5.2	4.8	10.0	6.8	8.1	7.4	9.7	5.4	5.0
London	6.3	4.9	6.1	6.0	10.3	8.2	9.5	9.0	10.7	6.4	5.9
South East	5.8	4.0	5.4	4.6	10.5	7.0	8.3	6.8	10.3	5.3	4.4
South West	6.3	4.0	5.0	4.8	10.8	6.9	7.9	7.3	10.4	5.8	4.8
England	6.5	4.3	5.7	5.3	11.7	7.6	8.9	8.2	10.9	6.3	5.6
Wales	7.3	4.5	5.8	5.4	14.1	7.0	8.3	8.0	12.6	5.5	5.7
Scotland	6.3	4.8	6.4	6.1	11.6	8.0	9.6	8.7	11.3	6.5	5.6
Northern Ireland	8.8	4.1	5.2	5.1	15.3	7.7	8.8	8.1	13.2	7.1	5.6

1 See Notes and Definitions for Population chapter.
2 Rate per 1,000 live and still births.
3 On 1 October 1992 the legal definition of a still birth was altered from a baby born dead after 28 completed weeks gestation or more to one born dead after 24 weeks gestation or more. Figures are given on both the old and new definitions for continuity/comparison.
4 Still births and deaths of infants under 1 week of age per 1,000 live and still births.
5 Deaths of infants under 1 year of age per 1,000 live births.

Source: Office for National Statistics; General Register Office for Scotland; Northern Ireland Statistics and Research Agency

7.3 Consultations with NHS general medical practitioners[1] and reports of limiting long-standing illness[2]: by age, 1998-99

Percentages

| | Persons who consulted an NHS general medical practitioner | | | | | | Persons who reported limiting long-standing illness | | | | | |
	0-4	5-15	16-44	45-64	Aged 65 or over	All ages	0-4	5-15	16-44	45-64	Aged 65 or over	All ages
United Kingdom	12		13	16	19	14	7		13	28	43	20
North East	16	10	14	19	18	15	5	10	15	36	44	23
North West	16	8	13	16	19	14	4	11	16	29	45	22
Yorkshire and the Humber	17	9	13	17	21	15	6	8	13	33	52	22
East Midlands	13	7	11	14	13	12	1	8	10	31	47	21
West Midlands	18	8	13	15	21	14	4	8	15	27	43	20
East	20	12	13	14	18	14	4	9	11	25	41	18
London	22	9	11	18	19	14	4	6	12	28	39	17
South East	16	6	13	15	19	14	4	7	11	22	37	17
South West	23	8	13	12	17	14	5	9	13	22	36	19
England	18	8	13	15	18	14	4	8	13	27	42	20
Wales	17	13	12	17	22	16	5	8	16	38	58	27
Scotland	17	11	15	19	22	17	5	6	12	31	39	19
Northern Ireland	10		13	18	22	15	8		12	30	48	20

1 In the 14 days before interview.
2 See Notes and Definitions.

Source: General Household Survey, Office for National Statistics; Continuous Household Survey, Northern Ireland Statistics and Research Agency

7.4 Prevalence of cardiovascular disease, ischaemic heart disease and stroke among people aged 35 and over[1]: by gender, 1998

Percentages

| | Males | | Females | |
	CVD	IHD or stroke	CVD	IHD or stroke
England				
Northern and Yorkshire	36.6	14.3	35.9	9.4
North West	32.9	11.5	37.7	9.9
Trent	35.4	12.0	36.0	9.9
West Midlands	33.4	11.5	34.9	11.2
Anglia and Oxford	33.2	10.8	36.9	9.2
North Thames	34.2	11.1	35.4	6.9
South Thames	35.3	9.7	35.0	8.9
South and West	35.3	10.9	34.2	8.1

1 Age-standardised. See Notes and Definitions.

Source: Health Survey for England, Department of Health

7.5 Contributions of selected foods to nutritional intakes (household food), 1998-1999[1]

| | Percentage of fat and energy derived from | | | | | | | | | | Total intake[2] per person per day | | Per-centage of food energy derived from fat[2] |
| | Liquid and processed milk and cream | | Meat and meat products | | All fats | | Fresh and processed fruit and vegetables | | Cereals including bread | | | | |
	Fat	Energy	Fat	Energy	Fat	Energy	Fat	Energy	Fat	Energy	Fat (grams)	Energy (Kcal)	
United Kingdom	11.1	10.4	22.3	14.6	28.1	10.9	8.3	15.4	17.4	35.4	73	1,714	39
North East	10.5	9.8	24.0	15.7	25.7	10.0	8.6	15.4	18.6	36.4	73	1,704	39
North West	12.5	11.4	23.8	15.6	26.5	10.4	7.6	14.6	17.2	35.1	70	1,635	39
Yorkshire and the Humber	11.3	10.5	22.9	15.1	28.8	11.4	7.2	14.2	17.3	35.6	73	1,694	38
East Midlands	11.3	10.6	22.1	14.2	29.2	11.3	8.2	15.2	16.7	35.1	74	1,744	39
West Midlands	11.3	10.5	21.0	13.5	29.5	11.4	8.4	15.4	16.8	35.1	73	1,732	38
East	10.1	9.8	22.6	14.5	26.1	10.0	9.5	16.5	18.1	35.8	70	1,671	37
London	10.3	9.6	20.2	13.2	33.1	12.8	8.3	15.8	15.7	36.0	73	1,724	39
South East	10.6	10.2	21.6	14.1	27.2	10.7	8.7	15.7	17.9	35.2	75	1,751	39
South West	10.4	10.1	21.7	14.2	26.8	10.5	9.0	16.3	18.1	35.2	76	1,778	39
England	10.9	10.3	22.1	14.4	28.2	11.0	8.4	15.5	17.4	35.4	73	1,715	39
Wales	11.0	10.2	23.4	15.5	29.8	11.7	7.6	15.2	16.8	34.3	76	1,763	39
Scotland	11.6	10.5	23.8	15.9	25.0	9.9	8.2	14.4	18.3	35.7	73	1,684	40
Northern Ireland	13.3	12.1	22.4	14.3	31.7	12.3	6.9	15.5	16.6	35.0	72	1,681	39

1 See Notes and Definitions.
2 Total intake from all household food, excluding household consumption of soft and alcoholic drinks and confectionery.

Source: National Food Survey, Ministry of Agriculture, Fisheries and Food

7.6 Cigarette smoking among people aged 16 or over: by gender, 1998-99

Percentages and numbers

| | Males | | | | | Females | | | | |
| | Proportion who | | | Proportion of smokers smoking 20 or more daily | Smokers' average weekly con-sumption (numbers) | Proportion who | | | Proportion of smokers smoking 20 or more daily | Smokers' average weekly con-sumption (numbers) |
	Have never smoked	Are ex-regular smokers	Smoke			Have never smoked	Are ex-regular smokers	Smoke		
United Kingdom	41	31	28	37	110	53	21	26	28	93
North East	46	28	26	51	132	50	20	30	35	101
North West	41	31	28	38	111	47	21	32	33	100
Yorkshire and the Humber	38	32	29	37	107	52	20	28	21	86
East Midlands	39	35	26	36	108	55	20	25	28	91
West Midlands	38	31	31	34	103	55	20	25	25	89
East	43	32	25	37	101	53	24	23	21	88
London	43	24	33	33	98	56	17	27	25	84
South East	41	33	26	36	107	55	24	21	29	92
South West	42	34	25	31	104	55	21	24	28	88
England	41	31	28	36	106	53	21	26	27	91
Wales	45	27	28	31	103	54	20	26	25	90
Scotland	41	26	33	46	129	52	20	29	34	107
Northern Ireland	35	37	28	52	129	49	22	29	37	108

Source: General Household Survey, Office for National Statistics; Continuous Household Survey, Northern Ireland Statistics and Research Agency

7.7 Alcohol consumption[1] among people aged 16 or over: by gender, 1998-99

Percentages

	Males				Females			
	Drank nothing last week	Drank up to 4 units last week[2]	Drank more than 4 and up to 8 units[2]	Drank more than 8 units last week[2]	Drank Nothing last week	Drank up to 3 units last week[2]	Drank more than 3 and up to 6 units[2]	Drank more than 6 units last week[2]
United Kingdom	26	36	17	21	41	38	13	8
North East	23	32	22	23	43	35	14	8
North West	23	33	18	27	38	37	15	11
Yorkshire and the Humber	27	34	15	24	44	36	13	7
East Midlands	24	35	21	20	40	40	12	7
West Midlands	26	36	17	21	42	38	11	9
East	23	44	19	14	42	42	10	6
London	33	36	13	18	49	34	11	6
South East	21	44	17	18	34	46	13	7
South West	24	41	17	18	38	42	12	8
England	25	38	17	20	41	39	12	8
Wales	32	29	17	22	46	33	12	9
Scotland	29	32	16	24	40	32	15	12
Northern Ireland	47	18	13	22	61	18	10	10

1 Comparative consumption levels are different for males and females. See Notes and Definitions.
2 On the heaviest drinking day last week.

Source: General Household Survey, Office for National Statistics; Continuous Household Survey, Northern Ireland Statistics and Research Agency

7.8 Drug use among 16-29 year olds[1], 1998[2]

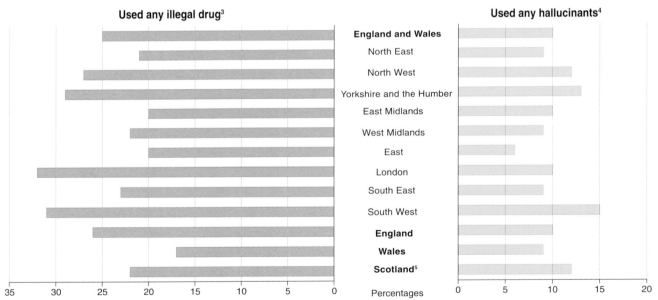

Used any illegal drug[3]   **Used any hallucinants[4]**

1 See Notes and Definitions.
2 Interviews were conducted between January and April, asking about drug use in the previous 12 months.
3 Amphetamines, cannabis, cocaine, crack, ecstasy, heroin, LSD, magic mushrooms, non-prescribed methadone, poppers, glue/solvents, steroids, non-prescribed tranquilisers.
4 Amphetamines, LSD, magic mushrooms, ecstasy, poppers.
5 Data for Scotland relate to 1996.

Source: British Crime Survey, Home Office; Scottish Crime Survey, Scottish Executive

7.9 Age adjusted mortality rates[1]: by cause[2] and gender, 1998

Rates per 100,000 population

	All circulatory diseases			All respiratory diseases			All injuries and poisonings				
	Total	Ischaemic heart disease	Cerebro-vascular disease	Total	Bronchitis and allied conditions	Cancer[3]	Total	Road traffic accidents	Suicides and open verdicts	All other causes	All causes[4]
Males											
United Kingdom	395	243	78	141	53	264	43	9	17	125	969
Northern and Yorkshire	426	272	84	155	61	288	44	10	18	126	1,039
North West	437	272	88	165	66	283	45	8	18	130	1,060
Trent	405	256	78	145	54	263	43	12	15	126	982
West Midlands	400	243	80	145	54	270	40	9	14	125	980
Anglia and Oxford	357	215	70	128	46	237	41	10	15	117	879
North Thames	354	219	63	144	52	252	38	7	14	124	912
South Thames	350	201	70	132	49	249	37	8	15	117	885
South and West	354	216	68	115	42	240	37	8	15	113	859
England	384	236	75	140	53	260	40	9	16	122	947
Wales	408	251	79	134	54	265	47	10	17	119	973
Scotland	472	287	103	142	57	300	59	11	26	158	1,132
Northern Ireland	431	281	83	159	53	270	49	12	14	107	1,017
Females											
United Kingdom	428	201	130	171	43	244	23	3	5	179	1,045
Northern and Yorkshire	456	227	136	191	56	257	22	3	5	191	1,117
North West	484	235	148	202	57	257	22	3	5	184	1,149
Trent	435	212	128	174	43	239	22	4	4	190	1,060
West Midlands	438	204	132	171	39	235	21	3	4	180	1,045
Anglia and Oxford	392	177	124	160	33	234	23	4	4	180	989
North Thames	373	175	106	173	38	237	21	3	5	168	971
South Thames	381	163	117	161	34	233	19	3	6	167	962
South and West	380	176	118	134	30	223	20	3	5	175	932
England	416	195	125	170	41	239	21	3	5	179	1,025
Wales	442	205	134	167	46	251	27	4	5	170	1,057
Scotland	526	247	170	174	54	281	35	5	9	191	1,207
Northern Ireland	458	231	140	204	41	235	23	4	4	133	1,052
All persons											
United Kingdom	415	224	104	158	48	256	33	6	11	153	1,014
Northern and Yorkshire	443	250	111	175	59	274	33	7	11	159	1,084
North West	462	254	118	185	62	271	33	5	12	157	1,108
Trent	425	237	104	162	50	254	32	8	9	159	1,031
West Midlands	423	226	107	160	47	255	31	6	9	153	1,021
Anglia and Oxford	380	200	98	147	40	239	31	7	10	149	946
North Thames	367	199	85	161	46	247	29	5	10	147	950
South Thames	368	183	94	148	42	242	28	5	10	143	929
South and West	371	198	94	127	37	234	28	6	10	145	904
England	403	218	101	157	48	252	31	6	10	151	994
Wales	428	230	107	152	50	259	37	7	11	145	1,022
Scotland	499	266	137	159	56	290	47	8	17	175	1,170
Northern Ireland	445	256	112	183	47	253	36	8	9	120	1,037

1 Rates standardised to the mid-1991 United Kingdom population for males and females separately. See Notes and Definitions.
2 Deaths at ages under 28 days occurring in England and Wales are not assigned an underlying cause.
3 Malignant neoplasms only.
4 Including deaths at ages under 28 days.

Source: Office for National Statistics; General Register Office for Scotland; Northern Ireland Statistics and Research Agency

7.10 Cancer – comparative incidence ratios[1] for selected sites: by gender, 1996

Lung

UK=100[2]

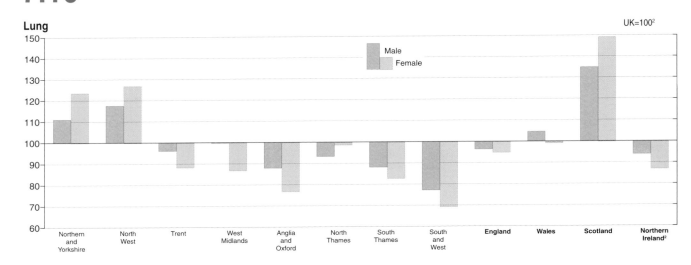

Colorectal

UK=100[2]

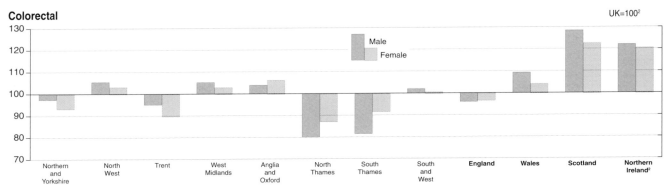

Breast

UK=100[2]

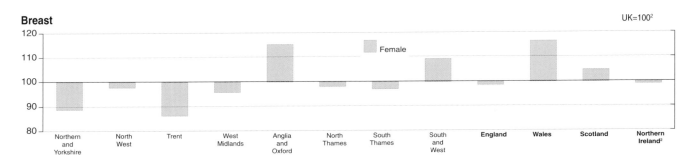

Prostate

UK=100[2]

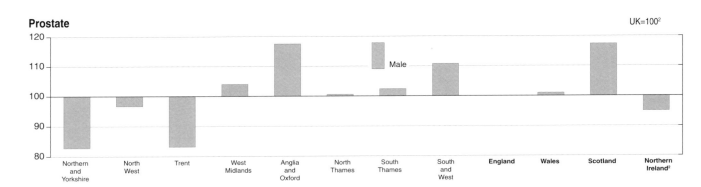

1 Comparative incidence ratio: the directly age-standardised incidence rate for each country and region as a percentage of the UK rate.
2 The UK age-standardised rates are given in the Notes and Definitions.

Source: Office for National Statistics; Information and Statistics Division, Scottish Health Service; Northern Ireland Cancer Registry

7.11 Cervical and breast cancer: screening and age-adjusted death rates

| | Cervical screening programme at 31 March 1999 | | | | | | | Breast screening programme at 31 March 1999 | | | | | | |
| | Percentage of target population screened: women aged[1] | | | | | Adequate test results[4] (thousands) | Per-centage recalled early[4,5] | Percentage of target population[6] screened: women aged | | | | Age-adjusted death rates[7], 1998 | |
	25-34[2,3]	35-44	45-54	55-64[3]	All aged 25-64[2,3]			50-54	55-59	60-64	All aged 50-64	Cervical cancer	Breast cancer
United Kingdom	..	..	..	..	..	..	..	59.4	74.6	72.9	68.1	5.8	56.5
Northern and Yorkshire	85.4	88.7	87.2	80.1	85.8	418.6	1.3	56.9	74.6	72.6	67.0	6.4	52.3
North West	83.9	86.3	84.5	76.9	83.5	424.3	1.5	57.9	74.6	72.4	67.4	7.2	54.0
Trent	87.2	89.8	88.3	81.9	87.2	322.0	1.3	58.7	76.8	75.5	69.2	5.0	54.0
West Midlands	83.3	87.5	86.7	80.5	84.8	350.5	1.7	60.4	77.7	75.5	70.3	5.5	57.5
Anglia and Oxford	83.6	88.6	88.7	82.7	86.1	343.8	1.4	63.0	79.5	78.8	72.5	4.8	62.8
North Thames	74.5	80.4	81.7	77.0	78.0	442.2	1.4	51.4	67.5	66.0	60.5	4.8	58.3
South Thames	81.5	86.4	86.7	80.9	84.0	477.9	1.5	56.4	73.4	72.8	66.3	4.6	55.8
South and West	84.3	88.1	86.8	80.9	85.4	378.0	1.3	59.1	76.2	75.2	69.0	5.9	57.4
England	82.3	86.6	86.2	80.0	84.0	3,157.2	1.4	57.8	74.8	73.4	67.6	5.5	56.4
Wales	77.8	87.0	86.1	78.1	81.9	208.8	1.7	60.7	76.6	75.0	69.9	6.3	58.8
Scotland	80.9	91.5	92.0	85.6	86.6	..	..	71.8	72.5	68.9	71.1	7.3	57.1
Northern Ireland	..	..	..	..	..	..	..	72.0	71.0	68.1	70.5	5.7	51.6

1 For England the target population relates to women aged 25-64, for Wales to women 20-64 and for Scotland to women 20-59 years screened in the previous 5 years (5.5 years in Scotland). Medically ineligible women (women who as a result of surgery etc do not require screening) in the target population are excluded from the figures.
2 For Wales the age groups are 20-34 and 20-64 respectively.
3 For Scotland the age groups are 20-34, 55-60 and 20-60 respectively.
4 Adequate test results and percentages recalled early relate to the year 1998-1999.
5 Women whose screening test results are borderline or show mild dyskaryosis are recalled for a repeat smear in approximately 6 months instead of the routine 5 years; if the condition persists they are referred to a gynaecologist.
6 Percentage of the target population - women aged 50-64 years - screened in the previous 3 years. Medically ineligible women (women who as a result of surgery etc do not require screening) in the target population are excluded from the figures. See Notes and Definitions.
7 Deaths per 100,000 women aged 20 or over. Standardised to mid-1991 UK population. See Notes and Definitions.

Source: Office for National Statistics; Department of Health; National Assembly for Wales; General Register Office for Scotland; Information and Statistics Division, NHS in Scotland; Northern Ireland Statistics and Research Agency; Department of Health, Social Services and Public Safety, Northern Ireland

7.12 Notification rates of tuberculosis

Rates per 100,000 population

	1989	1990	1991	1992	1993	1994	1995	1996	1997	1998	1999
United Kingdom	10.6	10.2	10.5	11.1	11.3	10.7	10.5	10.6	10.8	11.2	..
Northern and Yorkshire	10.2	9.9	11.1	10.1	11.0	8.9	9.8	9.5	9.7	10.1	9.6
North West	10.5	11.4	10.1	11.7	12.1	10.0	9.6	8.8	9.4	10.3	10.5
Trent	10.1	9.3	9.3	8.6	10.5	9.2	10.2	10.6	9.4	9.6	9.6
West Midlands	16.1	13.9	15.5	16.5	14.9	13.8	12.3	12.3	11.5	12.6	13.4
Eastern	5.3	5.1	4.9	5.6	5.2	4.7	5.1	4.9	4.3	5.0	4.2
London	24.2	23.4	26.0	29.0	28.3	29.8	29.1	31.0	33.9	34.0	34.7
South East	5.2	4.9	4.9	4.7	5.6	5.5	5.7	5.4	5.6	5.8	5.7
South West	4.4	3.7	3.5	4.2	4.4	4.2	4.2	4.2	4.3	4.3	4.3
England	10.9	10.4	10.9	11.6	11.8	11.1	11.1	11.2	11.5	12.0	12.0
Wales	7.3	6.7	5.7	6.9	6.8	6.2	6.2	5.5	6.7	5.9	7.0
Scotland	10.5	11.0	10.7	10.9	10.8	10.6	9.3	9.9	8.5	8.9	..
Northern Ireland	5.9	8.2	6.0	5.2	5.5	5.6	5.5	4.5	4.5	3.6	3.6

Source: Public Health Laboratory Service, Communicable Disease Surveillance Centre; Scottish Centre for Infection and Environmental Health; Department of Health, Social Services and Public Safety, Northern Ireland

7.13 Diagnosed HIV-infected patients: by probable route of HIV infection and region of residence when last seen for care, 1999[1]

Numbers

	Homo/ bisexual	Injecting drug use	Heterosexual	Blood/blood products	Mother to infant	Other/ Not known	Total
United Kingdom	10,911	1,335	5,699	525	616	951	20,037
Northern and Yorkshire	371	29	184	51	10	19	664
North West	909	78	248	73	22	35	1,365
Trent	293	66	178	32	12	10	591
West Midlands	420	22	205	26	6	20	699
Eastern	327	64	209	35	18	30	683
London	6,417	478	3,593	137	464	716	11,805
South East	1,075	110	486	90	40	57	1,858
South West	436	50	177	12	14	34	723
England[2]	10,252	897	5,280	456	586	921	18,392
Wales	179	17	77	32	3	6	314
Scotland	422	415	313	35	27	24	1,236
Northern Ireland	58	6	29	2	0	0	95

1 Patients seen for statutory medical HIV related care at services in England, Wales and Northern Ireland in 1999 (includes 212 children born to HIV infected mothers in 1999 whose HIV infection status had not been confirmed: two resident in Northern and Yorkshire, four in North West, three in Trent, three in West Midlands, four in Eastern, 172 in London, 14 in South East, one in South West, one in Wales and eight in Scotland).
2 Includes four patients whose region of residence was not known.

Source: Public Health Laboratory Service, Communicable Disease Surveillance Centre; Institute of Child Health; Scottish Centre for Infection and Environmental Health.

7.14 NHS hospital waiting lists: by patients' region of residence, at 31 March 2000

	NHS hospital waiting lists[1]					
	Percentage waiting:			Total waiting (=100%) (thousands)	Mean waiting time (months)[2]	Median waiting time (months)[2]
	Less than 6 months	6 months but less than 12	12 months or longer			
Northern and Yorkshire	78.6	20.6	0.8	124.6	3.8	2.8
North West	74.8	20.5	4.7	160.1	4.2	2.9
Trent	76.4	19.9	3.8	102.6	4.1	2.8
West Midlands	79.4	17.7	3.0	86.0	3.8	2.7
Eastern	70.8	22.5	6.6	118.8	4.7	3.4
London	72.0	22.1	5.9	139.9	4.5	3.1
South East	70.6	23.2	6.2	193.4	4.7	3.4
South West	75.2	19.6	5.2	99.3	4.3	3.0
England	74.2	21.1	4.7	1,024.7	4.3	3.0
Wales	85.8		14.2	79.8	..	..
Scotland	85.4	13.2	1.4	82.3	3.0	1.9
Northern Ireland	60.9	19.1	20.0	47.3	..	..

1 The figures relate to people on the waiting lists on 31 March 2000 who were waiting for admission as either an in-patient or a day case and the length of time they had waited to date. Figures for Northern Ireland included all patients waiting for treatment at Northern Ireland Trusts including private patients and patients from outside Northern Ireland. Patients undergoing a series of repeat admissions and those who were temporarily suspended from the waiting list for medical or social reasons are excluded. There are differences between countries in the ways that waiting times are counted; comparisons between countries should be made with caution.
2 Average time patients had been waiting at 31 March 2000. The mean and median are different types of 'average'. See Notes and Definitions. Figures for Northern Ireland are available only in quarterly time bands.

Source: Department of Health; National Assembly for Wales; Information and Statistics Division, NHS in Scotland; Department of Health, Social Services and Public Safety, Northern Ireland

7.15 Total waiting on NHS hospital waiting lists[1]: by patients' region of residence, March 2000

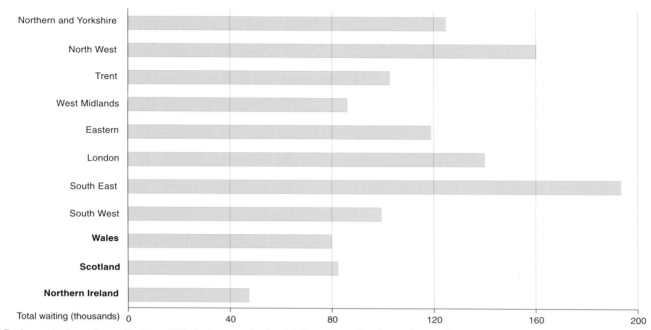

Total waiting (thousands)

1 The figures relate to people on waiting lists on 31 March who were waiting for admission as either an in-patient or a day case. There are differences between countries in the ways that waiting lists are counted; comparisons between countries should be made with caution. See Notes and Definitions.

Source: Department of Health; National Assembly for Wales; Information and Statistics Division, NHS in Scotland; Department of Health, Social Services and Public Safety, Northern Ireland

7.16 NHS hospital activity[1], 1998-99

	In-patients (all specialties)				Average length of stay in hospital for non-psychiatric specialties (mean)(days)	Day cases (thousands)	Total accident & emergency attend-ances (thousands)	Consultant out-patient attendances	
	Average daily available beds[2] per 1,000 population	Cases[3] treated per available bed[2]	Cases[3] treated per 1,000 population	Finished consultant episodes/ discharges and deaths[3] (thousands)				Total (thousands)	Of which new[4] (percentages)
United Kingdom	4.2	41.8	177	10,406	..	4,298	17,491	51,118	27.8
Northern and Yorkshire	4.3	43.3	185	1,175	5.9	475	1,900	5,566	26.1
North West	4.1	48.7	202	1,331	6.0	636	2,185	6,294	26.5
Trent	3.8	47.5	181	931	5.6	345	1,324	4,415	28.5
West Midlands	3.7	47.3	174	926	6.0	341	1,689	4,495	27.8
Eastern	3.4	43.0	147	788	6.2	290	1,132	3,939	28.4
London	4.3	39.6	171	1,227	6.8	501	2,495	7,602	27.3
South East	3.2	46.6	149	1,282	5.9	465	2,146	6,167	29.8
South West	4.0	46.2	184	901	6.1	368	1,409	3,676	30.4
England	3.8	45.1	173	8,563	6.1	3,421	14,280	42,154	27.9
Wales	5.1	38.6	196	575	6.6	328	982	2,678	25.8
Scotland	7.1	27.0	192	982	7.6	438	1,566	4,858	27.5
Northern Ireland	5.2	38.0	198	335	5.7	112	663	1,428	28.0

1 See Notes and Definitions.
2 Excluding cots for healthy new-born babies except in Northern Ireland.
3 Finished consultant episodes in England and Wales. Data for Wales relate to discharges and deaths. Data for Scotland relate to discharges and deaths and transfers to other specialties and hospitals. Data for Northern Ireland relate to discharges and deaths and transfers to another hospital. Healthy new-born babies are included for Northern Ireland but excluded for the other countries.
4 In Northern Ireland, data refer to GP referrals, not first attendances.

Source: Department of Health; National Assembly for Wales; Information and Statistics Division, NHS in Scotland; Department of Health, Social Services and Public Safety, Northern Ireland

7.17 Prescriptions dispensed, 1999[1]

	Prescription items dispensed (millions)[2]	Percentage of prescription items exempt from charge[3,4]	Percentage of prescription items[3,5] that were for		Number of prescription items per person	Average net ingredient cost[6]	
			Children	People aged 60 or over		£ per person	£ per prescription item
Northern and Yorkshire	74.8	86.4	8.9	52.4	11.8	111.0	9.4
North West	83.7	86.7	8.7	51.5	12.7	121.6	9.6
Trent	59.2	85.3	8.1	54.3	11.5	109.1	9.5
West Midlands	58.5	86.0	10.3	51.0	11.0	106.4	9.7
Anglia and Oxford	52.6	81.9	9.0	52.1	9.6	99.2	10.4
North Thames	65.9	84.8	11.2	47.1	9.2	97.8	10.6
South Thames	65.5	84.1	9.0	53.8	9.4	102.6	10.9
South and West	69.5	84.4	7.5	57.2	10.3	103.7	10.0
England	529.8	85.1	9.1	52.3	10.6	106.4	10.0
Wales[1]	40.0	87.9	8.6	51.6	13.6	118.6	8.7
Scotland	60.4	89.7	..	..	11.7	122.8	10.5
Northern Ireland[1]	22.2	94.7	..	..	13.0	121.8	9.4

1 For Wales and Northern Ireland, data relate to 1998-99.
2 Figures relate to NHS prescription items dispensed by community pharmacies, appliance contractors (appliance suppliers in Scotland), and dispensing doctors, and prescriptions submitted by prescribing doctors for items personally administered, known as stock orders in Scotland.
3 For England, figures relate to items dispensed by community pharmacists and appliance contractors only. Items dispensed by dispensing doctors and personal administration are not analysed into exempt, non-exempt or other categories and are therefore excluded. Personally administered items are free of charge.
4 Figures for the English regions, England and Wales exclude prescriptions for which prepayment certificates have been purchased. For Scotland and Northern Ireland they are included.
5 Items for children includes those for young adults aged 16 to 18 who are in full-time education. Data for Wales are calculated from a 5 per cent sample of prescriptions. Age specfic data are not available in Scotland.
6 Net ingredient cost is the cost of medicines before any discounts and does not include any dispensing costs or fees. This is known as Gross Ingredient Cost in Scotland.

Source: Department of Health; National Assembly for Wales; Information and Statistics Division, NHS in Scotland; Central Services Agency, Northern Ireland

7.18 NHS Hospital and Community Health Service staff: by type of staff[1], 30 September 1999

Percentages and thousands

	Direct care staff				Management and support staff			
	Medical and dental	Nursing, midwifery and health visiting[2]	Scientific, therapeutic and technical	All direct care staff	Adminis- tration and estates	Other	All management and support staff	Total staff[1] (=100%) (thousands)
Northern and Yorkshire	7.2	46.6	13.0	66.8	20.7	12.5	33.2	104.3
North West	7.6	47.9	13.6	69.1	20.7	10.2	30.9	112.1
Trent	7.5	47.2	13.4	68.0	21.0	11.0	32.0	81.8
West Midlands	7.4	46.9	13.1	67.5	21.4	11.1	32.5	83.5
Eastern	8.1	48.7	13.2	70.1	20.9	9.0	29.9	66.6
London	10.0	44.9	14.9	69.8	24.1	6.1	30.2	123.6
South East	7.7	46.6	12.9	67.2	21.8	11.1	32.9	115.7
South West	7.0	48.0	13.9	69.0	20.1	11.0	31.0	74.1
England[3]	7.7	46.3	13.7	67.7	22.1	10.2	32.3	782.1
Wales	7.0	44.3	12.8	64.1	20.9	15.0	35.9	54.1
Scotland	8.1	51.4	7.2	66.7	19.2	21.0	40.3	107.0
Northern Ireland	6.7	40.1	10.9	57.6	23.7	18.7	42.4	36.5

1 Directly employed whole-time equivalents. See Notes and Definitions.
2 Nursing, midwifery and health visiting staff includes learners and healthcare assistants.
3 The England totals include staff in special health authorities and other statutory authorities which are not assigned to a specific region.

Source: Department of Health; National Assembly for Wales; Information and Statistics Division, NHS in Scotland; Department of Health, Social Services and Public Safety, Northern Ireland

7.19 General practitioners, dentists and opticians[1], at 1 October 1999

Numbers and percentages

	General medical services						General dental services[1,2]			
	Number of practices	Number of general medical practitioners (GPs)[1]	Percentage who were female GPs	Average list size per GP	Number of practice staff[3] (whole-time equivalents)	Percentage who were direct care practice staff[3,4]	Number of dentists	Persons registered with a dentist as a percentage of the population[5]	Average regis-trations per dentist	Number of opticians[6]
United Kingdom	10,888	34,129	32	1,787	74,286	19.5	21,410	48	1,331	..
Northern and Yorkshire	1,054	3,592	30	1,794	8,046	19.2	2,074	52	1,586	1,040
North West	1,349	3,605	31	1,865	8,217	17.3	2,283	52	1,512	1,055
Trent	877	2,778	29	1,873	6,635	22.3	1,558	52	1,705	936
West Midlands	1,048	2,893	28	1,917	6,448	20.4	1,619	47	1,538	931
Eastern	821	2,956	30	1,850	6,976	23.2	1,898	48	1,370	1,300
London	1,711	3,947	40	1,931	8,772	18.6	3,087	40	933	1,496
South East	1,317	4,779	32	1,860	11,792	19.0	3,300	44	1,139	2,041
South West	767	3,041	30	1,644	6,201	21.2	1,896	48	1,237	873
England	8,944	27,591	32	1,845	63,087	19.9	17,715	47	1,325	7,517
Wales	525	1,761	28	1,694	3,868	20.2	1,004	50	1,457	582
Scotland	1,054	3,696	36	1,441	7,324	15.9	2,031	53	1,328	1,313
Northern Ireland	365	1,054	30	1,679	..	..	660	51	1,315	364

1 See Notes and Definitions.
2 At 30 September 1999 for general dental practitioners. Dentists are assigned to the region where they carry out their main work.
3 Other than GPs. Figure for the United Kingdom relates to Great Britain as figures for Northern Ireland are not held centrally.
4 Figures relate to practice nurses, physiotherapists, chiropodists, counsellors, dispensers and complementary therapists.
5 Registrations with dentists practising in each region.
6 Optometrists and ophthalmic medical practitioners contracted to perform NHS sight tests at 31 December 1999 (31 March 1999 for optometrists in Scotland). As some practitioners have contracts in more than one region, the sum of the regions does not equal the England total. Similarly, as some practitioners have contracts in more than one country, it is not possible to calculate a United Kingdom figure.

Source: Department of Health; National Assembly for Wales; Information and Statistics Division, NHS in Scotland; Central Services Agency, Northern Ireland

7.20 Places available in residential care homes[1]: by type of care home, at 31 March 1999

Percentages and thousands

	Percentage of places available in			Percentage of places available in homes for				Total number of places available (=100%) (thousands)
	Local authority homes[2]	Registered homes		Older people	People with physical, sensory or learning disabilities	People with mental health problems	Other people	
		Voluntary homes[3]	Other[4]					
United Kingdom	19	20	61	71	17	10	2	394
North East	21	11	68	72	15	12	1	20
North West	17	17	66	77	12	11	1	51
Yorkshire and the Humber	21	13	66	61	16	22	1	37
East Midlands	18	11	71	73	16	10	-	30
West Midlands	22	17	60	74	20	6	1	33
East	18	21	60	74	18	7	-	34
London	23	36	41	63	19	16	2	30
South East	12	20	68	68	20	11	1	64
South West	11	18	71	73	18	7	1	46
England	17	19	64	71	17	11	1	345
Wales	32	11	57	84	12	4	1	17
Scotland	33	42	25	67	22	6	5	24
Northern Ireland	35	28	37	69	18	9	5	7

1 The figures for percentage of places available in Wales and Scotland exclude children's homes. The figures for England and Scotland include residential places in homes registered as both residential and nursing. See Notes and Definitions.
2 For England, figures relate to local authority staffed homes. For Northern Ireland, figures relate to places available in statutory homes operated by the Health and Social Services Trusts.
3 For England, figures include dual registered voluntary homes.
4 The figures for England include independent small homes (fewer than 4 places) and private homes. The figures for Wales and Scotland relate to private homes only. The figures for Northern Ireland relate to all private homes regardless of size.

Source: Department of Health; National Assembly for Wales; Scottish Executive; Department of Health, Social Services and Public Safety, Northern Ireland

7.21 Children looked after by local authorities, year ending 31 March 1999[1]

	Total children looked after per thousand resident population[2]			Manner of accommodation (percentages)			Number of children looked after (=100%)
	Children admitted	Ceased to be looked after	Looked after	Foster homes	Community homes[3]	Other[4]	
North East	3.1	2.9	5.5	63.3	12.0	24.7	3,300
North West	2.8	2.6	5.9	61.7	11.2	27.1	9,500
Yorkshire and the Humber	2.5	2.5	5.5	62.3	10.6	27.1	6,400
East Midlands	2.3	2.2	4.2	66.0	8.8	25.3	4,000
West Midlands	2.5	2.4	4.9	66.2	9.2	24.7	6,200
East	1.9	1.9	4.0	66.9	6.9	26.3	4,800
London	3.1	2.9	5.8	67.3	6.7	26.1	9,500
South East	2.0	2.0	3.9	65.1	7.5	27.4	7,000
South West	2.5	2.6	4.4	73.4	5.9	20.7	4,600
England	2.5	2.4	4.9	65.5	8.7	25.8	55,300
Wales1	2.6	2.6	4.9	76.0	7.3	16.8	3,300
Scotland	..	..	9.8	28.2	15.9	55.9	11,200
Northern Ireland	2.2	2.2	5.0	64.8	11.2	24.0	2,300

1 Data for Wales relate to year ending 31 March 1997.
2 Rates are based on mid-1998 estimates of population aged under 18.
3 Scottish figures relate to residential care homes for children.
4 Includes children looked after at home. The total number of children looked after in Scotland includes 5309 children (47 per cent of the total) who were looked after at home.

Source: Department of Health; National Assembly for Wales; Scottish Executive; Department of Health, Social Services and Public Safety, Northern Ireland

7.22 Children and young people on child protection registers: by age and category[1], at 31 March 1999

	Percentage aged					Number of children on registers[3] (=100%)	Rate per 10,000 children aged under 18[4]	Percentage of children in each category of abuse					
	Under 1	1-4	5-9[2]	10-15[2]	16 or over			Neglect	Physical injury	Sexual abuse	Emotional abuse	Multiple categories	Other[5]
North East	10	33	29	26	2	2,166	37	32	21	13	13	17	4
North West	11	31	31	25	1	4,205	26	37	23	13	14	10	4
Yorkshire and the Humber	11	30	30	26	2	4,014	34	33	22	16	14	11	4
East Midlands	9	32	29	28	2	3,088	33	30	21	19	13	16	1
West Midlands	8	30	30	28	3	3,759	30	34	21	15	21	9	-
East	9	31	31	26	2	3,082	25	33	20	15	18	13	-
London	9	30	30	29	2	4,901	30	44	17	12	18	9	1
South East	9	30	29	27	2	3,817	21	38	20	16	19	8	-
South West	8	30	32	28	2	2,844	27	26	18	20	21	11	4
England	9	30	30	27	2	31,900	28	35	20	15	17	11	2
Wales	8	31	31	27	2	2,673	40	34	26	12	18	10	-
Northern Ireland	7	28	42	21	3	1,463	32	38	21	17	14	10	-

1 Data for Scotland are not available in the same form.
2 Age bands for Northern Ireland are 5 to 11 and 12 to 15.
3 Includes a number of unborn children not included elsewhere in this table.
4 Figure for Northern Ireland calculated using the mid-1999 population estimate.
5 For England and Wales data relate to children or young people on the child protection registers who have not been allocated a specific category.

Source: Department of Health; National Assembly for Wales; Department of Health, Social Services and Public Safety, Northern Ireland

8 Income and Lifestyles

Household income

During the period 1996-97 to 1998-99, households in London had the highest average gross weekly income at £523, while those in Northern Ireland had the lowest with £347 per week.

(Table 8.1)

During the period 1996-97 to 1998-99, the proportion of households with a weekly income of less than £100 ranged from a tenth in the East of England and the South East to almost two tenths of households in the North East region.

(Table 8.2)

Individual income

In 1998-99, individuals living in the North East and in Yorkshire and the Humber were over represented at the lower end of the distribution and under represented at the top on a before housing cost basis, excluding the self employed. For these regions 26 per cent and 25 per cent of individuals had incomes in the bottom quintile respectively and around 13 per cent had incomes in the top quintile.

(Table 8.3)

Savings

In 1998-99, almost two fifths of households in the South East owned stocks and shares, compared to around a fifth of households in the North East.

(Table 8.4)

Around a fifth of households in Scotland, the North East and the West Midlands do not have a current account.

(Table 8.4)

Income tax

In 1997-98, the percentage of individuals with an income liable to assessment for tax in excess of £50,000 ranged from 1.4 per cent in the North East and in Wales, to 5 per cent in London.

(Table 8.5)

Receipt of benefit

In many of the large cities in the UK over a quarter of the population of working age were claiming key social security benefits in November 1999.

(Map 8.7)

In 1998-99, around a quarter of households in the South West and Scotland were in receipt of Child Benefit compared with around three in ten households in all other regions of Great Britain.

(Table 8.8)

Children's spending

Children in Scotland spent on average £12.40 per head per week over the period 1997-98 to 1998-99, the highest amount in Great Britain; children in the South East and West Midlands spent nearly 30 per cent less than this, an average of £8.80 per child per week.

(Chart 8.9)

Charitable giving

Charitable giving by households over the period 1996-97 to 1998-99 ranged from an average of 80 pence per household per week in the North East to £3.90 per household per week in Northern Ireland.

(Chart 8.10)

Household expenditure

Average household expenditure in London and the South East over the period 1996-97 to 1998-99, at around £375 per week, was the highest of any regions, and around £40 higher than the UK average.

(Table 8.11)

In Northern Ireland over the period 1996-97 to 1998-99, households spent on average 10 per cent of their total weekly expenditure on housing, a lower proportion than in any other UK region.

(Table 8.11)

In 1998-1999, households in Wales bought the most meat and meat products, while households in Northern Ireland bought the most liquid and processed milk and cream, and Londoners bought the most fruit.

(Table 8.13)

Consumption

In 1998-1999, households in Northern Ireland consumed around a quarter less fruit than the GB average of 1.04 kg per person per week, while people in London bought over a fifth more fruit than the GB average.

(Table 8.13)

Consumer goods

In 1998-99, around nine in ten households in the United Kingdom owned a washing machine and a similar proportion owned a deep freezer. In contrast, fewer than one in four households in the United Kingdom owned a dishwasher.

(Table 8.14)

Leisure

In 1999, the average weekly television viewing ranged from 27.6 hours for men in Scotland to 21.4 hours for men living in the East of England; for women the range was from 31.6 hours in Scotland to 25.1 hours in the South, South East and Channel Islands.

(Table 8.15)

In 1999, children in East of England, the Midlands and the South, South East and the Channel Islands watched the least television at around 16 hours per week.

(Table 8.15)

In 1998, the proportion of adults taking a holiday abroad ranged from around two fifths of adults from the South East, North and North West regions to under three tenths of those in East Anglia.

(Table 8.16)

Voluntary work

In 1996-97, 18 per cent of householders in the South West had done some voluntary work in the last year.

(Table 8.17)

National Lottery

In the period 1996-97 to 1998-99, three fifths of households in the UK participated in the National Lottery in the two weeks preceding interview, with the lowest proportion being households in London and the South West.

(Chart 8.18)

In 1996-97 to 1998-99, on average, households in the North East and West Midlands spent around £4.10 a week on the National Lottery compared with households in the South West; who spent around £3.40 a week.

(Chart 8.18)

8.1 Household income: by source, 1996-1999[1]

Percentages and £

	Percentage of average gross weekly household income						Average gross weekly household income[3] (=100%) (£)
	Wages and salaries	Self-employ-ment	Invest-ments	Annuities and pensions[2]	Social security benefits[3]	Other income	
United Kingdom	67	8	4	7	13	1	430
North East	63	6	4	8	18	1	357
North West	66	9	3	6	15	1	401
Yorkshire and the Humber	66	7	4	7	15	1	390
East Midlands	67	8	4	7	13	1	405
West Midlands	67	9	4	6	14	1	408
East	69	8	5	7	11	1	451
London	70	9	5	6	9	1	523
South East	68	9	6	7	9	1	511
South West	61	10	7	8	12	2	411
England	67	8	5	7	12	1	441
Wales	60	8	4	7	19	2	359
Scotland	66	7	3	7	16	1	387
Northern Ireland[4]	62	9	2	5	21	1	347

1 See Notes and Definitions. Combined data from the 1996-97, 1997-98 and 1998-99 surveys. Data are for adults only.
2 Other than social security benefits.
3 Excluding Housing Benefit and Council Tax Benefit (rates rebate in Northern Ireland).
4 Northern Ireland data are calculated from an enhanced sample, but the United Kingdom figures are calculated from the main Family Expenditure Survey sample.

Source: Family Expenditure Survey, Office for National Statistics and Northern Ireland Statistics and Research Agency

8.2 Distribution of household income, 1996-1999[1]

Percentages and £

	Percentage of households in each weekly income group								Average gross weekly income[2] (£)	
	Under £100	£100 but under £150	£150 but under £250	£250 but under £350	£350 but under £450	£450 but under £600	£600 but under £750	£750 or over	Per house-hold	Per person
United Kingdom	12	10	16	13	11	14	9	14	430	181
North East	18	13	19	15	8	10	7	11	357	158
North West	14	10	17	13	11	14	9	11	401	167
Yorkshire and the Humber	13	11	18	12	12	13	9	12	390	164
East Midlands	11	11	17	13	11	16	9	12	405	169
West Midlands	14	10	16	14	11	14	8	13	408	166
East	10	10	16	13	10	15	10	16	451	190
London	12	9	13	12	10	13	10	21	523	219
South East	10	9	13	12	11	13	10	21	511	223
South West	11	11	16	14	14	15	8	12	411	177
England	12	10	16	13	11	14	9	15	441	187
Wales	14	13	19	14	11	13	7	10	359	149
Scotland	14	10	18	13	13	12	8	11	387	166
Northern Ireland[3]	14	14	21	13	11	11	8	8	347	129

1 See Notes and Definitions. Combined data from the 1996-97, 1997-98 and 1998-99 surveys. Data are for adults only.
2 Excluding Housing Benefit and Council Tax Benefit (rates rebate in Northern Ireland).
3 Northern Ireland data are calculated from an enhanced sample, but the United Kingdom figures are calculated from the main Family Expenditure Survey sample.

Source: Family Expenditure Survey, Office for National Statistics and Northern Ireland Statistics and Research Agency

8.3 Income distribution of individuals, 1998-99[1]

Percentages

	Quintile groups of individuals ranked by net equivalised household income[2]				
	Bottom fifth	Next fifth	Middle fifth	Next fifth	Top fifth
Great Britain	20	20	20	20	20
North East	26	26	20	16	13
North West	24	20	19	20	17
Yorkshire and the Humber	25	23	21	17	14
East Midlands	22	21	23	19	15
West Midlands	19	23	22	20	15
East	16	20	19	20	25
London	19	18	15	19	30
South East	13	15	19	24	29
South West	19	20	21	23	17
England	20	20	20	20	21
Wales	23	22	24	19	12
Scotland	21	19	23	20	18

1 See Notes and Definitions.
2 Income before housing costs.

Source: Department of Social Security from Households Below Average Income

8.4 Households[1] with different types of saving, 1998-99

Percentages[2]

	Accounts				Other savings					
	Current[3]	Post Office	TESSA	Other bank/ building society[4]	Gilts or unit trusts	Stocks and shares	National Savings	Save As You Earn	Premium Bonds	PEPs
Great Britain	85	11	14	64	6	28	8	2	27	15
North East	79	6	10	54	4	19	4	1	16	10
North West	83	11	13	58	6	26	7	2	23	13
Yorkshire and the Humber	85	9	12	60	5	25	7	1	22	11
East Midlands	87	12	16	67	5	30	8	2	26	14
West Midlands	80	10	14	68	6	24	8	1	26	13
East	89	13	17	72	6	32	9	2	33	17
London	85	10	14	64	7	28	7	2	28	16
South East	92	14	17	75	9	38	11	2	37	20
South West	91	17	15	68	8	31	11	2	34	17
England	86	12	14	66	7	29	8	2	28	15
Wales	84	12	13	56	4	22	7	1	23	12
Scotland	76	7	11	55	6	24	6	1	17	13

1 Households in which at least one member has an account. See Notes and Definitions.
2 As a percentage of all households.
3 A current account may be either a bank account or a building society account.
4 All bank/building society accounts excluding current accounts and TESSAs plus other accounts yielding interest.

Source: Family Resources Survey, Department of Social Security

8.5 Distribution of income liable to assessment for tax, 1997-98[1]

Percentages and thousands

	Percentage of individuals in each income range								Individuals with incomes of £4,045 or more (=100%) (thousands)
	£4,045-£4,999	£5,000-£7,499	£7,500-£9,999	£10,000-£14,999	£15,000-£19,999	£20,000-£29,999	£30,000-£49,999	£50,000 and over	
United Kingdom[2]	6.9	17.0	13.9	22.8	15.3	15.3	6.2	2.8	27,933
North East	9.7	18.5	15.4	22.2	14.8	14.0	4.0	1.4	1,108
North West	7.5	18.2	14.1	23.8	14.6	14.9	5.2	1.8	3,070
Yorkshire and the Humber	8.1	18.7	14.1	25.1	14.6	13.1	4.6	1.8	2,313
East Midlands	7.2	17.4	14.7	24.7	15.3	13.5	5.1	2.1	2,042
West Midlands	6.8	17.6	14.7	24.2	15.5	14.7	4.8	1.7	2,364
East	5.4	15.3	14.7	21.1	15.4	16.6	7.8	3.7	2,659
London	5.2	13.9	11.1	20.8	17.3	18.2	8.5	5.0	3,246
South East	6.2	16.0	12.4	21.4	14.4	16.5	8.6	4.6	4,075
South West	7.7	19.4	13.7	23.6	14.8	13.2	5.8	1.9	2,477
England	6.8	17.0	13.6	22.8	15.2	15.3	6.4	2.9	23,353
Wales	8.5	18.1	16.7	23.1	14.0	14.1	4.1	1.4	1,283
Scotland	5.8	16.4	14.8	22.7	16.8	16.2	5.4	1.9	2,500
Northern Ireland	8.8	19.1	14.6	24.9	14.4	12.3	4.4	1.6	611

1 See Notes and Definitions.
2 Figures for United Kingdom include members of HM Forces and others who are liable to some UK tax but reside overseas on a long-term basis. In addition, the United Kingdom total includes a very small number of individuals who could not be allocated to a region.

Source: Survey of Personal Incomes, Board of Inland Revenue

8.6 Average total income[1] and average income tax payable[2]: by gender, 1997-98[3]

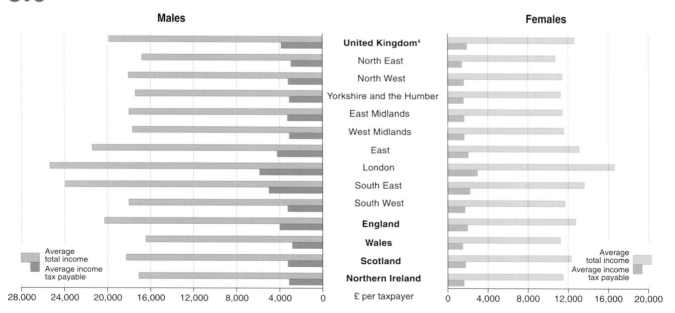

1 Figures are based on individuals with total income above the single person's allowance (£4,045 in 1997-98).
2 Figures relate to taxpayers only.
3 See Notes and Definitions.
4 Figures for United Kingdom include members of HM forces and others who are liable to some tax but reside overseas on a long-term basis. In addition, the United Kingdom total includes a very small number of individuals that could not be allocated to a region.

Source: Survey of Personal Incomes, Board of Inland Revenue

8.7

Percentage of the population of working age claiming a key social security benefit[1]: by local authority, November 1999

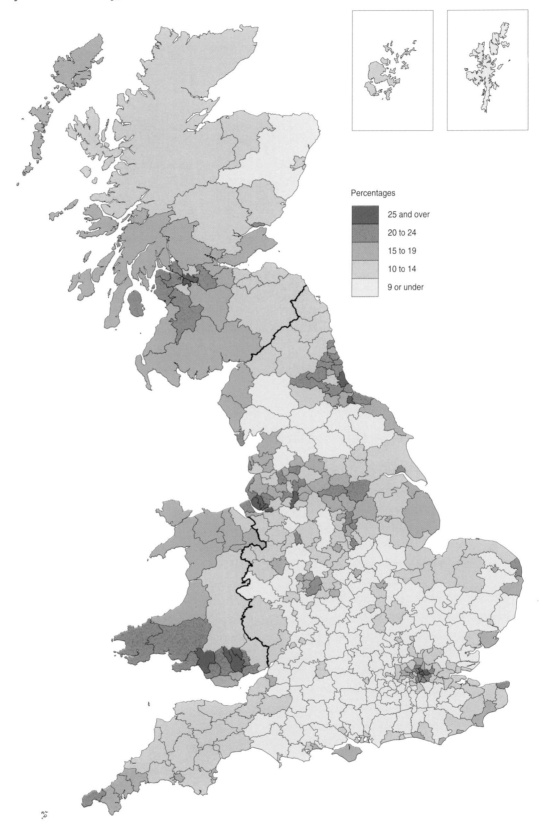

Percentages

- 25 and over
- 20 to 24
- 15 to 19
- 10 to 14
- 9 or under

1 Key benefits are Jobseeker's Allowance, Incapacity Benefit, National Insurance Credits (only through Jobseeker's Allowance and Incapacity Benefit), Severe Disablement Allowance, Disability Living Allowance and Income Support.

Source: Department of Social Security

8.8 Households in receipt of benefit[1]: by type of benefit, 1998-99

Percentages[2]

	Family Credit or Income Support	Housing Benefit	Council Tax Benefit	Jobseeker's Allowance	Retirement Pension	Incapacity or Disablement Benefits[3]	Child Benefit	Any benefit
Great Britain	15	18	23	4	30	16	29	70
North East	20	26	33	7	29	25	29	75
North West	18	19	26	4	29	20	30	74
Yorkshire and the Humber	18	21	27	5	30	18	30	72
East Midlands	16	15	21	4	30	16	28	70
West Midlands	17	19	25	3	32	17	29	73
East	11	15	18	3	29	12	29	66
London	17	22	25	4	26	10	29	65
South East	10	11	14	2	29	10	29	65
South West	12	13	18	3	33	15	26	69
England	15	17	22	4	29	15	29	69
Wales	18	17	26	5	34	24	29	77
Scotland	17	24	29	5	30	19	26	71

1 Households in which at least one member is in receipt of benefit. See Notes and Definitions.
2 As a percentage of all households.
3 Incapacity Benefit, Disability Living Allowance (Care and Mobility components), Severe Disablement Allowance, Disability Working Allowance, Industrial Injuries Disablement Benefit, War Disablement Pension and Attendance Allowance.

Source: Family Resources Survey, Department of Social Security

8.9 Children's spending[1], 1997-99[2]

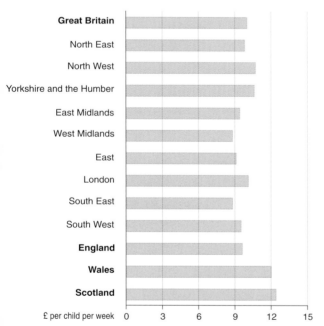

£ per child per week

1 Children aged 7 to 15. See Notes and Definitions.
2 Combined data from the 1997-98, and 1998-99 surveys.

Source: Family Expenditure Survey, Office for National Statistics

8.10 Charitable giving[1], 1996-99[2]

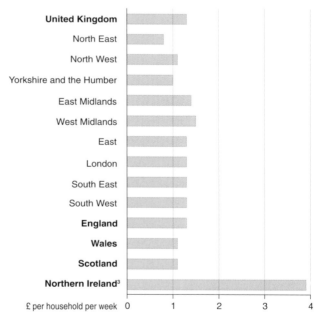

£ per household per week

1 See Notes and Definitions.
2 Combined data from the 1996-97, 1997-98 and 1998-99 surveys.
3 Northern Ireland data are obtained from an enhanced sample but the United Kingdom figures are obtained from the main Family Expenditure Survey Sample.

Source: Family Expenditure Survey, Office for National Statistics and Northern Ireland Statistics and Research Agency

8.11 Household expenditure[1]: by commodity and service, 1996-1999

£ per week and percentages

	Housing	Fuel, light and power	Food	Alcohol and tobacco	Clothing and footwear	House-hold goods and services	Motoring and fares	Leisure goods and services	Miscellan-eous and personal goods and services	Average house-hold expend-iture	Average expend-iture per person
£ per week											
United Kingdom	52.50	12.50	57.40	19.90	20.20	45.30	55.20	55.80	13.70	332.60	140.20
North East	43.80	12.40	49.70	20.70	19.30	36.70	42.30	47.20	10.70	282.90	125.70
North West	46.70	12.70	55.30	22.90	21.20	39.50	54.20	53.90	13.10	319.50	132.70
Yorkshire and the Humber	46.60	12.40	54.10	21.70	20.60	45.60	51.20	55.20	12.90	320.40	134.50
East Midlands	48.30	12.30	56.40	19.40	18.80	45.60	53.00	52.80	13.10	319.80	133.70
West Midlands	47.20	12.90	55.70	19.70	19.20	43.40	52.90	53.40	11.90	316.30	129.00
East	54.50	12.10	59.20	16.90	21.40	49.50	59.70	57.90	14.60	345.70	145.90
London	69.30	11.70	63.10	20.10	22.80	50.80	56.40	65.40	16.90	376.40	157.70
South East	65.40	12.00	59.60	18.20	20.10	52.30	67.40	65.10	16.00	376.10	164.00
South West	51.90	12.00	55.00	17.50	17.30	44.10	54.10	50.00	13.70	315.50	135.80
England	54.60	12.20	57.20	19.70	20.30	46.20	56.20	57.10	14.10	337.70	142.80
Wales	45.30	13.50	54.90	19.80	18.30	42.60	48.90	51.50	12.70	307.50	127.80
Scotland	43.50	13.60	57.80	21.80	19.90	39.80	50.50	48.00	10.90	305.70	131.60
Northern Ireland[2]	30.60	15.50	62.40	20.40	24.40	40.90	49.40	44.30	13.60	301.40	112.50
As a percentage of average weekly household expenditure											
United Kingdom	16	4	17	6	6	14	17	17	4	100	
North East	15	4	18	7	7	13	15	17	4	100	
North West	15	4	17	7	7	12	17	17	4	100	
Yorkshire and the Humber	15	4	17	7	6	14	16	17	4	100	
East Midlands	15	4	18	6	6	14	17	17	4	100	
West Midlands	15	4	18	6	6	14	17	17	4	100	
East	16	3	17	5	6	14	17	17	4	100	
London	18	3	17	5	6	13	15	17	4	100	
South East	17	3	16	5	5	14	18	17	4	100	
South West	16	4	17	6	5	14	17	16	4	100	
England	16	4	17	6	6	14	17	17	4	100	
Wales	15	4	18	6	6	14	16	17	4	100	
Scotland	14	4	19	7	6	13	17	16	4	100	
Northern Ireland[2]	10	5	21	7	8	14	16	15	5	100	

1 See Notes and Definitions. Combined data from the 1996-97, 1997-98 and 1998-99 surveys.
2 Northern Ireland data are calculated from an enhanced sample, but the United Kingdom figures are calculated from the main Family Expenditure Survey sample. The data from the main Family Expenditure Survey include expenditure by children; the Northern Ireland data relate to adults only.

Source: Family Expenditure Survey, Office for National Statistics and Northern Ireland Statistics and Research Agency

8.12 Expenditure on selected foods bought for household consumption and expenditure on eating out, 1998-1999[1]

£ per person per week

	Liquid and processed milk and cream	Cheese	Uncooked carcass meat and poultry	Other meat and meat products	Fish	Vegetables and vegetable products[2]	Fresh and other fruit	Bread	Cereals other than bread	Drinks and confectionary	Total household food and drink	Eating out[3]
United Kingdom[4]	1.34	0.52	1.86	1.98	0.78	2.30	1.32	0.72	1.97	2.12	16.86	6.91
North East	1.21	0.41	1.66	2.06	0.77	2.06	1.00	0.71	2.00	1.91	15.57	7.90
North West	1.36	0.44	1.90	1.98	0.71	2.07	1.14	0.72	1.81	1.83	15.73	6.54
Yorkshire and the Humber	1.31	0.41	1.76	1.86	0.80	2.01	1.11	0.70	1.86	1.76	15.37	5.70
East Midlands	1.33	0.52	1.66	1.83	0.71	2.15	1.11	0.71	1.81	1.99	15.61	6.23
West Midlands	1.32	0.50	1.72	1.80	0.74	2.29	1.14	0.68	1.82	1.87	15.84	6.11
East	1.31	0.52	1.93	2.01	0.86	2.37	1.51	0.67	2.00	2.06	17.20	6.97
London	1.31	0.53	1.96	1.86	0.93	2.73	1.68	0.71	2.11	2.09	17.99	9.63
South East	1.42	0.65	1.98	2.12	0.89	2.56	1.60	0.72	2.17	2.60	18.88	7.37
South West	1.42	0.63	1.89	1.90	0.71	2.47	1.48	0.76	2.04	2.20	17.58	6.04
England	1.34	0.52	1.85	1.95	0.80	2.33	1.35	0.71	1.97	2.07	16.85	7.01
Wales	1.27	0.47	1.79	2.17	0.78	2.20	1.24	0.72	1.83	2.22	16.69	6.18
Scotland	1.24	0.53	1.91	2.25	0.72	2.14	1.20	0.79	2.08	2.78	17.59	6.36
Northern Ireland	1.51	0.34	2.01	1.91	0.50	2.00	0.97	0.80	1.87	1.42	14.99	..

1 See Notes and Definitions.
2 Including tomatoes, fresh potatoes and potato products.
3 Individual expenditure on all food and drink consumed outside the home and not obtained from household stocks, whether consumed by the purchaser or others or both. Expenditure which is to be reclaimed as business expenses is not included.
4 Figure relating to eating out is for Great Britain only.

Source: National Food Survey, Ministry of Agriculture, Fisheries and Food

8.13 Household consumption of selected foods, 1991-1992 and 1998-1999[1]

Kilograms per person per week[2]

	Liquid and processed milk and cream		Meat and meat products		Fish		Vegetables and vegetable products[3]		Fresh and other fruit		Cereals including bread	
	1991-1992	1998-1999	1991-1992	1998-1999	1991-1992	1998-1999	1991-1992	1998-1999	1991-1992	1998-1999	1991-1992	1998-1999
Great Britain	2.15	2.06	0.97	0.92	0.14	0.14	2.24	2.01	0.92	1.04	1.46	1.47
North East	2.08	1.93	1.02	0.99	0.15	0.15	2.44	1.97	0.83	0.91	1.56	1.53
North West	2.11	2.09	0.98	0.95	0.14	0.14	2.23	1.84	0.81	0.94	1.42	1.41
Yorkshire and the Humber	2.13	2.04	0.97	0.94	0.16	0.15	2.21	1.88	0.85	0.98	1.46	1.46
East Midlands	2.35	2.13	0.91	0.89	0.13	0.14	2.26	2.06	0.95	0.97	1.42	1.50
West Midlands	2.02	2.09	0.99	0.87	0.13	0.15	2.33	2.04	0.70	0.99	1.52	1.48
East	2.13	1.91	0.97	0.89	0.15	0.15	2.21	1.98	1.06	1.19	1.44	1.44
London	2.02	1.90	0.99	0.86	0.16	0.17	2.15	2.08	1.15	1.26	1.41	1.48
South East	2.20	2.05	0.90	0.91	0.13	0.15	2.15	2.02	1.10	1.23	1.41	1.48
South West	2.25	2.07	0.96	0.94	0.12	0.14	2.45	2.18	0.98	1.24	1.46	1.50
England	2.13	2.03	0.97	0.92	0.14	0.15	2.26	2.00	0.95	1.10	1.45	1.47
Wales	2.29	2.08	1.02	1.02	0.14	0.15	2.41	2.14	0.74	0.99	1.52	1.45
Scotland	2.20	1.98	0.92	0.97	0.13	0.13	2.02	1.17	0.82	0.94	1.52	1.47
Northern Ireland	..	2.36	..	0.87	..	0.09	..	2.15	..	0.76	..	1.48

1 See Notes and Definitions.
2 Except equivalent litres of milk and cream.
3 Including tomatoes, fresh potatoes and potato products.

Source: National Food Survey, Ministry of Agriculture, Fisheries and Food

8.14 Households with selected durable goods, 1998-99[1]

Percentages

	Micro-wave oven	Wash-ing machine	Dish-washer	Deep freezer[2]	Video	Compact-disc player	Satellite dish	Home comp-uter
United Kingdom	78	91	23	92	84	66	29	31
North East	83	93	15	91	85	64	32	23
North West	80	91	19	92	85	66	31	30
Yorkshire and the Humber	81	93	18	92	83	63	27	26
East Midlands	81	93	22	93	84	66	28	30
West Midlands	79	92	20	91	83	63	30	30
East	80	93	28	93	86	71	31	36
London	68	85	23	90	79	66	30	36
South East	78	91	31	93	86	73	27	40
South West	76	90	27	93	84	67	23	32
England	78	91	23	92	84	67	29	33
Wales	80	90	18	91	82	61	28	24
Scotland	78	93	22	89	85	67	26	27
Northern Ireland	79	92	25	85	80	49	27	22

1 See Notes and Definitions.
2 Includes fridge freezers.

Source: Family Resources Survey, Department of Social Security; Continuous Household Survey, Northern Ireland Statistics and Research Agency

8.15 Average weekly television viewing[1]: by audience

Hours[2]

	Men[3]			Women[3]			Children[3]		
	1995	1997	1999	1995	1997	1999	1995	1997	1999
ITV Viewing regions									
United Kingdom[4]	24.9	25.0	25.2	28.3	28.2	28.7	18.1	17.7	18.2
North East	24.0	25.7	27.0	27.9	28.3	28.6	16.7	17.7	18.0
North West	25.5	25.1	25.4	28.7	29.8	30.3	18.3	17.4	18.2
Yorkshire	25.3	25.3	25.1	28.6	28.6	28.5	17.9	18.5	18.3
Midlands	22.9	23.5	24.7	27.4	27.1	27.5	17.6	16.1	16.4
East of England	21.3	21.7	21.4	24.9	25.0	25.8	17.8	16.3	15.9
London	22.5	23.1	23.2	24.6	25.0	25.4	16.1	16.2	17.3
South, South East and Channel Islands	23.0	23.2	21.6	26.6	25.6	25.1	16.2	17.2	16.6
South West	24.6	24.5	24.2	25.8	26.4	27.3	18.0	16.0	17.7
Wales and West	23.2	23.3	23.2	27.4	26.9	26.4	17.0	16.5	17.5
Border	24.0	24.8	25.6	28.2	29.5	28.5	17.5	19.2	19.0
Scotland	26.7	27.3	27.6	29.1	30.0	31.6	18.0	17.9	18.4
Ulster	25.9	25.9	26.6	29.6	28.6	30.4	19.0	18.1	19.6

1 Including timeshift, i.e. viewing of broadcast material recorded at home and played back within seven days of recording.
2 Per person in UK private households containing a television set in working order.
3 Men and women are defined as individuals aged 16+ and children are defined as individuals aged 4 to 15 years old.
4 Figures for the regions exclude viewing of other regions' broadcasts, whereas figures for the United Kingdom include all viewing, and are therefore higher.

Source: Broadcasters' Audience Research Board; RSMB Television Research Limited

8.16 Adults taking a holiday[1]: by region of domicile, 1998

Percentages

	A holiday abroad	A holiday in Great Britain	Any holiday
Standard Statistical Regions			
Great Britain	38	31	59
North	40	26	54
North West (SSR)	43	26	59
Yorkshire and Humberside	38	37	64
East Midlands	39	36	64
West Midlands	36	29	56
East Anglia	29	33	50
South East (SSR)	43	29	60
Greater London	41	20	56
Rest of South East	43	34	63
South West	32	37	61
England	39	31	59
Wales	34	36	55
Scotland	35	32	59

Percentage of adults taking

1 Defined as four or more nights away.

Source: British Tourist Authority

8.17 Participation in local voluntary work, 1996-97[1]

Percentages

	Still volunteering	No longer volunteering	Interested in volunteering	Not interested in volunteering
North East	8	1	32	59
North West	11	2	36	52
Yorkshire and the Humber	8	1	30	61
East Midlands	10	2	39	49
West Midlands	11	2	35	53
East	14	1	39	46
London	10	2	43	45
South East	14	2	40	45
South West	16	2	40	43
England	11	2	37	50

1 Participation by householders in voluntary work to benefit their local area in the 12 months before interview. See Introduction and Notes and Definitions.

Source: Survey of English Housing, Department of the Environment, Transport and the Regions

8.18 Participation in the National Lottery[1], 1996-99[2]

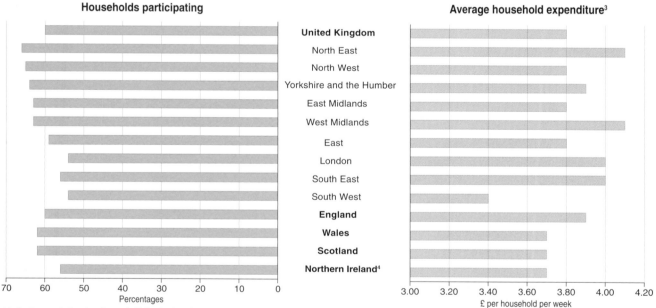

Households participating   **Average household expenditure[3]**

United Kingdom, North East, North West, Yorkshire and the Humber, East Midlands, West Midlands, East, London, South East, South West, **England**, **Wales**, **Scotland**, **Northern Ireland[4]**

Percentages (70 to 0) — £ per household per week (3.00 to 4.20)

1 In the two-week diary keeping period following interview; including scratchcards.
2 Combined data from the 1996-97, 1997-98 and 1998-99 surveys.
3 Average weekly expenditure of participating households.
4 Northern Ireland data are calculated from an enhanced sample, but the United Kingdom figures are calculated from the main Family Expenditure Survey sample. The data from the main Family Expenditure Survey include expenditure by children; the Northern Ireland data relate to adults only.

Source: Family Expenditure Survey, Office for National Statistics; Northern Ireland Statistics and Research Agency

8.19 The National Lottery grants: totals from 1996 to 1999

Numbers and £ million

	Number of grants awarded from the start of National Lottery to:				Total value of grants (£ million)			
	end 1996	end 1997	end 1998	end 1999	end 1996	end 1997	end 1998	end 1999
United Kingdom[1]	11,277	25,281	33,669	52,102	2,684.2	5,067.9	6,120.2	7,495.0
North East	440	1,246	1,839	2,870	75.8	196.5	229.2	338.2
North West	1,200	2,115	2,603	3,961	178.5	380.0	514.9	693.7
Yorkshire and the Humber	863	1,827	2,220	3,491	134.5	249.8	307.7	364.4
East Midlands	631	1,383	1,742	3,864	73.3	152.1	195.9	245.9
West Midlands	853	2,090	2,469	3,936	147.1	234.6	316.0	396.4
East	653	1,392	1,742	2,967	110.3	200.2	263.1	316.5
London	1,117	2,535	3,059	4,607	557.3	957.3	1,113.6	1,225.6
South East	1,114	2,390	2,906	4,333	157.3	272.8	388.5	467.8
South West	858	2,902	3,828	5,498	107.1	205.1	277.7	342.3
England[2]	8,085	18,490	23,346	37,233	2,049.9	3,595.6	4,423.8	5,522.0
Wales	1,226	2,294	3,481	4,863	166.7	254.2	304.1	388.5
Scotland	1,405	3,489	5,513	8,156	303.1	484.0	604.5	709.1
Northern Ireland	378	640	904	1,201	48.4	126.1	160.4	197.2

1 Includes grants made UK-wide or to institutions of national significance. Further grants have been made overseas. See Notes and Definitions.
2 Includes grants not allocated to a specific English region. See Notes and Definitions.

Source: Department of Culture, Media and Sport

9 Crime and Justice

Crime rates

1998-99, London had the highest recorded crime rate in England and Wales as a proportion of its population, at more than 12 crimes per 100 people. The East of England had the lowest recorded crime rate, at just over 7 crimes per 100 people.

(Table 9.1)

London and Yorkshire and the Humber had the highest recorded rates of theft and handling stolen goods, around 20 per cent higher than the average rate for England and Wales in 1998-99.

(Table 9.1)

The East of England had the lowest recorded rates for violence against the person in 1998-99, over a third lower than comparable figures for England and Wales.

(Table 9.1)

Victims

Around a third of households in the Yorkshire and Humber region suffered at least one crime against their property in 1997.

(Table 9.2)

Just over a quarter of households that owned a vehicle in Yorkshire and the Humber suffered a vehicle theft in 1997, compared to less than 1 in 6 of households in the South West and South East.

(Table 9.2)

In 1998, 18 per cent of women surveyed in England and Wales felt 'very' unsafe when walking alone at night, compared to 3 per cent of men.

(Table 9.13)

Clear up rates

The clear-up rate for recorded crimes in England and Wales in 1998-99 was highest in Wales, at 46 per cent, compared with 22 per cent in London.

(Table 9.3)

Firearms

In 1998-99, Wales had the lowest number of recorded offences in which firearms were reported to have been used; Scotland had the lowest number of operations in which firearms were issued to the police.

(Table 9.4)

Drug seizures

In 1998, Northern Ireland had the lowest number of drugs seizures for class A, B and C drugs of any region of the United Kingdom.

(Table 9.5)

Over half of the total for class A drug seizures were made in four United Kingdom regions: London, Scotland, the North West and Yorkshire and the Humber.

(Table 9.5)

Cautions

In 1998, just under 20 per cent of people aged 18 and over, found guilty or cautioned for an indictable offence in Yorkshire and the Humber were cautioned, one of the lowest rates for cautioning in England and Wales. London had the highest cautioning rates for people aged 18 and over found guilty or cautioned for an indictable offence, at around 40 per cent.

(Table 9.7)

Offenders

The North East had the highest rates of young males found guilty or cautioned for an offence in England and Wales in 1998, at around 6 per thousand of the population.

(Table 9.8)

In the North East in 1998, 11 per cent of males aged 21 or over found guilty of offences received an absolute or conditional discharge; for females, the proportion was 13 per cent.

(Table 9.9)

In 1998, in England and Wales, Cleveland was the police force area with the highest rates of persons found guilty or cautioned for indictable offences; Thames Valley had the lowest rates. For summary offences; the area with the highest rates for persons found guilty or cautioned was Gwent; the area with the lowest rates was Surrey.

(Table 9.10)

Prisoners

Of the total number of women aged 21 or over who were sentenced to immediate imprisonment in 1998, the South East had the highest proportion of females sentenced to four years or more.

(Table 9.11)

In April 2000 the Yorkshire and Humberside prison service region had the largest prison population in England and Wales, accomodating more than one in ten of all prisoners. Wales had the smallest prison population at less than one in thirty of the total prison population.

(Chart 9.12)

Police

Almost a fifth of ordinary duty police officers in the West Midlands in March 1999 were women, compared with around a seventh of those in Wales and the East Midlands.

(Table 9.6)

In 1998, around 6 in 7 adults in the East of England and London said that their local Police "did a very or fairly good job" compared with around 5 in 7 in the North West, Scotland and Northern Ireland.

(Table 9.14)

9.1 Recorded crimes[1]: by offence group, 1998-99

Rates per 100,000 population

	Violence against the person	Sexual offences	Burglary	Robbery	Theft and handling stolen goods	Fraud and forgery	Criminal damage	Drugs	Other	Total
England and Wales	963	69	1,826	128	4,197	535	1,685	260	122	9,785
North East	786	60	2,246	94	4,409	330	2,060	248	125	10,359
North West	1,025	71	2,162	169	4,321	417	1,974	280	138	10,556
Yorkshire and the Humber	744	69	2,927	100	5,084	411	2,051	276	109	11,770
East Midlands	963	71	2,049	81	4,411	561	1,754	184	157	10,231
West Midlands	943	63	2,137	157	4,096	486	1,660	229	130	9,901
East	602	56	1,167	41	3,222	348	1,314	175	91	7,017
London	1,733	107	1,632	346	5,091	1,068	1,800	429	148	12,354
South East	712	57	1,319	49	3,642	434	1,373	204	106	7,897
South West	708	57	1,486	61	3,817	570	1,227	194	81	8,201
England	955	70	1,849	134	4,236	543	1,671	256	121	9,835
Wales	1,097	62	1,433	29	3,539	410	1,916	333	133	8,951
Scotland[2,3]	359	76	1,051	99	3,819	472	1,554	622	457	8,510
Northern Ireland[2]	1,093	95	917	83	2,094	405	1,638	83	51	6,458

1 Due to a change in the recording of offences the figures for England, Wales and Northern Ireland are not comparable to those shown in previous editions of *Regional Trends*. See Notes and Definitions.
2 Figures for Scotland and Northern Ireland are not comparable with those for England and Wales, nor with each other, because of the differences in the legal systems, recording practises and classifications.
3 Figures for Scotland relate to the calendar year 1999.

Source: Home Office; Scottish Executive; Royal Ulster Constabulary

9.2 Offences committed against households[1], 1997

Rates per 10,000 households[2] and percentages

	Offences per 10,000 households[2]				Percentage of households[2] victimised at least once			
	Vandalism	Burglary[3]	Vehicle thefts[4]	All household offences[5]	Vandalism	Burglary[3]	Vehicle thefts[4]	All household offences[5]
England and Wales	1,345	756	2,122	4,914	8.2	5.6	15.7	27.7
North East	888	1,049	2,158	4,794	4.6	8.6	15.4	28.2
North West	1,755	977	2,521	6,021	9.6	6.8	18.6	31.7
Yorkshire and the Humber	1,597	1,085	2,704	6,263	9.1	8.3	20.0	32.5
East Midlands	1,236	735	1,961	4,564	8.2	5.6	14.7	26.6
West Midlands	1,108	735	2,092	4,363	7.1	5.9	16.0	27.0
East	1,291	428	1,911	4,249	7.9	3.1	15.2	26.7
London	1,135	710	2,372	4,279	7.9	5.7	15.6	25.0
South East	1,586	649	1,889	5,016	9.6	4.3	13.8	27.6
South West	1,043	558	1,785	4,537	7.2	4.4	13.5	25.7
England	1,333	753	2,137	4,918	8.2	5.7	15.8	27.9
Wales	1,548	812	1,869	4,861	8.7	4.8	14.2	25.0
Scotland[6]	1,105	386	1,739	3,211	6.4	3.0	12.6	18.6
Northern Ireland	863	301	1,163	2,112	6.9	2.4	8.4	15.4

1 See Notes and Definitions.
2 The vehicle theft risks are based on vehicle-owning households only.
3 The term used in Scotland is housebreaking. The figures include attempts at burglary/housebreaking.
4 Comprises theft of vehicles, thefts from vehicles and associated attempts.
5 Comprises the three individual categories plus thefts of bicycles and other household thefts.
6 Data for Scotland relate to 1995.

Source: British Crime Survey, Home Office; Scottish Crime Survey, Scottish Executive; Northern Ireland Crime Survey, Northern Ireland Office

9.3 Recorded crimes cleared up by the police[1]: by offence group, 1998-99[2]

Percentages

	Violence against the person	Sexual offences	Burglary	Robbery	Theft and handling stolen goods	Fraud and forgery	Criminal damage[3]	Drugs	Other[3]	Total[3]
England and Wales[4]	71	68	19	23	22	36	17	97	78	29
North East	81	73	15	27	24	56	16	97	86	29
North West	78	77	16	22	22	51	16	99	85	30
Yorkshire and the Humber	83	75	20	29	22	52	14	97	84	28
East Midlands	79	69	21	30	24	44	19	95	83	31
West Midlands	79	69	20	24	24	37	18	97	79	31
East	82	70	25	31	26	43	18	92	73	33
London	44	42	13	16	14	15	13	97	58	22
South East	79	80	20	34	22	36	19	97	79	30
South West	83	78	19	28	21	46	20	95	78	31
England	70	67	18	22	21	35	17	97	77	28
Wales	90	90	41	50	38	58	23	98	92	46
Scotland[5,6]	82	77	23	38	33	76	24	99	97	43
Northern Ireland[5]	57	76	17	19	23	44	14	90	71	29

1 See Notes and Definitions.
2 Some offences cleared up may have been initially recorded in an earlier year.
3 The Northern Ireland figure excludes Offences against the State.
4 Due to a change in the recording of offences the figures for England and Wales are not comparable with those shown in previous editions. See Notes and Definitions.
5 Figures for Scotland and Northern Ireland are not compatible with those for England, Wales and Northern Ireland, nor with each other, because of the differences in the legal systems, recording practises and classifications.
6 Figures for Scotland relate to the calendar year 1999.

Source: Home Office; Scottish Executive; Royal Ulster Constabulary

9.4 Firearms

Numbers

	Offences recorded[1] by the police in which firearms were reported[2] to have been used					Operations in which firearms were issued to the police[3,4]				
	1994	1995	1996	1997	1998-99[5]	1994-95[6]	1995-96[6]	1996-97[6]	1997-98[6]	1998-99[5,6]
United Kingdom	15,985	15,730	16,174	14,422	15,784	5,960	8,671	12,649	12,134	11,184
North East	767	723	681	486	727	800	1,050	2,517	1,029	832
North West	2,044	2,308	2,426	1,751	2,308	420	922	1,578	1,462	1,611
Yorkshire and the Humber	2,264	2,270	2,175	1,968	2,079	427	1,026	1,128	1,506	1,183
East Midlands	970	1,014	1,187	1,140	1,407	283	346	470	671	659
West Midlands	1,394	1,510	1,570	1,251	1,092	237	420	730	751	935
East	808	771	730	607	761	620	871	1,172	1,327	1,327
London	2,376	2,248	2,605	2,930	3,005	1,812	2,203	2,747	2,885	2,889
South East	1,526	1,367	1,232	1,123	1,276	790	883	1,064	1,284	562
South West	569	588	608	560	628	284	511	575	403	294
England	12,718	12,799	13,214	11,816	13,283	5,673	8,232	11,981	11,318	10,292
Wales	449	635	662	594	591	151	244	398	524	636
Scotland	1,788	1,721	1,650	1,187	985	136	195	270	292	256
Northern Ireland	1,030	575	651	827	925	.	.	.	.	.

1 See Notes and Definitions.
2 Alleged' in Scotland.
3 In England and Wales, police shots were fired in 6 operations in 1994-95, 5 in both 1995-96 and 1996-97, 3 in 1997-98 and 5 in 1998-99. In Scotland, police shots were fired in 3 operations in 1994, 4 in 1995-96, 9 in 1996-97, 1 in 1997-98 and 8 in 1998-99. In Northern Ireland, police officers are armed at all times.
4 Figures for the United Kingdom relate to Great Britain only.
5 The collection of recorded crime data in England and Wales changed to a financial year basis from 1 April 1998, which coincided with a change in the counting rules for recorded crime. Due to this, the data shown for 1998-99 are not comparable with those shown for previous years. See Notes and Definitions.
6 In England and Wales the data collection changed from a calendar year to a financial year basis from 1 April 1994. In Scotland data collection changed to financial year basis from 1 April 1995.

Source: Home Office; Scottish Executive Justice Department; Royal Ulster Constabulary

9.5 Seizures of controlled drugs[1]: by type of drug, 1998

Number of seizures

	Class A drugs						Class B drugs			All class C drugs[2,4]
	Heroin	Cocaine	Crack	LSD	Ecstasy type	All class A drugs[2]	Cannabis	Amphetamines	All class B drugs[2]	
United Kingdom[3]	15,095	4,993	2,440	613	4,790	28,472	114,163	18,535	126,989	3,112
North East	639	107	23	38	161	1,019	5,171	1,208	6,105	286
North West	2,036	326	195	36	441	3,012	11,335	2,025	12,749	146
Yorkshire and the Humber	2,453	142	203	43	397	3,294	6,324	1,717	7,631	232
East Midlands	694	57	79	25	238	1,086	4,599	1,442	5,602	193
West Midlands	918	97	96	39	222	1,414	6,067	1,037	6,751	34
East	698	299	63	37	213	1,374	6,833	1,049	7,488	77
London	2,789	2,374	1,534	143	1,188	7,946	27,212	2,968	29,351	417
South East	760	281	68	42	414	1,639	9,450	1,621	10,465	109
South West	1,004	128	123	37	351	1,752	7,190	1,237	7,951	151
England	11,991	3,811	2,384	440	3,625	22,536	84,181	14,304	94,093	1,645
Wales	380	62	10	62	311	951	7,541	1,818	8,845	165
Scotland	2,410	244	18	82	459	3,231	13,747	2,091	15,124	1,207
Northern Ireland	30	10	4	6	275	320	1,257	102	1,312	17
Regional Crime Squads/ National Crime Squad[3]	44	14	8	4	16	79	299	30	321	5
British Transport Police[3]	93	19	10	3	15	151	768	54	810	9
Customs and Excise[3]	147	833	6	16	89	1,204	6,370	136	6,484	64

1 See Notes and Definitions.
2 Since a seizure may involve drugs other than those listed, figures for individual drugs cannot be added together to produce totals.
3 Figures for the Regional Crime Squads/National Crime Squad, the British Transport Police and the Customs and Excise cannot be split by region or country, but are included in the UK totals.
4 Class C drugs include benezodiazepines (including temazepam) and anabolic steroids.

Source: Home Office

9.6 Police manpower: by type, March 1999[1]

	Police officers on ordinary duty[2]				Special constables and civilian staff (rates per 1,000 officers on ordinary duty)		Traffic wardens (numbers)
		Percentage of which		Population per officer[3]	Special constables[4]	Civilian staff[5]	
	Number	Ethnic minorities	Women officers				
United Kingdom[6]	150,009	1.7	15.6	393	129	400	3,912
North East	6,824	0.7	15.8	380	92	374	109
North West	17,462	1.7	16.9	394	117	367	421
Yorkshire and the Humber	11,461	1.9	15.7	439	118	425	269
East Midlands	8,253	2.7	14.4	504	187	456	229
West Midlands	12,491	3.1	19.4	426	158	403	264
East	9,502	1.5	16.3	538	211	465	285
London	26,851	3.4	15.2	284	45	431	945
South East	14,931	1.3	16.4	515	166	461	327
South West	9,420	0.7	15.2	518	232	467	324
England	117,195	2.1	16.2	421	131	426	3,173
Wales	6,646	0.8	13.9	440	164	385	169
Scotland[7]	14,810	0.4	15.3	346	111	287	405
Northern Ireland[7]	11,358	..	10.7	149	107	257	165

1 Full-time equivalents as at 31 March 1999 for England and Wales and for Scotland. Actual numbers (whether full or part-time) as at 31 March 1999 for Northern Ireland.
2 Includes full-time Reserves in Northern Ireland.
3 Based on mid-1997 population estimates for England and Wales and 1998 population estimates for Scotland and Northern Ireland.
4 Part-time Reserves in Northern Ireland.
5 Excludes traffic wardens.
6 Great Britain for ethnic minorities.
7 For civilian staff and traffic wardens, part-time staff are counted as half full-time.

Source: Home Office; Scottish Executive; Royal Ulster Constabulary

9.7 Persons given a police caution[1]: by type of offence and age, 1998

Percentages and thousands

	Those cautioned as a percentage of persons found guilty or cautioned for each offence category										All persons found guilty or cautioned (thousands)	
	Violence against the person	Sexual offences	Burglary	Robbery	Theft and handling stolen goods	Fraud and forgery	Criminal damage	Other indict-able off-ences	Total indict-able off-ences	Sum-mary off-ences[2]	Indictable offences	Summary offences[2]
Persons aged 10-17												
England and Wales	61	56	44	20	67	63	42	26	61	56	126.5	58.1
North East	55	55	35	13	61	58	44	23	55	55	10.5	5.5
North West	60	46	45	18	68	63	30	24	61	62	20.2	10.9
Yorkshire and the Humber	58	44	38	12	61	51	37	21	54	56	12.9	5.5
East Midlands	57	63	39	13	62	58	33	22	56	54	9.0	4.3
West Midlands	64	56	44	15	67	65	28	22	60	61	14.6	5.5
East	65	71	48	20	70	60	48	30	65	55	9.9	4.2
London	64	54	53	28	74	72	31	35	68	44	19.7	8.7
South East	62	62	41	21	63	63	51	30	60	56	14.3	5.9
South West	66	71	51	22	72	67	52	36	68	60	7.4	3.1
England	61	56	43	20	67	63	39	27	61	56	117.8	53.6
Wales	58	57	48	8	66	60	62	22	61	60	8.7	4.5
Persons aged 18 or over												
England and Wales	31	22	7	3	28	23	11	7	29	13	396.6	498.6
North East	27	31	6	1	29	28	10	7	29	22	22.5	30.7
North West	27	18	4	0	23	23	6	4	25	21	64.8	74.6
Yorkshire and the Humber	23	16	4	1	20	18	5	3	19	10	39.5	48.1
East Midlands	27	22	7	0	27	23	12	7	26	10	26.6	42.7
West Midlands	34	24	8	2	29	25	3	8	30	12	39.9	52.0
East	37	30	11	4	28	21	12	7	29	9	28.3	44.1
London	42	24	13	8	38	31	4	15	41	11	78.2	84.2
South East	33	18	6	2	24	22	18	9	25	11	42.3	45.7
South West	28	20	7	1	25	19	17	7	28	10	28.1	40.5
England	32	22	7	3	28	24	9	8	29	13	370.3	462.7
Wales	18	17	6	0	25	17	26	5	27	11	26.3	35.8

1 Persons committing an offence who on admission of guilt were given a formal oral caution by the police as a proportion of those found guilty or cautioned. See Notes and Definitions.
2 Excludes motoring offences for which written warnings were issued.

Source: Home Office

9.8 Persons found guilty or cautioned[1]: by type of offence and age, 1998

Rates per 100,000 population in the relevant age group

	Persons aged 10-17						Persons aged 18 or over					
	Violence against the person plus common assault[2]	Sexual off-ences	Burglary, robbery and theft[3]	Drugs off-ences	Other indict-able off-ences[4]	All indictable offences plus common assault[2]	Violence against the person plus common assault[2]	Sexual off-ences	Burglary, robbery and theft[3]	Drugs off-ences	Other indict-able off-ences[4]	All indictable offences plus common assault[2]
Males												
England and Wales	663	39	2,327	475	385	3,890	323	26	681	426	347	1,803
North East	839	52	4,018	572	489	5,970	357	34	856	391	355	1,993
North West	818	41	2,643	615	426	4,544	362	32	856	489	459	2,198
Yorkshire and the Humber	636	50	2,410	352	390	3,837	304	28	752	382	400	1,866
East Midlands	772	55	2,206	263	325	3,620	356	28	615	278	290	1,568
West Midlands	743	55	2,475	450	482	4,205	323	27	675	383	397	1,805
East	550	35	1,853	361	347	3,147	269	22	503	297	248	1,339
London	660	25	2,449	890	405	4,428	372	24	806	779	412	2,392
South East	543	23	1,923	307	294	3,089	265	22	523	284	248	1,342
South West	474	27	1,476	315	221	2,513	270	23	556	300	237	1,385
England	662	38	2,306	473	372	3,851	319	26	679	418	342	1,783
Wales	682	54	2,674	509	601	4,519	393	31	710	572	437	2,142
Females												
England and Wales	220	1	845	42	80	1,188	47	0	183	51	67	348
North East	330	1	1,525	49	99	2,004	60	1	242	57	79	438
North West	302	0	849	41	94	1,286	54	0	250	56	90	450
Yorkshire and the Humber	235	2	815	44	75	1,171	49	0	182	56	69	356
East Midlands	280	0	742	24	82	1,128	62	1	161	37	62	324
West Midlands	240	1	928	40	92	1,301	46	0	165	35	63	310
East	183	0	731	37	73	1,024	39	0	139	40	41	260
London	143	0	1,050	60	59	1,311	46	1	245	69	93	453
South East	158	1	631	29	64	883	34	0	126	39	44	243
South West	155	2	588	31	61	836	34	1	136	47	46	264
England	216	1	840	40	76	1,173	46	0	183	49	66	344
Wales	281	0	926	74	138	1,419	70	0	185	79	82	417
All persons												
England and Wales	448	21	1,606	264	237	2,575	181	13	425	233	203	1,055
North East	591	27	2,804	317	299	4,039	203	17	538	218	212	1,188
North West	566	21	1,767	335	264	2,953	203	16	543	265	269	1,295
Yorkshire and the Humber	441	26	1,634	202	236	2,540	173	14	460	215	230	1,091
East Midlands	533	28	1,496	147	207	2,411	206	14	383	155	174	932
West Midlands	498	29	1,723	251	293	2,794	182	13	414	205	226	1,041
East	370	18	1,305	203	213	2,110	151	11	317	165	142	787
London	408	13	1,768	486	237	2,912	204	12	517	414	248	1,396
South East	356	12	1,296	172	182	2,018	146	11	318	158	143	775
South West	319	15	1,045	177	143	1,700	148	11	339	169	138	805
England	445	20	1,593	262	228	2,549	178	13	424	228	200	1,043
Wales	487	27	1,820	297	375	3,006	226	15	438	317	253	1,250

1 See Notes and Definitions.
2 Following the introduction of a charging standard on 31 August 1994, some people who would have been charged with an indictable offence are now charged with common assault, a
    summary offence. Common assaults have therefore been included for comparability with figures in previous editions of *Regional Trends*.
3 Includes handling stolen goods.
4 Includes criminal damage and fraud and forgery.

Source: Home Office

9.9 Persons aged 21 or over found guilty of offences[1]: by gender and type of sentence, 1998

	Result as a percentage of number of persons sentenced						All sentenced	
	Absolute or condit- ional discharge	Fine	All community penalties	Fully sus- pended sentence[2]	Immed- iate custodial sentence[3]	Otherwise dealt with	(=100%) (numbers)	Rates[4]
Males								
England and Wales	7	76	8	-	7	1	997,918	54
North East	11	69	9	-	8	2	42,451	46
North West	7	76	8	-	8	1	163,616	67
Yorkshire and the Humber	7	72	10	-	8	3	97,805	55
East Midlands	7	75	9	-	7	2	76,760	51
West Midlands	6	81	7	-	6	1	116,753	62
East	5	79	8	-	6	2	88,744	48
London	5	77	7	-	9	1	145,332	54
South East	7	76	9	-	6	2	117,204	43
South West	8	77	8	-	6	1	83,347	47
England	7	76	8	-	7	1	932,012	53
Wales	7	78	7	-	6	1	65,906	64
Scotland[5]	8	74	6	.	11	1	88,206	49
Northern Ireland[6]	5	72	2	8	7	6	19,416	35
Females								
England and Wales	9	80	7	-	3	1	205,442	10
North East	13	75	8	-	2	1	9,786	10
North West	11	76	9	-	4	1	32,542	13
Yorkshire and the Humber	10	77	8	-	2	2	20,617	11
East Midlands	9	81	6	1	2	2	16,161	10
West Midlands	7	85	6	-	2	1	21,247	11
East	8	82	7	-	2	1	18,119	9
London	7	82	7	-	4	1	31,966	11
South East	10	78	8	1	2	1	21,776	7
South West	10	81	7	-	2	1	18,405	10
England	9	80	7	-	3	1	190,619	10
Wales	11	80	6	-	2	1	14,823	13
Scotland[5]	18	68	9	.	4	1	14,643	7
Northern Ireland[6]	11	70	1	6	2	11	2,586	4

1 See Notes and Definitions. The coverage of the table is all offences, including motoring offences. A defendant is recorded only once for each set of court proceedings, against the
   principal offence.
2 Fully suspended sentences are not available to courts in Scotland.
3 Includes custodial sentences imposed following a sentence deferred for good behaviour in Scotland.
4 Rates per 1,000 population aged 21 or over.
5 To improve comparability, this table excludes breaches of probation and community service orders normally included in Scottish figures.
6 Northern Ireland figures relate to 1997.

Source: Home Office; Scottish Executive; Royal Ulster Constabulary

9.10 Persons found guilty or cautioned: by police force area[1], 1998

Indictable offences

Rates per 100,000 population
- 1,000 or over
- 800 to 999
- 700 to 799
- 699 or under

Summary offences

Rates per 100,000 population
- 2,800 or over
- 2,500 to 2,799
- 2,000 to 2,499
- 1,999 or under

1 Metropolitan Police Force area includes the City of London. See map on page 240.

Source: Home Office

9.11 Persons aged 21 or over sentenced to immediate imprisonment: by gender and length of sentence imposed for principal[1] offence, 1998

Percentages and numbers

	Males				Females			
	Length of sentence (percentages)			Total sentenced to immediate imprisonment (=100%) (numbers)	Length of sentence (percentages)			Total sentenced to immediate imprisonment[2] (=100%) (numbers)
	One year or less	Over one year but less than four years	Four years or over		One year or less	Over one year but less than four years	Four years or over	
Great Britain	75	18	7	80,972	84	11	5	5,990
North East	69	22	9	3,324	89	8	4	221
North West	75	19	6	12,590	87	10	3	1,154
Yorkshire and the Humber	72	22	6	7,652	85	13	2	512
East Midlands	75	19	6	5,531	83	14	3	326
West Midlands	71	22	7	6,839	78	17	5	372
East	76	18	6	5,610	83	13	4	428
London	75	17	8	13,096	81	11	8	1,288
South East	71	20	9	7,534	76	13	11	502
South West	75	19	6	4,736	86	12	2	323
England	74	19	7	66,912	83	12	5	5,126
Wales	75	20	5	4,054	87	12	1	254
Scotland[3]	87	8	5	10,006	92	5	2	610
Northern Ireland[4]	21	44	35	333	82	9	9	11

1 Figures for Scotland are for the length of sentence in total given for all offences and not just for the principal offence. Figures on sentence lengths for principal offences only are not available for Scotland.

2 For Scotland, the custodial total for females includes one female who was given a custodial sentence where the sentence length was not known.

3 To improve comparability, this table excludes breaches of probation and community service orders normally included in Scottish figures.

4 Figures for Northern Ireland are not comparable with those for Great Britain as they relate to Crown Court only.

Source: Home Office; Scottish Executive; Northern Ireland Office

9.12 Prison population in England and Wales: by prison service region[1], April 2000

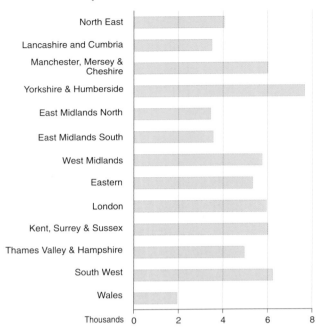

North East
Lancashire and Cumbria
Manchester, Mersey & Cheshire
Yorkshire & Humberside
East Midlands North
East Midlands South
West Midlands
Eastern
London
Kent, Surrey & Sussex
Thames Valley & Hampshire
South West
Wales

Thousands 0   2   4   6   8

1 People in prison establishments in the region. See map on page 240.

Source: Home Office

9.13 Feelings of insecurity[1]: by gender, 1998[2]

Percentages

| | Percentage feeling 'very' unsafe at night when: | | | |
| | Alone at home | | Walking alone[3] | |
	Males	Females	Males	Females
England and Wales	-	3	3	18
North East	-	4	5	22
North West	2	3	4	23
Yorkshire and the Humber	1	2	3	22
East Midlands	1	3	3	16
West Midlands	1	3	5	17
East	-	2	2	17
London	-	3	4	17
South East	-	2	2	16
South West	1	2	2	17
England	1	3	3	18
Wales	-	5	5	15
Scotland	1	3	6	23
Northern Ireland	3	12	11	34

1 People aged 16 or over.
2 Data for Scotland relate to 1996. See Notes and Definitions.
3 For Northern Ireland the question relates to fear of 'walking in the dark' (ie alone or with others); the figures also include those people who never go out.

Source: British Crime Survey, Home Office; Scottish Crime Survey, Scottish Executive; Northern Ireland Crime Survey, Northern Ireland Office

9.14 Adults' assessment of the local police: by age, 1998[1]

Percentages

| | Percentage in each age group saying police do a very or fairly good job | | | |
	16-29[2]	30-59	60 or over	All aged 16 or over
North East	..	75	82	76
North West	66	73	82	74
Yorkshire and the Humber	75	78	71	76
East Midlands	71	79	86	79
West Midlands	74	79	86	81
East	79	88	91	87
London	83	89	90	87
South East	77	84	89	84
South West	..	84	82	83
England	76	82	84	81
Wales	..	85	86	83
Scotland[1]	69	74	74	73
Northern Ireland	60	73	79	71

1 Data for Scotland relate to 1996. See Notes and Defintions.
2 For some regions, sample sizes are too small to provide a reliable estimate.

Source: British Crime Survey, Home Office; Scottish Crime Survey, Scottish Executive; Community Attitudes Survey, Northern Ireland Statistics and Research Agency

10 Transport

Cars

Between 1995 and 1998, the number of cars licensed increased within all regions of the United Kingdom.

(Table 10.1)

One in six cars licensed in the West Midlands in 1998 was a company car compared with one in 20 in the North East region.

(Table 10.1)

In 1998, the proportion of households with two or more cars ranged from over a third of households in the South West and South East regions to around one fifth of households in London, the North East and Scotland.

(Table 10.2)

In 1998, the percentage of households with no car ranged from 19 per cent in the South East of England to 38 per cent in Scotland.

(Table 10.2)

On average, households in Scotland and the North West region have the newest cars, (under three years old), while those in the South West have the oldest cars (seven years or older).

(Table 10.3)

Travel

On average people in the North East, the South East and the East of England make more journeys per year than those in all other regions, people in Wales and London make the least number of journeys.

(Table 10.4)

In 1996-1998, Londoners spent on average over half an hour travelling to work compared with between 19 and 24 minutes in all other parts of Great Britain.

(Chart 10.5)

People living in the East of England travelled the longest distances to get to work; 16.7 km on average, compared with an average of 13 km for Great Britain as a whole in 1996-1998.

(Chart 10.5)

Londoners and people living in Scotland travel, on average, more than a thousand miles per year by public transport, while those living in the South West and Wales travel only around 500 miles by that mode. Those in the South East travelled the greatest number of miles by car.

(Table 10.6)

In 1996-1998, the distance travelled by pedal cycle ranged from an average of around 60 miles per year for people living in the East of England to around 20 miles for people living in Wales and the North East region.

(Table 10.6)

Traffic

The average daily motor vehicle traffic flow on roads in London is around three times that on roads in Wales and in Scotland.

(Table 10.7)

Average daily motor vehicle flows on motorways in Great Britain in 1998 ranged from 93.8 thousand vehicles per day in London to 35.9 thousand vehicles in Scotland.

(Table 10.7)

In 1998, over 40 per cent of the motor vehicle traffic in the North West was on motorways compared with less than ten per cent in the North East.

(Table 10.8)

Accidents

In 1998, the fatal or serious accident rate per vehicle kilometre on major roads in London was nearly three times that for any other region in England and that for Scotland.

(Table 10.9)

In 1998, the fatal or serious accident rate per 100 million vehicle kilometres on major roads in Great Britain ranged from just over 4 accidents in the North East and the South West to almost 20 accidents in London.

(Table 10.9)

In 1998, the rate of fatal and serious road accidents per 100,000 population fell in all regions compared with the 1981-1985 average. The greatest fall was from 158 to 50 in the South West.

(Table 10.9)

In 1998, just over one in seven road casualties in London was a motor cyclist compared with about one in 40 in Northern Ireland. Almost one in five road casualties in London and Scotland in 1998 were pedestrians compared with around one in ten in the East of England, Northern Ireland and the South East region.

(Table 10.10)

Breath tests

In 1998, motor vehicle drivers from the South East region had the highest breath test failure rate (or refused to provide a specimen) for all ages from 17 to 69 years.

(Table 10.11)

Airports

London's Heathrow and Gatwick airports accounted for more than a half of all the air passenger traffic, and over two thirds of all freight handled at UK airports in 1999.

(Table 10.12)

Over half of all international non-scheduled flights were from Gatwick and Manchester airports in 1999.

(Table 10.12)

Sea ports

Dover handled almost three-fifths of all the UK's international sea passengers in 1998.

(Table 10.13)

Almost half of all the freight handled by UK ports is handled by East coast ports.

(Table 10.13)

10.1 Motor cars currently licensed and new registrations[1]

Thousands and percentages

	Currently licensed[2]				Percentage company cars	New registrations			Percentage company cars
	1995	1996	1997	1998	1998	1996	1997	1998	1998[3,4]
United Kingdom	21,917	22,784	23,408	23,878	10	2,077	2,251	2,366	52
North East	753	783	800	824	5	74	76	78	36
North West	2,396	2,501	2,589	2,647	13	235	255	265	49
Yorkshire and the Humber	1,653	1,707	1,764	1,808	9	138	151	157	44
East Midlands	1,550	1,609	1,655	1,698	9	140	152	173	61
West Midlands	2,102	2,183	2,283	2,290	17	275	292	288	60
East	2,213	2,295	2,311	2,429	9	200	223	232	48
London	2,294	2,362	2,359	2,369	11	277	278	270	61
South East	3,329	3,469	3,652	3,709	10	292	315	361	55
South West	1,995	2,109	2,159	2,230	10	130	132	138	45
England	18,285	19,018	19,574	20,006	11	1,762	1,873	1,961	53
Wales	1,021	1,067	1,100	1,129	7	73	76	83	38
Scotland	1,598	1,674	1,727	1,775	9	154	173	175	43
Northern Ireland	523	546	576	585	9	55	61	71	..

1 Figures for United Kingdom include motor vehicles where the country of the registered keeper is unknown.
2 At 31 December.
3 Figure for the United Kingdom relates to Great Britain.
4 Within the Private and light goods tax class only.

Source: Annual Vehicle Census/Vehicle Information Database, Department of the Environment, Transport and the Regions; Department of the Environment, Northern Ireland

10.2 Households with regular use of a car[1], 1998

Percentages

	Percentage of households with			
	No car	One car	Two cars	Three or more cars
United Kingdom	28	45	23	5
North East	35	46	17	2
North West	32	44	21	4
Yorkshire and the Humber	31	45	20	4
East Midlands	24	48	23	4
West Midlands	28	41	24	6
East	21	46	26	7
London	36	46	16	3
South East	19	44	28	8
South West	21	43	29	6
England	27	45	23	5
Wales	30	43	23	4
Scotland	38	41	17	4
Northern Ireland	30	48	19	3

1 Includes cars and light vans normally available to the household.

Source: General Household Survey and Family Expenditure Survey, Office for National Statistics; National Travel Survey, Department of the Environment, Transport and the Regions; Continuous Household Survey, Northern Ireland Statistics and Research Agency

10.3 Age of household cars, 1996-1998

Percentages

	Age of car[1]		
	Under 3 years old	3-6 years old	7 years or more
Great Britain	26	30	44
North East	24	34	42
North West	31	30	39
Yorkshire and the Humber	27	29	44
East Midlands	25	30	45
West Midlands	26	31	43
East	25	29	45
London	23	28	49
South East	26	30	44
South West	19	27	54
England	25	30	45
Wales	25	31	44
Scotland	33	30	37

1 Age of main or only car or light van normally available to the household. See Notes and Definitions.

Source: National Travel Survey, Department of the Environment, Transport and the Regions

10.4 Journeys per person per year[1]: by journey purpose and gender, 1996-1998

Percentages and numbers

	Commuting	Business	Education	Shopping	Other personal business	Leisure	Average number of journeys (=100%)
Males							
Great Britain	18	5	7	19	20	31	1,070
North East	17	4	9	21	18	33	1,162
North West	17	5	7	20	21	31	1,057
Yorkshire and the Humber	19	4	6	20	18	32	1,030
East Midlands	20	5	7	17	21	30	1,087
West Midlands	20	4	7	19	21	29	1,032
East	20	6	5	18	19	32	1,140
London	18	6	8	19	21	28	1,020
South East	19	7	5	17	22	31	1,118
South West	19	5	5	19	18	34	1,120
England	19	5	7	19	20	31	1,079
Wales	16	6	7	19	19	34	978
Scotland	18	6	6	19	20	32	1,036
Females							
Great Britain	13	2	6	23	25	30	1,034
North East	12	1	7	25	22	33	1,093
North West	13	1	7	24	25	30	999
Yorkshire and the Humber	13	1	6	25	23	31	1,001
East Midlands	14	2	6	23	26	29	1,073
West Midlands	13	2	8	24	24	29	1,024
East	14	1	6	22	24	33	1,076
London	13	2	7	23	28	27	946
South East	14	2	6	22	27	30	1,110
South West	13	2	5	23	24	32	1,079
England	13	2	6	23	25	30	1,040
Wales	12	2	6	24	24	32	961
Scotland	13	2	7	24	23	31	1,022
All persons							
Great Britain	16	3	7	21	22	31	1,051
North East	14	2	8	23	20	33	1,125
North West	15	3	7	22	23	30	1,026
Yorkshire and the Humber	16	2	6	23	21	32	1,015
East Midlands	17	3	6	20	23	30	1,080
West Midlands	16	3	8	21	22	29	1,028
East	17	4	6	20	21	32	1,107
London	15	4	8	21	24	27	981
South East	16	4	5	19	24	30	1,114
South West	16	4	5	21	21	33	1,099
England	16	3	7	21	23	30	1,059
Wales	14	4	6	21	22	33	970
Scotland	16	4	6	21	21	32	1,029

1 Within Great Britain only. Figures relate to region of residence of the traveller and include journeys undertaken outside of their region. They include journeys of less than one mile; these were excluded from the table in *Regional Trends 32* and earlier editions. See Notes and Definitions.

Source: National Travel Survey, Department of the Environment, Transport and the Regions

10.5 Travel to work[1]: distance travelled and average time taken, 1996-1998

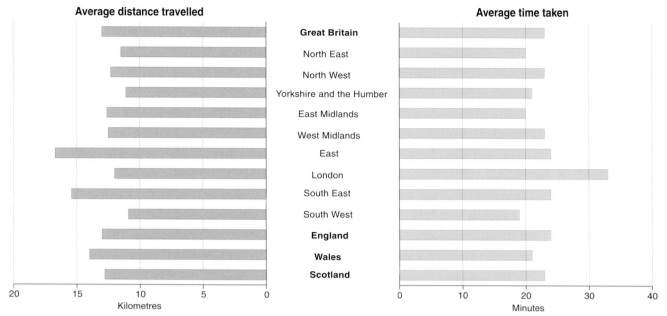

1 By region of residence. See Notes and Definitions.

Source: National Travel Survey, Department of the Environment, Transport and the Regions

10.6 Distance travelled per person per year[1]: by mode of transport, 1996-1998

Miles

	Walk	Pedal cycle	Cars and other private road vehicles	Public transport	All modes of transport
Great Britain	193	38	5,704	793	6,728
North East	235	22	4,710	759	5,726
North West	183	27	5,166	603	5,979
Yorkshire and the Humber	195	32	5,304	747	6,278
East Midlands	192	57	6,245	660	7,154
West Midlands	169	25	5,593	636	6,424
East	179	62	6,920	855	8,016
London	219	38	3,633	1,318	5,207
South East	187	50	7,219	760	8,216
South West	205	51	6,419	500	7,175
England	194	41	5,707	775	6,716
Wales	165	22	6,244	509	6,940
Scotland	200	26	5,342	1,151	6,720

1 Within Great Britain only. Figures relate to region of residence of the traveller and include journeys undertaken outside of this region. They include journeys of less than one mile; these were excluded from the table in *Regional Trends 32* and earlier editions. See Notes and Definitions.

Source: National Travel Survey, Department of the Environment, Transport and the Regions

10.7 Average daily motor vehicle flows[1]: by road class, 1998

Thousand vehicles per day

		Major roads		Minor roads		All roads
	Motorway	Non built-up	Built-up	Non built-up	Built-up	
Great Britain	67.3	10.7	15.1	0.8	2.1	3.4
North East	44.9	14.3	13.5	0.6	2.2	3.1
North West	66.4	11.8	15.8	0.9	1.9	4.1
Yorkshire and the Humber	58.9	12.2	15.7	0.9	2.0	3.5
East Midlands	76.6	12.0	13.9	0.8	1.7	3.3
West Midlands	80.1	11.1	16.7	0.8	2.5	3.9
East	80.8	17.2	13.8	1.0	2.2	3.8
London	93.8	53.4	23.4	.	2.3	6.0
South East	79.3	17.4	15.5	1.4	2.4	5.0
South West	59.9	10.2	13.5	0.6	1.8	2.5
England	71.9	13.7	16.2	0.9	2.1	3.8
Wales	50.3	7.8	9.8	0.5	1.7	2.0
Scotland	35.9	4.2	10.8	0.6	2.0	2.1

1 Average daily flow is annual traffic divided by road length divided by 365. See Notes and Definitions.

Source: National Road Traffic Survey, Department of the Environment, Transport and the Regions

10.8 Road traffic and distribution of accidents on major roads, 1998

	Motor vehicle traffic on major roads (percentages)[1]			All major roads (=100%) (billion vehicle kilometres)[1]	All roads (billion vehicle kilometres)[1]	Distribution of accidents (percentages)			Total accidents	
									On major roads (=100%) (numbers)	On all roads (numbers)[2]
	Motorway	Built-up 'A'	Non built-up 'A'			Motorway	Built-up 'A'	Non built-up 'A'		
United Kingdom	27.7	27.0	45.3	292.8	459.4	7.4	63.0	29.6	122,091	246,410
North East	9.4	23.6	67.0	10.6	18.8	2.9	54.5	42.5	3,701	8,700
North West	41.1	30.7	28.2	34.8	51.9	11.9	69.5	18.7	16,140	31,937
Yorkshire and the Humber	29.6	31.3	39.1	24.3	38.7	7.0	65.0	28.0	9,667	21,591
East Midlands	22.6	20.9	56.4	23.4	34.7	6.4	50.1	43.4	8,394	16,987
West Midlands	38.8	26.7	34.5	28.1	44.2	9.8	62.6	27.7	9,884	20,879
East	22.9	17.3	59.8	32.3	51.4	10.3	44.9	44.8	10,464	22,288
London	11.3	69.1	19.6	19.4	29.0	1.8	95.2	3.0	23,609	38,258
South East	36.2	20.3	43.6	52.8	79.9	11.9	49.7	38.3	16,252	33,350
South West	24.8	22.3	52.9	27.4	44.0	6.8	46.4	46.7	8,247	18,395
England	29.3	27.2	43.5	253.2	392.9	7.7	64.9	27.4	106,358	212,385
Wales	15.7	24.8	59.5	15.3	25.3	5.6	45.8	48.7	4,482	10,026
Scotland	18.6	26.0	55.4	24.2	41.2	5.1	48.4	46.4	7,828	16,512
Northern Ireland	..	..	..	..	..	5.1	59.4	35.5	3,423	7,487

1 Figures for United Kingdom relate to Great Britain.
2 Includes B,C and unclassified roads. See Notes and Definitions.

Source: Department of the Environment, Transport and the Regions; Royal Ulster Constabulary

10.9 Fatal and serious road accidents[1]

Numbers and rates

	Fatal and serious accidents on all roads						Fatal and serious accidents on major roads[2]			
	Numbers			Rates per 100,000 population			Numbers		Rates per 100 million vehicle kms	
	1981-1985 average[3]	1991	1998	1981-1985 average[3]	1991	1998	1991	1998	1991	1998
Great Britain	67,843	47,931	37,770	124	85	66	24,344	19,349	9.4	6.6
North East	2,255	1,769	1,117	86	68	43	734	475	5.9	4.3
North West	6,178	4,914	3,965	90	71	58	2,506	2,046	9.6	6.1
Yorkshire and the Humber	5,713	4,352	3,287	117	87	65	2,084	1,513	10.1	6.3
East Midlands	5,333	3,451	3,194	138	86	77	1,796	1,599	9.3	6.8
West Midlands	6,526	4,447	3,653	126	84	69	2,055	1,726	8.5	6.4
East	6,885	4,802	3,971	140	93	74	2,264	1,850	6.5	5.5
London	7,588	7,279	6,286	112	105	87	4,399	3,871	23.7	19.8
South East	10,169	5,843	4,809	139	76	60	2,882	2,447	6.9	4.6
South West	6,697	3,793	2,471	158	80	50	1,833	1,214	7.1	4.4
England	57,348	40,650	32,753	123	84	66	20,553	16,741	9.2	6.6
Wales	3,083	2,112	1,365	107	73	47	1,139	741	8.9	4.8
Scotland	7,412	5,169	3,652	144	101	71	2,652	1,867	12.1	7.3
Northern Ireland	..	1,381	1,246	..	85	74	643	529	9.5	..

1 See Notes and Definitions.
2 Motorways, A(M) roads and A roads.
3 Used as a basis for the government targets for reducing road casualties in Great Britain, and fatal and serious road casualties in Northern Ireland, by a third by the year 2000.

Source: Department of the Environment, Transport and the Regions; Royal Ulster Constabulary

10.10 Road casualties[1]: by age and type of road user, 1998

Percentages and numbers

	Percentage of all road casualties								All road casualties (=100%) (numbers)	Percentage change over 1981-85 average[4]
	Who were aged[2]			Type of road user						
	0-15	16-59	60 or over	Pedes- trians	Pedal cyclists	Motor cyclists	Car occupants[3]	Other road users		
United Kingdom	13.4	75.2	9.7	13.6	6.9	7.4	65.3	6.9	338,614	2.6
North East	16.7	74.4	8.9	15.8	5.7	3.4	66.8	8.2	12,310	10.8
North West	15.3	75.4	9.2	13.9	6.0	4.3	68.5	7.2	45,815	26.3
Yorkshire and the Humber	14.8	74.7	10.0	13.7	6.3	5.6	66.5	7.9	30,639	18.2
East Midlands	13.4	74.7	9.3	11.9	6.8	6.9	67.4	7.0	24,087	4.4
West Midlands	14.5	74.6	9.7	14.2	6.4	5.9	67.0	6.5	28,766	3.9
East	11.4	76.1	9.7	9.1	8.1	7.8	68.7	6.3	30,821	1.8
London	11.0	75.7	8.3	19.8	9.5	15.0	48.1	7.7	45,679	-15.7
South East	11.4	75.3	10.1	10.4	8.0	8.7	67.5	5.4	45,135	-0.8
South West	12.1	75.8	11.7	11.3	7.4	8.7	67.0	5.5	24,964	-5.3
England	13.1	75.3	9.6	13.5	7.3	7.9	64.5	6.8	288,216	2.8
Wales	15.0	74.0	10.9	13.4	4.4	5.1	70.9	6.2	14,540	1.0
Scotland	15.7	73.1	11.1	18.2	5.1	4.3	63.9	8.5	22,456	-17.2
Northern Ireland	14.2	78.3	7.6	9.1	2.4	2.5	78.5	7.4	13,402	63.4

1 See Notes and Definitions.
2 Excludes age not reported.
3 Includes occupants of taxis and minibuses.
4 Used as a basis for the government targets for reducing road casualties in Great Britain, and fatal and serious road casualties in Northern Ireland, by a third by the year 2000.

Source: Department of the Environment, Transport and the Regions; Royal Ulster Constabulary

10.11 Breath tests and breath test failures: by motor vehicle driver age, 1998

Numbers and percentages

	Number of motor vehicle drivers tested who were aged						Motor vehicle drivers that failed[1] as a percentage of those tested who were aged					
	17-24	25-34	35-44	45-54	55-69	70 and over	17-24	25-34	35-44	45-54	55-69	70 and over
Great Britain	45,272	60,933	43,595	30,453	20,579	6,678	5.1	3.9	3.3	2.4	1.9	1.3
North East	1,757	2,399	1,686	1,233	726	210	6.3	3.1	2.7	1.8	1.5	1.4
North West	5,694	7,693	5,456	3,840	2,409	811	4.8	3.4	3.4	1.9	2.0	1.2
Yorkshire and the Humber	4,342	5,717	4,190	2,971	1,881	622	5.2	4.1	3.3	2.7	1.8	1.9
East Midlands	3,675	4,644	3,304	2,409	1,653	549	5.0	4.0	2.8	1.9	1.5	0.4
West Midlands	4,562	6,151	4,191	3,121	2,103	624	5.0	4.3	3.7	2.3	2.3	1.4
East	4,228	4,831	3,440	2,563	1,847	697	5.7	4.6	3.8	3.0	1.7	1.1
London	5,494	10,469	6,970	3,982	2,565	526	2.9	2.6	2.5	2.0	2.1	1.5
South East	5,777	6,743	5,033	3,602	2,504	976	6.6	5.3	4.6	3.3	2.7	1.6
South West	3,553	4,206	3,099	2,451	1,802	743	5.7	4.9	3.1	2.6	1.6	1.3
England	39,082	52,853	37,369	26,172	17,490	5,758	5.1	3.9	3.3	2.4	2.0	1.4
Wales	2,465	2,868	2,203	1,608	1,200	425	5.0	4.4	2.9	2.3	1.8	0.9
Scotland	3,725	5,212	4,023	2,673	1,889	495	4.5	3.0	3.2	1.9	1.3	0.6

1 Failed or refused to provide a specimen of breath.

Source: Department for the Environment, Transport and the Regions

10.12 Activity at major airports[1], 1999

	Air passengers (thousands)[2]				Freight handled[3] (thousands tonnes)
		International			
	Domestic[3]	Scheduled	Non-scheduled	Total	
All UK Airports[4]	35,041	96,708	36,728	168,478	2,201
Newcastle	886	562	1,486	2,934	1
Manchester	2,683	5,390	9,345	17,419	108
Leeds/Bradford	413	496	543	1,451	-
East Midlands	362	446	1,409	2,217	129
Birmingham	1,160	3,035	2,741	6,936	29
Luton	1,313	2,569	1,370	5,251	27
Stansted	1,456	6,630	1,325	9,411	175
Heathrow	7,141	54,727	111	61,979	1,265
Gatwick	2,775	16,698	10,937	30,410	294
Bristol	383	484	1,099	1,966	-
Cardiff	99	225	979	1,303	-
Aberdeen	1,641	346	469	2,456	5
Edinburgh	3,762	978	351	5,090	18
Glasgow	3,507	1,015	2,238	6,759	9
Belfast City	1,272	9	1	1,282	1
Belfast International	2,061	183	768	3,012	26
Other UK airports	4,127	2,916	1,557	8,601	113

1 Airports handling one million passengers or more in 1999. Passengers are recorded at both airport of departure and arrival. Includes British Government/armed forces on official business and travel to/from oil rigs.
2 Arrivals and departures.
3 Domestic traffic is counted at airports on arrival and departure.
4 Including airports handling fewer than one million passengers.

Source: Civil Aviation Authority

10.13 Activity at major seaports[1], 1998

Millions and million tonnes

	International sea passenger movements (millions)	Freight handled (million tonnes)
All UK ports	33.3	568.5
All East coast ports	3.0	278.1
Sullom Voe	0.0	31.1
Forth	0.0	44.4
Tees and Hartlepool	-	51.5
Hull	1.0	10.2
Grimsby and Immingham	-	48.4
Harwich	1.4	3.3
All Thames and Kent ports	20.4	93.9
London	-	57.3
Ramsgate	0.2	1.9
Dover	19.3	17.7
All South coast ports	5.3	49.2
Portsmouth	3.5	4.5
Southampton	-	34.3
All West coast ports	4.6	127.2
Milford Haven	0.5	28.8
Holyhead	2.8	3.4
Liverpool	0.3	30.4
All Northern Ireland ports	0.0	20.1

1 Individual ports handling one million passengers or more in 1998 and/or 25 million tonnes of freight. See Notes and Definitions.

Source: Department of the Environment, Transport and the Regions

11 Environment

Sunshine	For 5 out of the 12 months between October 1998 and September 1999, the average hours of sunshine per day was higher in the South East of England than all other regions. In contrast, for 9 out of the 12 months, the North of Scotland had the lowest daily mean hours of sunshine. *(Table 11.1)*
Rainfall	In 1998, the summer half-year was notably wet, the Northumbrian region recorded its highest April-September rainfall for over 30 years. *(Table11.2)*
Atmospheric pollution	Black smoke and sulphur dioxide concentrations have fallen considerably in most parts of the UK since the mid 1970s, but there are still significant local variations. *(Table 11.3)*
Radon	In the UK, the South West is the region most affected by high levels of radon. *(Map 11.4)*
Water consumption	Water consumption among unmetered households in 1998-99 ranged from 158 litres per head per day in the Southern water and sewerage company region to 134 litres in the Yorkshire region; among metered households the range was from 154 litres in the Thames region to 121 litres in Yorkshire. *(Table 11.5)*
Water abstraction	The Thames region abstracts more water from ground water than any other region in England. *(Table 11.6)*
Water quality	In 1995, the biological quality of rivers and canals in the UK ranged from 100 per cent of those in Northern Ireland being of good or fair quality to 78 per cent of those in the North West region. *(Table 11.7)*
	In 1996-98, the chemical quality of rivers and canals in the UK ranged from 99 per cent of those in Scotland and Wales being of good or fair quality to 82 per cent of those in the Anglian region. *(Table 11.7)*
	In 1998, the Midlands had the highest number of substantiated water pollution incidents in the UK. *(Table 11.8)*
	In 1999, around nine in ten bathing waters in the UK complied with the EC Bathing Water Directive. Compliance was highest in Northern Ireland and the Thames region with all bathing water meeting the standard, and in Wales, where 99 per cent did so. *(Table 11.9)*
	The lowest proportion of bathing waters complying with the EC Bathing Water Directive is in the North West. This region has also experienced the greatest rise in the percentage complying between 1995 and 1999, increasing from 44 per cent in 1995 to 68 per cent in 1999. *(Table 11.9)*

Prosecutions for pollution

In 1999, the highest number of companies and individuals combined prosecuted for pollution incidents was in the Midlands region, accounting for almost a fifth of all prosecutions in England and Wales.

(Table 11.10)

Land

A fifth of Yorkshire and the Humber and of Wales is within National Parks and three tenths of the South East and South West regions are designated Areas of Outstanding Natural Beauty.
(Table 11.12)

More than a fifth of London and the West Midlands is Green Belt land.
(Table 11.12)

More than 600km of the South West and nearly 500km of Wales are designated as Heritage Coasts.
(Table 11.12)

In 1994 over half of land changing to urban use in the East Midlands, East of England and the South West was previously in rural use.
(Table 11.13)

The South West has the highest projected rate of urban growth between 1991 and 2016 at 18 per cent, followed by the East Midlands and the East of England at around 15 per cent.
(Map 11.14)

Recycling

The South East and South West recycle a greater weight of waste per household than any other region in England and Wales.
(Table 11.17)

Paper and card are the most common type of household waste that is recycled by weight.
(Table 11.17)

Waste disposal

Landfill is the most common method of waste disposal within England and Wales although there are variations between regions: 95 per cent of waste in the North West region was landfilled in 1997-98 compared with 71 per cent in the West Midlands.
(Table 11.18)

In 1997-98, around a fifth of all waste in the West Midlands and London regions was incinerated with energy recovery, while in the North West, Wales, the South West and the South East regions this practice did not occur at all.
(Table 11.18)

11.1 Average daily sunshine

Hours per day

	Oct 1998	Nov 1998	Dec 1998	Jan 1999	Feb 1999	Mar 1999	Apr 1999	May 1999	June 1999	July 1999	Aug 1999	Sept 1999
England and Wales	3.17	2.47	1.26	1.93	2.85	3.53	5.23	5.27	6.44	7.59	5.07	5.26
England NW & Wales N	2.88	1.96	1.10	1.50	2.42	3.15	4.92	4.62	6.16	6.76	4.37	4.87
England SW & Wales S	3.17	2.64	1.29	1.80	2.28	3.67	5.25	4.71	6.29	7.69	5.34	5.36
England E & NE	3.61	2.29	1.31	2.05	3.53	3.49	5.04	5.30	5.43	6.90	4.59	5.33
Midlands	3.12	2.20	1.23	1.96	2.88	3.37	4.89	4.84	6.20	7.59	4.66	5.35
England SE	3.01	3.03	1.30	2.03	3.17	3.92	5.46	6.21	7.30	8.41	6.09	5.32
East Anglia	3.16	2.66	1.33	2.25	3.25	3.37	5.89	6.50	7.18	8.12	5.27	5.08
Scotland	3.24	1.77	0.85	1.13	2.58	3.50	4.62	5.20	4.69	5.36	5.07	3.55
Scotland N	2.81	1.40	0.53	0.79	1.83	3.20	4.48	4.86	4.01	5.06	5.25	3.30
Scotland E	3.64	2.14	1.14	1.45	3.33	3.66	4.53	5.39	4.78	5.47	4.84	3.74
Scotland W	3.19	1.74	0.84	1.09	2.61	3.57	4.77	5.25	5.12	5.53	5.00	3.51
Northern Ireland	3.31	2.00	0.90	1.34	1.93	3.89	4.96	4.61	4.20	5.75	3.81	3.37

Source: Meteorological Office

11.2 Winter and summer half-year rainfall[1,2]

Percentages and millimetres

	Rainfall average 1961-1990 (millimetres)		Rainfall as a percentage of the 1961-1990 winter and summer rainfall averages					
			1995-1996		1996-1997		1997-1998	
	Winter	Summer	Winter	Summer	Winter	Summer	Winter	Summer
United Kingdom	609	471	86	76	102	98	109	115
North West	669	534	62	75	99	89	103	114
Northumbria	456	397	92	73	100	100	102	136
Severn Trent	397	357	85	70	85	111	107	116
Yorkshire	441	380	72	72	93	103	104	117
Anglian	298	298	77	63	84	106	106	125
Thames	362	327	95	63	78	93	102	124
Southern	444	335	83	71	82	92	121	114
Wessex	477	361	110	81	88	110	113	128
South West	718	456	99	86	86	111	110	133
England	444	379	84	70	89	100	107	118
Wales[3]	796	560	83	83	90	101	110	117
Scotland	836	601	84	77	117	94	108	108
Northern Ireland	597	462	111	102	95	103	103	124

1 Winter rainfall is the October-March accumulation; summer rainfall is the April-September accumulation.
2 The regions of England shown in this table correspond to the original nine regions of the National Rivers Authority; the NRA became part of the Environment Agency upon its creation in April 1996. See Notes and Definitions.
3 The figures in this table relate to the country of Wales; not to the Environment Agency Welsh Region.

Source: Meteorological Office; Centre for Ecology and Hydrology, Wallingford

11.3 Atmospheric pollution[1,2]

Micrograms per cubic metre and percentages

	Black smoke[3]				Sulphur dioxide			
	Micrograms per cubic metre			Percentage change 1977-78 to 1997-98	Micrograms per cubic metre			Percentage change 1977-78 to 1997-98
	1977-78	1987-88	1997-98		1977-78	1987-88	1997-98	
Newcastle	149	124	41	-72	252	163	79	-69
Manchester	188	78	53	-72	266	127	38	-86
Barnsley	301	122	58	-81	301	170	161	-47
Mansfield Woodhouse	311	154	60	-81	257	168	93	-64
Stoke-on-Trent	256	94	56	-78	274	121	70	-74
Norwich	93	61	40	-57	89	39	20	-78
Stepney	98	40	38	-61	311	178	89	-71
Slough	85	72	42	-51	175	87	24	-86
Swindon	82	61	39	-52	106	59	20	-81
Cardiff	148	52	51	-66	175	64	108	-38
Glasgow	146	93	44	-70	229	67	49	-79
Belfast	304	133	53	-83	142	201	141	-1

1 One site chosen for each UK Statistical Region.
2 Figures shown are for 98th percentile daily mean concentration ie the level which is exceeded by the highest 2 per cent of daily mean concentrations during the year.
3 Measured in OECD units; measurements in British Standard units are equivalent to 0.85 x OECD units.

Source: National Environmental Technology Centre

11.4 Radon levels[1], 1999

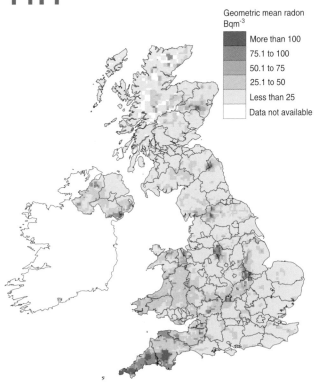

Geometric mean radon Bqm⁻³

- More than 100
- 75.1 to 100
- 50.1 to 75
- 25.1 to 50
- Less than 25
- Data not available

1 Levels of natural radioactivity from Radon. See Notes and Definitions.

Source: National Radiological Protection Board

11.5 Estimated household water consumption[1]

Litres per head per day and percentages

	Unmetered households			Metered households			Percentage of billed households that are metered		
	1996-97	1997-98[2]	1998-99[2]	1996-97	1997-98[2]	1998-99[2]	1996-97	1997-98[3]	1998-99[3]
Water and sewerage companies									
England and Wales[3]	149	150	148	134	137	136	*8*	*11*	*14*
North West	138	141	138	109	134	132	*6*	*7*	*8*
Northumbrian	144	144	147	122	119	132	*2*	*3*	*3*
Yorkshire	132	136	134	118	125	121	*8*	*12*	*16*
Severn Trent	137	137	138	130	130	131	*10*	*13*	*15*
Anglian	153	153	150	141	141	134	*21*	*31*	*38*
Thames	159	161	156	151	155	154	*5*	*8*	*12*
Southern	160	161	158	130	138	138	*12*	*14*	*16*
Wessex	145	141	138	124	124	124	*12*	*14*	*18*
South West	153	155	156	138	123	129	*10*	*14*	*18*
Welsh	146	146	144	136	132	132	*4*	*5*	*6*
Scotland[4]	154	155	156	..	..	..	..	..	..

1 Excluding underground supply pipe leakage.
2 OFWAT does not consider South East Water's water balances for 1997-98 and 1998-99 to be robust. Therefore industry weighted averages for England and Wales for these years include South East Water's 1996-97 data.
3 Figures for England and Wales are industry weighted averages; these include both the ten major water and sewerage companies and 16 smaller water companies.
4 Figures given are estimates taken from the projections provided in the publication Public Water Supply in Scotland an Assessment of Demands and Resources at 1994 produced by the Scottish Office Agriculture, Environment and Fisheries Department. They cover the North of Scotland, the East of Scotland and the West of Scotland Water Authorities.

Source: OFWAT; Scottish Executive Water Services Unit

11.6 Estimated abstractions from groundwaters: by purpose[1], 1998

Megalitres per day

	Public water supply	Spray irrigation	Agriculture (excluding spraying)	Electricity supply	Other industry	Mineral washing	Fish farming, etc	Private water supply[2]	Other[3]	Total
Environment Agency Regions[4]										
England and Wales	5,042	137	100	25	669	187	313	86	94	6,653
North East	349	20	9	10	88	1	2	12	39	530
North West	161	0	5	0	106	28	3	0	1	304
Midlands	890	20	8	8	139	0	4	4	0	1,072
Anglian	724	61	23	0	95	90	3	32	2	1,032
Thames	1,571	21	7	1	105	47	45	24	0	1,821
Southern	1,003	11	7	2	87	18	133	5	12	1,279
South West	256	2	35	1	24	2	121	7	39	489
England	4,954	135	94	23	643	187	311	85	94	6,526
Wales[4]	88	1	6	3	26	0	2	2	0	127

1 Some regions report licensed and actual abstractions for financial rather than calendar years. As figures represent an average for the whole year expressed in daily amounts, differences between amounts reported for financial and calendar years are small.
2 Private abstractions for domestic use by individual households.
3 'Other' includes some private domestic water supply wells and boreholes, public water supply transfer licenses and frost protection use. For Midlands region it also includes mineral washing.
4 The boundaries of the Environment Agency Regions are based on river catchment areas and not county borders. In particular, the figures shown for Wales are for the Environment Agency Welsh Region, the boundary of which does not coincide with the boundary of Wales. See map on page 239 and Notes and Definitions.

Source: Environment Agency

11.7 Rivers and canals: by biological[1] and chemical quality[2]

Percentages and kilometres

| | Biological quality (percentages) | | | | Total length surveyed (=100%) (kms) 1990 | Total length surveyed (=100%) (kms) 1995 | Chemical quality (percentages) | | | | Total length surveyed (=100%) (kms) 1988-90[4] | Total length surveyed (=100%) (kms) 1996-98 |
| | 1990 | | 1995 | | | | 1988-90 | | 1996-98 | | | |
	Good/ Fair	Poor/ Bad	Good/ Fair	Poor/ Bad			Good/ Fair	Poor/ Bad	Good/ Fair	Poor/ Bad		
Environment Agency Regions[3]												
North East	83	17	86	14	4,130	5,460	81	19	88	12	4,350	6,360
North West	63	37	78	22	4,020	4,970	73	27	87	13	3,180	5,750
Midlands	82	18	92	8	3,810	5,840	81	19	90	10	5,640	6,680
Anglian	93	7	97	3	4,170	4,730	81	20	82	18	4,560	4,800
Thames	89	11	95	5	3,090	3,570	81	19	85	15	3,560	3,790
Southern	97	3	98	2	1,420	2,190	88	12	89	11	2,180	2,200
South West	96	4	99	1	5,550	5,940	94	6	97	3	6,770	6,320
England	86	14	92	8	26,190	32,690	84	16	89	11	30,250	35,900
Wales	97	3	99	1	3,810	4,860	96	4	99	1	4,030	5,040
Scotland[2]	97	3	98	2	10,870	16,710	97	3	99	3	49,050	50,050
Northern Ireland[4]	100	-	100	-	2,190	2,330	94	5	96	4	1,690	2,360

1 Classification based on the River Invertebrate Prediction and Classification System (RIVPACS). See Notes and Definitions.
2 Based on the chemical quality grade of the General Quality Assessment (GQA) scheme. See Notes and Definitions. The chemical quality classification for Scotland changed in 1996.
3 In England and Wales. The boundaries of the Environment Agency Regions are based on river catchment areas and not county borders. In particular, the figures shown for Wales are for the Environment Agency Welsh Region, the boundary of which does not correspond with the boundary of Wales. See map on page 239 and Notes and Definitions.
4 Biological quality surveys refer to 1991 and 1997; Chemical quality surveys refer to 1989-91 and 1996-98.

Source: Environment Agency; Department of the Environment, Northern Ireland; Scottish Environment Protection Agency

11.8 Water pollution incidents: by source, 1998[1]

Numbers

| | Industrial | | Sewage and water related | | Agricultural | | Other | | Total | | Number of prose-cutions[3] |
	All	Major[2]	All	Major[2]	All	Major[2]	All	Major[2]	All	Major[2]	
Environment Agency Regions[4]											
United Kingdom	4,675	148	5,086	81	2,873	122	9,199	92	21,833	343	310
North East	398	4	645	6	190	5	760	7	1,993	22	9
North West	533	8	342	5	321	4	1,005	8	2,201	25	33
Midlands	807	8	986	0	418	2	1,850	7	4,061	17	20
Anglian	371	0	473	0	189	0	1,130	7	2,163	7	13
Thames	369	4	437	1	70	2	943	5	1,819	12	34
Southern	199	8	244	2	115	2	580	5	1,138	17	16
South West	459	5	628	1	483	4	1,033	2	2,603	12	34
England	3,136	37	3,755	15	1,786	19	7,301	41	15,978	112	159
Wales[4]	464	5	498	2	264	3	659	6	1,885	16	32
Scotland[1]	640	93	557	58	356	69	776	41	2,329	261	25
Northern Ireland	435	13	276	6	467	31	463	4	1,641	54	94

1 Data relate to substantiated reports of pollution only. Figures for Scotland relate to the financial year 1998-99.
2 Major incidents are those corresponding to Category 1 in the Environment Agency's pollution incidents classification scheme. For Scotland the term 'serious' incidents' is used and compares broadly with all of Category 1 and most of Category 2 used by the Environment Agency. In Northern Ireland the term 'high severity' is used, this compares broadly with all of Category 1 used by the Environment Agency. See Notes and Definitions.
3 For England and Wales total prosecutions include cases concluded and prosecutions outstanding. Prosecutions concluded relate to cases which had been brought to court by 31 March 1998. In Scotland, this figure relates only to legal proceedings which resulted in a conviction during 1998-99. In Northern Ireland total prosecutions include cases concluded and prosecutions outstanding for incidents which took place in 1998.
4 In England and Wales. The boundaries of the Environment Agency Regions are based on river catchment areas and not county borders. In particular, the figures shown for Wales are for the Environment Agency Welsh Region, the boundary of which does not coincide with the boundary of Wales. See map on page 239 and Notes and Definitions.

Source: Environment Agency; Scottish Environment Protection Agency; Department of the Environment, Northern Ireland

11.9 Bathing water – compliance with EC Bathing Water Directive[1] coliform standards[2]: by coastal region

Numbers and percentages

	Identified bathing waters (numbers)					Percentage complying during the bathing season[3]				
	1995	1996	1997	1998	1999	1995	1996	1997	1998	1999
Environment Agency Regions[4]										
United Kingdom	464	472	486	496	537	89	90	88	89	91
North East	56	56	56	56	55	95	88	91	84	95
North West[5]	34	34	34	34	34	44	59	50	62	68
Anglian	34	35	35	36	36	88	97	100	100	94
Thames	3	3	3	3	3	100	67	100	100	100
Southern	67	69	75	77	79	93	90	89	97	94
South West	176	180	180	183	184	95	93	91	91	91
England	370	377	383	389	391	89	89	88	90	90
Wales[4]	55	56	64	68	70	89	93	94	94	99
Scotland	23	23	23	23	60	83	91	78	52	88
Northern Ireland	16	16	16	16	16	94	100	88	94	100

1 76/160/EEC.
2 At least 95 per cent of samples must have counts not exceeding the mandatory limit values for total faecal coliforms.
3 The bathing season is from mid-May to end-September in England and Wales, but is shorter in Scotland and Northern Ireland.
4 In England and Wales. The boundaries of the Environment Agency Regions are based on river catchment areas and not county borders. In particular, the figures shown for Wales are for the Environment Agency Welsh Region, the boundary of which does not coincide with the boundary of Wales. See map on page 239 and Notes and Definitions.
5 In 1997 West Kirby was reclassified from the Welsh region to the North West region. West Kirby data are presented in the North West region for all years for consistency.

Source: Environment Agency; Scottish Environment Protection Agency; Environment and Heritage Service, Northern Ireland

11.10 Prosecutions[1] for pollution incidents, 1999

Numbers

	Waste	Water pollution	Integrated pollution control	Radioactive substances	Water abstraction	All
Environment Agency Regions[2]						
North East	67	15	3	1	0	86
North West	43	32	0	0	2	77
Midlands	74	26	1	2	2	105
Anglian	28	40	2	1	3	74
Thames	40	22	2	1	5	70
Southern	33	26	0	0	1	60
South West	14	28	1	0	0	43
England	299	189	9	5	13	515
Wales[2]	23	30	3	0	0	56

1 Figures are for the total numbers of defendants (companies and individuals) prosecuted in 1999 by type of prosecution.
2 In England and Wales. The boundaries of the Environment Agency Regions are based on river catchment areas and not county borders. In particular, the figures shown for Wales are for the Environment Agency Welsh Region, the boundary of which does not coincide with the boundary of Wales. See map on page 239 and Notes and Definitions.

Source: Environment Agency

11.11 Protected Areas[1], as at 1 April 2000

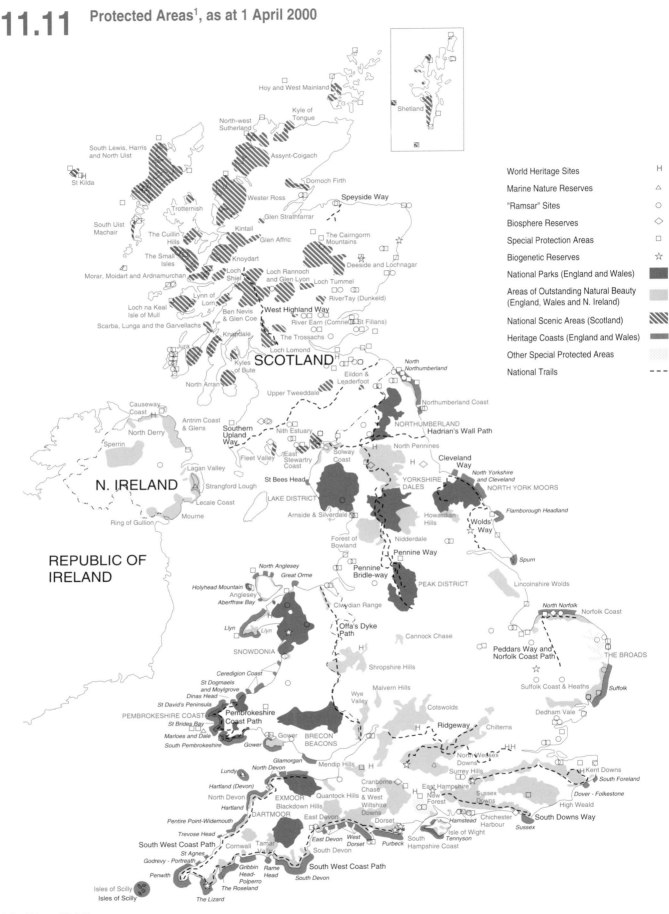

Legend:

H	World Heritage Sites
△	Marine Nature Reserves
○	"Ramsar" Sites
◇	Biosphere Reserves
□	Special Protection Areas
☆	Biogenetic Reserves
▓	National Parks (England and Wales)
░	Areas of Outstanding Natural Beauty (England, Wales and N. Ireland)
▨	National Scenic Areas (Scotland)
▬	Heritage Coasts (England and Wales)
▦	Other Special Protected Areas
- - -	National Trails

1 See Notes and Definitions.

Source: *Countryside Commission; English Nature; Department of Culture, Media and Sport; Institute of Terrestrial Ecology; Department of the Environment, Transport and the Regions; Countryside Council for Wales; Scottish National Heritage; Department of the Environment, Northern Ireland*

11.12 Designated areas[1], 1999[2]

	National Parks		Areas of Outstanding Natural Beauty[3]		Green Belt land[4]		Designated Heritage Coasts length (km)
	Area (sq km)	Percentage of total area in region	Area (sq km)	Percentage of total area in region	Area (sq km)	Percentage of total area in region	
North East	1,112	13	1,465	17	530	6	122
North West	2,607	18	1,570	11	2,519	18	6
Yorkshire and the Humber	3,146	21	921	6	2,637	17	82
East Midlands	917	6	519	3	799	5	0
West Midlands	202	2	1,269	10	2,674	21	.
East	303	2	1,122	6	2,369	12	121
London	0	0	..	..	355	22	.
South East	0	0	6,406	31	3,557	19	72
South West	1,647	7	7,121	30	1,056	4	638
England	9,934	8	20,393	16	16,500	13	1,041
Wales	4,077	20	844	4	-	-	496
Scotland	.	.	10,020	13	1,550	2	.
Northern Ireland	.	.	2,850	20	2,266	16	.

1 See Notes and Definitions. Some areas may be in more than one category.
2 At 1 April 1999 except for Green Belt land which relates to 1 January 1998.
3 National Scenic Areas in Scotland. The South East includes London.
4 Based on a new methodology in which the extent of Green Belt Land is captured in digital form. This approach provides much more reliable figures than those previously published in earlier years and therefore represents new baseline data.

Source: Department of the Environment, Transport and the Regions

11.13 Previous use of land changing to urban use in 1994[1]

Hectares and percentages

	Land changing to urban uses		Percentage previously in rural use
	Total hectares	Hectares per 100,000 population[2]	
North East	740	28	39
North West	1,805	26	27
Yorkshire and the Humber	1,570	31	44
East Midlands	1,335	33	51
West Midlands	1,270	24	35
East	1,900	36	55
London	630	9	14
South East	2,490	32	48
South West	1,425	30	60
England	13,165	27	44

1 The information relates only to map changes recorded by the Ordnance Survey between 1994 and 1998 for which the year of change is judged to be 1994. See Notes and Definitions.
2 Based on mid-1994 population estimates.

Source: Department of the Environment, Transport and the Regions

11.14 Projections of urban growth[1], 1991-2016[2]

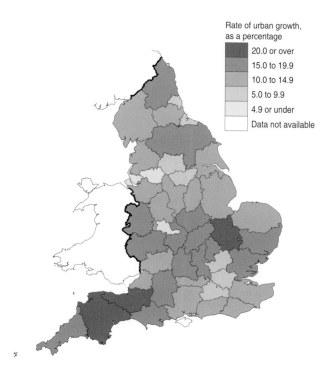

Rate of urban growth, as a percentage
- 20.0 or over
- 15.0 to 19.9
- 10.0 to 14.9
- 5.0 to 9.9
- 4.9 or under
- Data not available

1 The area projected to change net from rural uses to urban uses expressed as a percentage of the area of land in urban uses in 1991.
2 Data relates to counties prior to local government reorganisation.

Source: Department of the Environment, Transport and the Regions

11.15 Household waste: by local authority[1], 1997-98

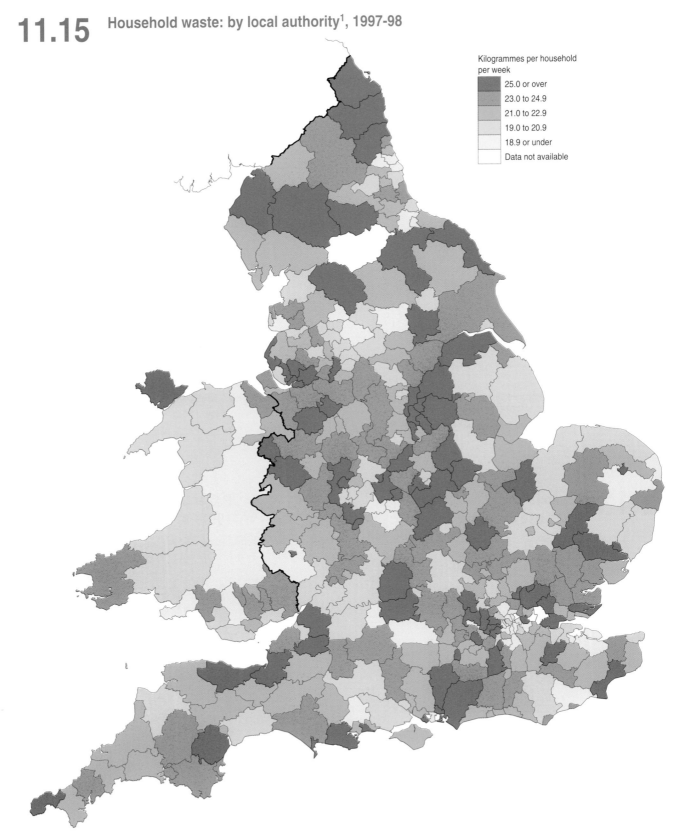

Kilogrammes per household
per week

- 25.0 or over
- 23.0 to 24.9
- 21.0 to 22.9
- 19.0 to 20.9
- 18.9 or under
- Data not available

1 Local government structure at 1 April 1997. See Notes and Definitions.

Source: Department of the Environment, Transport and the Regions; National Assembly for Wales

11.16 Recycling of household waste: by local authority[1], 1997-98

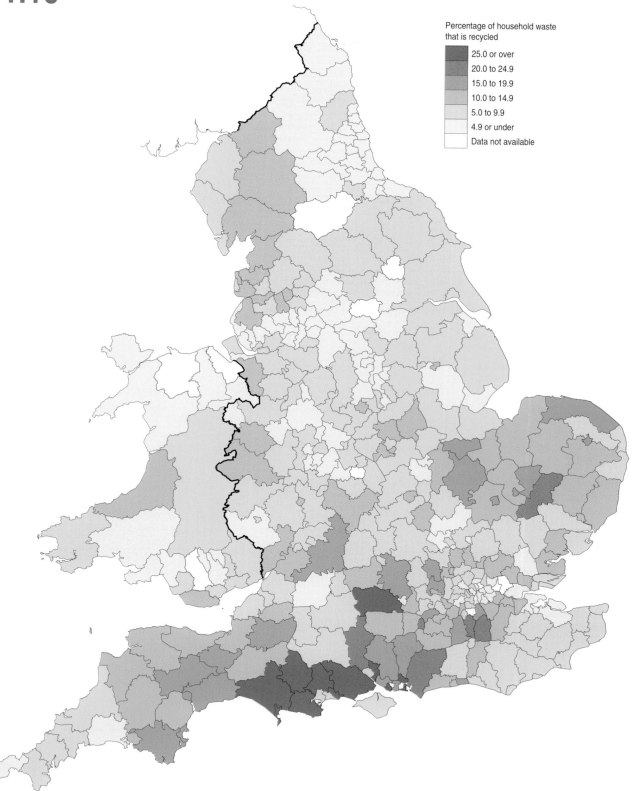

Percentage of household waste
that is recycled

- 25.0 or over
- 20.0 to 24.9
- 15.0 to 19.9
- 10.0 to 14.9
- 5.0 to 9.9
- 4.9 or under
- Data not available

1 Local government structure at 1 April 1997. See Notes and Definitions.

Source: Department of the Environment, Transport and the Regions; National Assembly for Wales

11.17 Recycling of household waste[1], 1997-98

Kilogrammes per household per year

	Glass	Total paper and card	Total cans	Plastics	Textiles	Scrap metal/ white goods	Compost	Other materials[2]	Total
England and Wales	16.2	33.3	0.9	0.3	1.8	11.1	18.0	7.8	89.4
North East	8.6	15.0	0.5	0.1	0.8	7.0	2.2	15.3	49.5
North West	11.6	23.2	0.4	0.3	1.3	7.1	15.1	5.9	64.9
Yorkshire and the Humber	11.8	21.6	0.6	0.4	0.9	6.7	10.2	4.0	56.2
East Midlands	13.9	29.3	1.0	0.4	1.7	13.4	15.8	15.0	90.5
West Midlands	13.7	25.3	0.3	0.0	1.2	9.8	12.0	1.5	63.8
East	21.1	46.1	1.0	0.3	4.3	13.5	40.6	3.3	130.2
London	14.3	33.0	0.6	0.0	1.0	8.1	6.0	4.4	67.4
South East	25.5	54.9	1.6	0.7	2.2	18.0	24.7	13.9	141.5
South West	23.3	42.1	1.7	0.7	2.9	16.7	41.0	16.5	144.9
Wales	10.9	22.3	0.9	0.1	1.7	6.6	3.5	1.1	47.1

1 Materials recycled by local authorities through civic amenity and bring/drop-off sites and kerbside collection schemes for household wastes.
2 Other materials includes oils, batteries, aluminium foil, books, shoes and co-mingled collections.

Source: Department of the Environment, Transport and the Regions; National Assembly for Wales

11.18 Waste disposal: by region and method, 1997-98

Percentages

	Landfill	Incin- eration without energy recovery	Incin- eration with energy recovery	RDF[1] manu- facture	Recycled/ composted	Other	All methods
North East	90	1	2	3	5	0	100
North West	95	0	-	0	5	0	100
Yorkshire and the Humber	94	0	1	0	5	0	100
East Midlands	84	0	7	0	9	0	100
West Midlands	71	2	20	0	6	1	100
East	89	0	-	0	11	0	100
London	76	0	19	0	6	0	100
South East	83	0	0	3	14	-	100
South West	86	-	0	0	14	0	100
England	85	0	6	8	1	0	100
Wales	92	0	0	0	5	3	100

1 Refuse derived fuel.

Source: Department of the Environment, Transport and the Regions

12 Regional Accounts

Gross domestic product (GDP)

In 1998, GDP for the South East and London were both around £116 billion, each accounting for about 16 per cent of total UK GDP. The region with the smallest share was Northern Ireland, at about 2 per cent (£16 billion).

(Table 12.1)

The contribution by Extra-regio to GDP fell by 33 per cent between 1997 and 1998.

(Table 12.1)

Compensation of employees (mainly wages and salaries) was the source of 67 per cent of GDP in the North East in 1998 compared with 60 per cent in the South West.

(Table 12.3)

GDP per head

In 1998, London had the highest level of GDP per head, over £16,200, followed by the South East and East of England at £14,500 and £14,200 respectively.

(Table 12.1)

Northern Ireland had the lowest regional GDP per head in 1998, at £9,400, followed by the North East at £9,800 and Wales at £9,900.

(Table 12.1)

Over the period 1990 to 1998, GDP per head has been consistently above the UK average for London, South East and the East of England.

(Chart 12.2)

GDP by industry

Thirty per cent of GDP in the East Midlands was derived from manufacturing in 1997 compared with around 11 per cent in London the same period. These represent the highest and lowest proportions respectively.

(Table 12.4)

More than two fifths of London's GDP in 1997 was generated by financial intermediation and real estate, renting and business services, compared with less than one fifth in the North East, East Midlands, Wales and Northern Ireland.

(Table 12.4)

Household income/ disposable income

Between 1996 and 1997, total household income per head grew most strongly in the South West and East Midlands, while London had the slowest growth

(Table 12.6)

Disposable household income per head in London in 1997 was 18 per cent higher than the UK average. In Wales and the North East it was 13 per cent and 14 per cent lower respectively.

(Table 12.6)

Individual consumption expenditure

London and the South East had the highest individual consumption expenditure per head in 1998, at over £10,000 per person, while the North East and Northern Ireland had the lowest, at less than £8,000.

(Table 12.7)

London and the East of England had the biggest increases in individual consumption expenditure per head relative to the UK average between 1996 and 1998; Wales had the largest decrease.

(Table 12.7)

Northern Ireland had the highest individual consumption expenditure on food drink and tobacco at 22 per cent in 1998, compared with the East of England, London and the South East at around 16 per cent.

(Table 12.8)

For all regions in the UK, individual consumption expenditure on recreation was between 10 and 12 per cent in 1998.

(Table 12.8)

12.1 Gross domestic product[1] at basic prices

	1990	1991	1992	1993	1994	1995	1996	1997	1998
£ million									
United Kingdom	499,742	521,547	543,904	571,838	604,163	634,067	672,570	713,615	747,544
North East	18,245	19,266	20,191	21,227	21,814	22,774	23,651	24,321	25,496
North West	53,389	55,657	57,517	60,265	63,602	65,806	68,776	72,475	75,834
Yorkshire and the Humber	37,383	39,271	40,302	42,393	44,366	46,837	49,852	53,002	55,232
East Midlands	32,500	33,919	35,120	36,860	38,801	40,786	44,024	47,289	49,260
West Midlands	41,789	43,216	45,236	47,491	50,137	52,781	55,134	58,053	60,927
East	49,411	50,798	53,680	55,757	59,589	62,151	66,191	72,229	76,308
London	74,674	78,700	82,713	87,043	91,635	94,399	99,903	108,645	116,444
South East	73,151	75,713	78,991	83,846	88,827	93,082	100,317	107,630	116,176
South West	36,600	38,167	40,143	42,302	44,527	47,373	50,164	53,453	56,068
England	417,143	434,706	453,893	477,185	503,299	525,991	558,013	597,096	631,746
Wales	20,353	21,518	22,154	23,195	24,405	25,860	26,886	27,912	29,027
Scotland	42,294	44,864	46,805	48,811	51,710	55,249	56,991	58,578	61,052
Northern Ireland	9,770	10,631	11,336	12,127	12,959	13,858	14,427	15,468	15,966
United Kingdom *less* Extra-Regio									
and statistical discrepancy	489,560	511,719	534,189	561,318	592,374	620,958	656,316	699,055	737,792
Extra-Regio	10,182	9,829	9,715	10,520	11,789	13,109	16,254	14,560	9,816
Statistical discrepancy (income adjustment)	-	-	-	-	-	-	-	-	-64
As a percentage of									
United Kingdom *less* Extra-Regio									
and statistical discrepancy									
United Kingdom	*100.0*	*100.0*	*100.0*	*100.0*	*100.0*	*100.0*	*100.0*	*100.0*	*100.0*
North East	*3.7*	*3.7*	*3.7*	*3.7*	*3.6*	*3.6*	*3.5*	*3.4*	*3.4*
North West	*10.7*	*10.7*	*10.6*	*10.5*	*10.5*	*10.4*	*10.2*	*10.2*	*10.1*
Yorkshire and the Humber	*7.5*	*7.5*	*7.4*	*7.4*	*7.3*	*7.4*	*7.4*	*7.4*	*7.4*
East Midlands	*6.5*	*6.5*	*6.5*	*6.4*	*6.4*	*6.4*	*6.5*	*6.6*	*6.6*
West Midlands	*8.4*	*8.3*	*8.3*	*8.3*	*8.3*	*8.3*	*8.2*	*8.1*	*8.2*
East	*9.9*	*9.7*	*9.9*	*9.8*	*9.9*	*9.8*	*9.8*	*10.1*	*10.2*
London	*14.9*	*15.1*	*15.2*	*15.2*	*15.2*	*14.9*	*14.9*	*15.2*	*15.6*
South East	*14.6*	*14.5*	*14.5*	*14.7*	*14.7*	*14.7*	*14.9*	*15.1*	*15.5*
South West	*7.3*	*7.3*	*7.4*	*7.4*	*7.4*	*7.5*	*7.5*	*7.5*	*7.5*
England	*83.5*	*83.3*	*83.5*	*83.4*	*83.3*	*83.0*	*83.0*	*83.7*	*84.5*
Wales	*4.1*	*4.1*	*4.1*	*4.1*	*4.0*	*4.1*	*4.0*	*3.9*	*3.9*
Scotland	*8.5*	*8.6*	*8.6*	*8.5*	*8.6*	*8.7*	*8.5*	*8.2*	*8.2*
Northern Ireland	*2.0*	*2.0*	*2.1*	*2.1*	*2.1*	*2.2*	*2.1*	*2.2*	*2.1*
£ per head									
United Kingdom *less* Extra-Regio									
and statistical discrepancy	8,505	8,852	9,209	9,646	10,144	10,595	11,162	11,847	12,455
North East	7,023	7,394	7,737	8,120	8,342	8,719	9,072	9,348	9,819
North West	7,775	8,078	8,338	8,727	9,200	9,519	9,958	10,504	10,990
Yorkshire and the Humber	7,533	7,882	8,060	8,453	8,825	9,301	9,890	10,506	10,939
East Midlands	8,097	8,411	8,661	9,039	9,466	9,899	10,635	11,378	11,812
West Midlands	7,960	8,203	8,568	8,976	9,459	9,940	10,363	10,896	11,417
East	9,664	9,880	10,382	10,740	11,424	11,840	12,528	13,570	14,222
London	10,897	11,431	11,974	12,563	13,164	13,487	14,167	15,280	16,245
South East	9,572	9,864	10,249	10,839	11,428	11,889	12,724	13,554	14,529
South West	7,798	8,094	8,470	8,880	9,295	9,827	10,360	10,983	11,448
England	8,692	9,020	9,384	9,834	10,336	10,759	11,371	12,119	12,768
Wales	7,072	7,445	7,641	7,980	8,374	8,856	9,196	9,530	9,888
Scotland	8,289	8,767	9,143	9,520	10,060	10,738	11,096	11,416	11,902
Northern Ireland	6,147	6,626	6,994	7,421	7,880	8,390	8,660	9,220	9,438

1 Based on the European System of Accounts 1995 (ESA95). See Notes and Definitions.

Source: Office for National Statistics

12.2 Gross domestic product per head[1] at basic prices

£ per head index, UK=100[2]

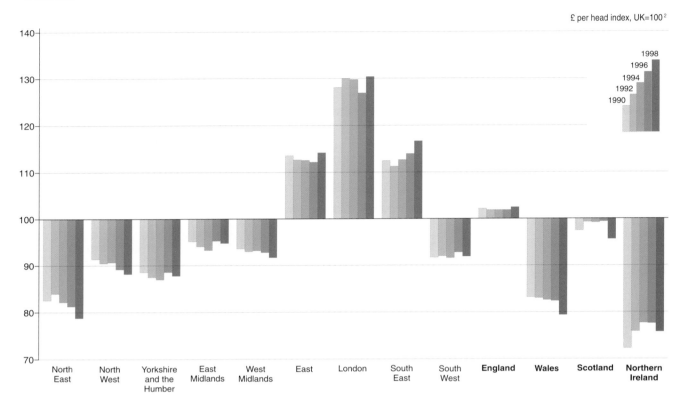

1 Based on the European System of Accounts 1995 (ESA95). See Notes and Definitions.
2 United Kingdom less Extra-Regio and statistical discrepancy.

Source: Office for National Statistics

12.3 Factor incomes in the gross domestic product[1] at basic prices, 1998

Percentages and £ million

	Income components as a percentage of total GDP		Gross domestic product (=100%) (£ million)
	Compensation of employees	Operating surplus/ mixed income	
United Kingdom	62	38	747,544
North East	67	33	25,496
North West	64	36	75,834
Yorkshire and the Humber	65	35	55,232
East Midlands	62	38	49,260
West Midlands	64	36	60,927
East	62	38	76,308
London	63	37	116,444
South East	61	39	116,176
South West	60	40	56,068
England	63	37	631,746
Wales	62	38	29,027
Scotland	64	36	61,052
Northern Ireland	62	38	15,966
Extra-Regio	13	87	9,816

1 Based on the European System of Accounts 1995 (ESA95). See Notes and Definitions.

Source: Office for National Statistics

12.4 Gross domestic product by industry groups[1,2] at basic prices

£ million

	1994	1995	1996	1997	1994	1995	1996	1997
	United Kingdom[3]				North East			
Agriculture, hunting, forestry and fishing	10,777	11,714	11,963	10,595	203	214	233	203
Mining, quarrying inc oil and gas extraction	2,459	2,581	2,470	2,394	123	115	136	106
Other mining and quarrying	1,250	1,437	1,621	1,609	86	76	91	83
Manufacturing	128,202	136,747	143,485	148,619	6,306	6,475	6,817	6,863
Electricity, gas and water supply	15,932	15,562	16,120	16,230	566	645	570	583
Construction	31,347	32,948	34,563	36,927	1,146	1,259	1,319	1,429
Wholesale and retail trade	71,201	74,148	78,698	85,865	2,412	2,399	2,494	2,563
Hotels and Restaurants	17,023	18,409	20,471	22,585	537	601	696	724
Transport and communication	50,708	52,297	53,994	57,916	1,497	1,538	1,559	1,599
Financial intermediation[4]	45,421	42,726	42,730	43,852	810	758	740	751
Real estate, renting and business activities	109,077	116,343	125,717	140,311	2,838	3,066	3,265	3,552
Public administration and defence[5]	37,945	38,186	38,284	38,101	1,471	1,465	1,500	1,433
Education	32,788	34,194	36,611	38,818	1,547	1,617	1,688	1,637
Health and social work	39,405	41,937	44,681	46,344	1,963	2,102	2,121	2,213
Other services	25,248	27,229	30,466	34,567	850	931	922	1,066
Adjustment for financial services (FISIM[6])	-26,410	-25,499	-25,557	-25,678	-543	-490	-499	-482
Total	592,374	620,958	656,316	699,055	21,814	22,774	23,651	24,321
	North West				Yorkshire and the Humber			
Agriculture, hunting, forestry and fishing	901	898	893	780	881	1,053	1,150	995
Mining, quarrying inc oil and gas extraction	22	31	28	28	197	207	165	243
Other mining and quarrying	87	107	111	216	135	151	235	141
Manufacturing	17,765	18,539	18,834	18,998	11,651	12,541	13,553	14,496
Electricity, gas and water supply	1,748	1,635	1,822	1,832	1,158	1,152	1,225	1,383
Construction	3,382	3,460	3,611	3,936	2,721	2,710	2,886	3,017
Wholesale and retail trade	7,570	7,854	8,303	8,874	5,541	5,757	6,084	6,418
Hotels and Restaurants	1,706	1,806	2,010	2,330	1,226	1,258	1,421	1,617
Transport and communication	5,436	5,503	5,723	5,957	3,531	3,764	3,826	4,179
Financial intermediation[4]	3,465	3,034	2,952	3,030	2,547	2,334	2,212	2,336
Real estate, renting and business activities	10,021	10,628	11,432	12,675	6,300	6,970	7,459	8,267
Public administration and defence[5]	3,233	3,291	3,265	3,373	2,659	2,792	2,748	2,541
Education	3,727	3,858	4,161	4,188	2,837	2,885	3,073	3,174
Health and social work	4,485	4,859	5,164	5,276	3,156	3,407	3,653	3,902
Other services	2,528	2,556	2,650	3,103	1,659	1,646	1,829	2,109
Adjustment for financial services (FISIM[6])	-2,474	-2,251	-2,184	-2,122	-1,831	-1,788	-1,668	-1,817
Total	63,602	65,806	68,776	72,475	44,366	46,837	49,852	53,002
	East Midlands				West Midlands			
Agriculture, hunting, forestry and fishing	1,094	1,160	1,191	1,031	976	1,054	1,110	986
Mining, quarrying inc oil and gas extraction	193	178	190	206	60	72	72	72
Other mining and quarrying	187	176	247	168	88	112	111	96
Manufacturing	11,623	12,241	13,332	14,144	14,315	15,466	16,279	17,104
Electricity, gas and water supply	1,146	1,054	1,214	1,109	1,420	1,438	1,381	1,452
Construction	2,234	2,249	2,242	2,405	2,706	2,741	2,888	3,043
Wholesale and retail trade	4,593	5,021	5,363	5,819	7,059	7,348	7,590	7,927
Hotels and Restaurants	907	1,019	1,189	1,314	1,245	1,370	1,447	1,462
Transport and communication	2,673	2,837	2,995	3,229	3,168	3,511	3,698	3,853
Financial intermediation[4]	1,615	1,471	1,390	1,430	2,387	2,410	2,398	2,429
Real estate, renting and business activities	5,792	6,012	6,745	7,781	8,383	8,649	9,188	9,916
Public administration and defence[5]	1,888	1,979	1,955	1,970	2,386	2,424	2,413	2,412
Education	2,109	2,271	2,417	2,739	2,763	2,854	3,030	3,208
Health and social work	2,530	2,785	2,930	2,875	3,266	3,382	3,368	3,464
Other services	1,273	1,327	1,558	1,995	1,610	1,672	1,878	2,274
Adjustment for financial services (FISIM[6])	-1,054	-994	-934	-925	-1,696	-1,724	-1,717	-1,645
Total	38,801	40,786	44,024	47,289	50,137	52,781	55,134	58,053

12.4 *(continued)*

£ million

	1994	1995	1996	1997	1994	1995	1996	1997
	East				London			
Agriculture, hunting, forestry and fishing	1,461	1,519	1,636	1,413	40	40	46	44
Mining, quarrying inc oil and gas extraction	201	256	222	210	194	209	162	126
Other mining and quarrying	32	43	67	70	40	45	64	77
Manufacturing	11,494	11,811	12,327	12,844	11,594	11,951	12,115	12,490
Electricity, gas and water supply	1,547	1,499	1,540	1,563	1,762	1,655	1,623	1,603
Construction	3,332	3,495	3,841	4,220	3,425	3,670	4,074	4,556
Wholesale and retail trade	7,115	7,528	7,978	9,209	10,881	11,494	12,455	13,945
Hotels and Restaurants	1,380	1,461	1,553	1,924	3,321	3,565	3,983	4,402
Transport and communication	6,203	6,344	6,410	6,811	9,743	9,657	9,816	10,995
Financial intermediation[4]	5,889	5,673	5,819	6,654	13,398	12,446	12,569	12,654
Real estate, renting and business activities	10,900	11,822	13,048	14,908	24,762	26,214	28,244	31,530
Public administration and defence[5]	3,386	3,333	3,320	3,422	5,257	5,306	5,201	5,070
Education	3,045	3,179	3,344	3,651	4,512	4,618	4,915	5,398
Health and social work	3,355	3,516	3,813	3,976	4,890	5,096	5,679	6,093
Other services	2,340	2,659	3,108	3,448	5,644	6,240	7,190	7,868
Adjustment for financial services (FISIM[6])	-2,090	-1,988	-1,834	-2,093	-7,828	-7,807	-8,234	-8,206
Total	59,589	62,151	66,191	72,229	91,635	94,399	99,903	108,645
	South East				South West			
Agriculture, hunting, forestry and fishing	1,048	1,128	1,144	996	1,518	1,765	1,740	1,554
Mining, quarrying inc oil and gas extraction	214	230	177	121	40	44	30	18
Other mining and quarrying	98	113	152	182	233	281	251	275
Manufacturing	14,696	15,916	16,694	17,391	8,866	9,435	10,264	10,609
Electricity, gas and water supply	1,995	1,848	1,834	1,982	1,782	1,634	1,716	1,844
Construction	4,508	4,871	5,381	5,698	2,482	2,685	2,623	2,789
Wholesale and retail trade	10,640	11,106	12,001	12,923	5,646	5,915	6,118	6,834
Hotels and Restaurants	2,361	2,558	2,911	3,160	1,448	1,586	1,793	2,012
Transport and communication	9,164	9,623	10,365	10,948	2,763	2,984	3,127	3,491
Financial intermediation[4]	7,207	6,810	7,128	7,147	3,321	3,163	3,017	2,862
Real estate, renting and business activities	20,278	21,676	23,661	26,926	7,732	8,362	9,044	9,884
Public administration and defence[5]	6,531	6,119	6,164	6,152	3,955	4,133	4,334	4,491
Education	4,136	4,328	4,720	5,340	2,279	2,318	2,498	2,775
Health and social work	5,419	5,685	6,428	6,577	3,096	3,280	3,436	3,520
Other services	3,960	4,333	4,989	5,518	1,641	1,960	2,305	2,487
Adjustment for financial services (FISIM[6])	-3,428	-3,261	-3,433	-3,431	-2,277	-2,169	-2,131	-1,991
Total	88,827	93,082	100,317	107,630	44,527	47,373	50,164	53,453

12.4 *(continued)*

£ million

	1994	1995	1996	1997	1994	1995	1996	1997
	England				Wales			
Agriculture, hunting, forestry and fishing	8,122	8,830	9,143	8,002	543	458	538	509
Mining, quarrying inc oil and gas extraction	1,245	1,341	1,181	1,129	87	108	114	70
Other mining and quarrying	986	1,104	1,328	1,308	97	131	99	99
Manufacturing	108,309	114,377	120,215	124,938	6,623	7,356	7,719	7,790
Electricity, gas and water supply	13,125	12,561	12,925	13,351	790	842	807	620
Construction	25,936	27,139	28,866	31,092	1,366	1,400	1,446	1,518
Wholesale and retail trade	61,459	64,423	68,385	74,512	2,495	2,524	2,605	2,919
Hotels and Restaurants	14,130	15,224	17,005	18,945	740	808	867	971
Transport and communication	44,179	45,760	47,519	51,062	1,596	1,637	1,586	1,699
Financial intermediation[4]	40,639	38,100	38,226	39,293	983	985	960	1,009
Real estate, renting and business activities	97,005	103,397	112,086	125,439	3,413	3,640	3,682	4,063
Public administration and defence[5]	30,766	30,842	30,901	30,864	1,733	1,813	1,896	1,775
Education	26,955	27,929	29,846	32,111	1,588	1,614	1,838	1,712
Health and social work	32,159	34,112	36,591	37,895	2,010	2,146	2,196	2,444
Other services	21,505	23,324	26,429	29,867	997	1,030	1,166	1,339
Adjustment for financial services (FISIM[6])	-23,220	-22,471	-22,634	-22,711	-657	-632	-634	-626
Total	503,299	525,991	558,013	597,096	24,405	25,860	26,886	27,912
	Scotland				Northern Ireland			
Agriculture, hunting, forestry and fishing	1,479	1,632	1,474	1,342	634	794	809	742
Mining, quarrying inc oil and gas extraction	1,117	1,121	1,165	1,186	10	11	9	9
Other mining and quarrying	97	115	129	131	70	86	65	71
Manufacturing	10,646	12,239	12,539	12,780	2,624	2,776	3,011	3,110
Electricity, gas and water supply	1,619	1,757	1,952	1,820	398	402	435	440
Construction	3,292	3,577	3,383	3,354	753	833	868	963
Wholesale and retail trade	5,770	5,590	6,001	6,494	1,477	1,611	1,707	1,940
Hotels and Restaurants	1,823	1,999	2,193	2,211	330	377	406	457
Transport and communication	4,212	4,131	4,095	4,274	721	769	795	882
Financial intermediation[4]	3,182	3,049	2,987	2,984	617	592	558	566
Real estate, renting and business activities	7,405	7,946	8,454	9,062	1,253	1,361	1,495	1,746
Public administration and defence[5]	3,760	3,866	3,875	3,841	1,685	1,666	1,612	1,620
Education	3,298	3,604	3,837	3,766	946	1,046	1,090	1,229
Health and social work	3,962	4,346	4,539	4,621	1,274	1,333	1,354	1,384
Other services	2,212	2,325	2,302	2,698	534	550	568	663
Adjustment for financial services (FISIM[6])	-2,165	-2,049	-1,934	-1,986	-367	-348	-355	-355
Total	51,710	55,249	56,991	58,578	12,959	13,858	14,427	15,468

1 Based on the European System of Accounts 1995 (ESA95). See Notes and Definitions.
2 Gross domestic product is shown for each industry, based on SIC 92, after deducting stock appreciation.
3 Excludes production from Extra-Regio.
4 Financial intermediation, real estate, renting, business activities, including rent on dwellings.
5 Public administration, national defence and compulsory social security.
6 FISIM – financial intermediation services indirectly measured.

Source: Office for National Statistics

12.5 Gross domestic product per head[1]: by sub-region (NUTS Level 2)[2], 1996

£ per head index
(UK=100)

125 or over
115.0-124.9
105.0-114.9
95.0-104.9
85.0-94.9
75.0-84.9
74.9 or under

1 Excluding the Continental Shelf and the statistical discrepancy of the income-based measure. The data plotted on this map are based on the old ESA 79 and are consistent with
    data published in *United Kingdom National Accounts 1997*. They can be found, together with reference maps for NUTS areas, in Chapters 14, 15, 16 and 17.
2 NUTS (Nomenclature of Units for Territorial Statistics) is a hierarchical classification of areas that provides a breakdown of the EU's economic territory. See Notes and Definitions.

Source: Office for National Statistics

12.6 Household income and disposable household income[1]

	1989	1990	1991	1992	1993	1994	1995	1996	1997
Household income									
£ million									
United Kingdom	474,620	528,626	573,576	613,469	639,781	666,938	705,589	749,007	788,914
North East	17,675	19,671	22,065	23,990	25,150	25,569	26,552	28,038	29,649
North West	51,869	57,799	62,909	67,181	69,852	72,497	75,954	80,126	84,533
Yorkshire and the Humber	36,703	40,688	44,566	48,058	50,455	52,378	55,434	58,271	61,508
East Midlands	31,551	34,534	37,553	39,997	42,032	44,125	46,735	48,897	52,444
West Midlands	39,641	44,151	48,397	52,099	53,998	56,421	59,382	62,964	65,639
East	46,813	51,366	55,157	58,856	60,603	63,690	68,050	73,158	77,276
London	70,129	79,742	84,976	89,199	93,046	96,991	102,631	111,071	115,210
South East	71,865	79,534	85,015	90,510	95,045	100,445	106,665	114,976	121,148
South West	38,483	42,574	46,396	49,762	51,252	53,046	56,759	59,344	64,134
England	404,729	450,059	487,033	519,654	541,432	565,163	598,162	636,846	671,540
Wales	19,911	21,963	24,315	26,173	27,175	28,343	30,148	31,551	33,317
Scotland	38,821	44,422	48,543	52,886	55,366	56,846	59,634	61,971	64,299
Northern Ireland	10,564	11,556	13,036	14,100	15,169	16,038	17,046	18,095	19,152
Extra-Regio	595	626	649	656	639	549	599	545	607
£ per head									
United Kingdom	8,275	9,184	9,942	10,592	11,010	11,439	12,060	12,759	13,392
United Kingdom *less* Extra-Regio	8,264	9,173	9,931	10,581	10,999	11,430	12,050	12,749	13,382
North East	6,814	7,572	8,486	9,207	9,634	9,793	10,183	10,772	11,415
North West	7,564	8,418	9,149	9,754	10,129	10,503	11,006	11,620	12,273
Yorkshire and the Humber	7,416	8,199	8,962	9,626	10,074	10,435	11,027	11,579	12,213
East Midlands	7,899	8,604	9,331	9,879	10,321	10,782	11,363	11,832	12,640
West Midlands	7,564	8,410	9,205	9,883	10,220	10,661	11,203	11,854	12,341
East	9,194	10,046	10,749	11,401	11,690	12,229	12,986	13,869	14,543
London	10,315	11,637	12,367	12,933	13,448	13,955	14,688	15,776	16,231
South East	9,447	10,408	11,098	11,761	12,304	12,943	13,647	14,607	15,283
South West	8,231	9,070	9,859	10,516	10,774	11,090	11,794	12,276	13,200
England	8,466	9,378	10,125	10,760	11,174	11,624	12,256	12,998	13,653
Wales	6,939	7,632	8,429	9,041	9,363	9,741	10,343	10,809	11,394
Scotland	7,617	8,706	9,505	10,347	10,813	11,076	11,610	12,085	12,552
Northern Ireland	6,674	7,270	8,141	8,712	9,296	9,768	10,337	10,879	11,435
Disposable household income									
£ per head									
United Kingdom	5,563	6,195	6,771	7,324	7,780	8,030	8,453	8,879	9,415
United Kingdom *less* Extra-Regio	5,553	6,184	6,760	7,313	7,769	8,020	8,443	8,870	9,405
North East	4,613	5,141	5,843	6,443	6,898	6,941	7,147	7,523	8,080
North West	5,114	5,712	6,292	6,839	7,251	7,439	7,783	8,157	8,703
Yorkshire and the Humber	5,011	5,559	6,150	6,705	7,174	7,387	7,808	8,140	8,676
East Midlands	5,305	5,806	6,360	6,849	7,293	7,541	7,931	8,195	8,926
West Midlands	5,059	5,646	6,298	6,879	7,260	7,502	7,828	8,240	8,640
East	6,128	6,727	7,242	7,816	8,215	8,539	9,090	9,740	10,371
London	6,922	7,776	8,254	8,733	9,305	9,667	10,147	10,776	11,084
South East	6,245	6,896	7,387	7,945	8,515	8,904	9,397	9,980	10,559
South West	5,643	6,196	6,827	7,378	7,719	7,923	8,446	8,704	9,543
England	5,683	6,305	6,868	7,413	7,872	8,140	8,572	9,027	9,585
Wales	4,712	5,214	5,942	6,445	6,798	7,018	7,441	7,702	8,217
Scotland	5,090	5,932	6,507	7,196	7,646	7,741	8,078	8,332	8,661
Northern Ireland	4,639	5,125	5,811	6,250	6,826	7,125	7,554	7,947	8,464
£ per head index, United Kingdom *less* Extra-Regio=100									
United Kingdom *less* Extra-Regio	100.0	100.0	100.0	100.0	100.0	100.0	100.0	100.0	100.0
North East	83.1	83.1	86.4	88.1	88.8	86.5	84.7	84.8	85.9
North West	92.1	92.4	93.1	93.5	93.3	92.8	92.2	92.0	92.5
Yorkshire and the Humber	90.2	89.9	91.0	91.7	92.3	92.1	92.5	91.8	92.2
East Midlands	95.5	93.9	94.1	93.7	93.9	94.0	93.9	92.4	94.9
West Midlands	91.1	91.3	93.2	94.1	93.4	93.5	92.7	92.9	91.9
East	110.4	108.8	107.1	106.9	105.7	106.5	107.7	109.8	110.3
London	124.7	125.7	122.1	119.4	119.8	120.5	120.2	121.5	117.9
South East	112.5	111.5	109.3	108.6	109.6	111.0	111.3	112.5	112.3
South West	101.6	100.2	101.0	100.9	99.4	98.8	100.0	98.1	101.5
England	102.3	101.9	101.6	101.4	101.3	101.5	101.5	101.8	101.9
Wales	84.9	84.3	87.9	88.1	87.5	87.5	88.1	86.8	87.4
Scotland	91.7	95.9	96.3	98.4	98.4	96.5	95.7	93.9	92.1
Northern Ireland	83.5	82.9	86.0	85.5	87.9	88.8	89.5	89.6	90.0

1 Based on the European System of Accounts 1995 (ESA95). See Notes and Definitions.

Source: Office for National Statistics

12.7 Individual consumption expenditure[1]

	Individual consumption expenditure (£ million)			Regional shares of the UK (percentages)			£ per head			£ per head index, UK = 100		
	1996	1997	1998	1996	1997	1998	1996	1997	1998	1996	1997	1998
United Kingdom	485,417	517,031	545,123	100.0	100.0	100.0	8,255	8,762	9,202	100.0	100.0	100.0
North East	19,122	20,122	20,416	3.9	3.9	3.7	7,335	7,734	7,862	88.9	88.3	85.4
North West	53,815	57,485	60,101	11.1	11.1	11.0	7,792	8,331	8,710	94.4	95.1	94.6
Yorkshire and the Humber	39,033	41,170	43,872	8.0	8.0	8.0	7,744	8,161	8,689	93.8	93.1	94.4
East Midlands	32,854	34,783	35,983	6.8	6.7	6.6	7,937	8,369	8,628	96.1	95.5	93.8
West Midlands	40,965	43,300	45,357	8.4	8.4	8.3	7,700	8,127	8,499	93.3	92.8	92.4
East	45,955	48,614	53,335	9.5	9.4	9.8	8,698	9,134	9,940	105.4	104.2	108.0
London	67,120	72,884	78,427	13.8	14.1	14.4	9,518	10,250	10,941	115.3	117.0	118.9
South East	72,297	77,599	82,638	14.9	15.0	15.2	9,170	9,772	10,335	111.1	111.5	112.3
South West	39,020	41,745	43,055	8.0	8.1	7.9	8,059	8,577	8,791	97.6	97.9	95.5
England	410,181	437,702	463,183	84.5	84.7	85.0	8,358	8,884	9,361	101.2	101.4	101.7
Wales	22,519	23,496	23,469	4.6	4.5	4.3	7,703	8,022	7,995	93.3	91.6	86.9
Scotland	40,856	43,445	45,634	8.4	8.4	8.4	7,955	8,467	8,896	96.4	96.6	96.7
Northern Ireland	11,860	12,388	12,837	2.4	2.4	2.4	7,119	7,384	7,588	86.2	84.3	82.5

1 Based on the European System of Accounts 1995 (ESA95). See Notes and Definitions.

Source: Office for National Statistics

12.8 Individual consumption expenditure[1]: by broad function, 1998

£ million

	Food, drink and tobacco	Clothing and footwear	Housing and fuel	House-hold goods and services	Vehicles, transport and commun-ications	Recreation	Other goods and services	Individual consumption expenditure in the UK[2]	Total individual consumption expenditure[3]
United Kingdom	95,690	32,479	94,341	31,999	89,640	58,485	120,735	523,368	545,123
North East	4,034	1,379	3,390	1,342	3,338	2,128	3,863	19,474	20,416
North West	11,249	3,533	9,652	3,448	9,351	6,632	12,399	56,264	60,101
Yorkshire and the Humber	8,083	2,654	6,917	2,784	6,453	5,007	9,355	41,252	43,872
East Midlands	6,476	1,894	6,089	2,452	5,539	4,061	7,316	33,826	35,983
West Midlands	7,996	2,512	7,662	2,862	7,600	4,913	9,468	43,012	45,357
East	8,380	2,930	9,533	3,077	9,996	5,514	11,484	50,914	53,335
London	12,625	5,698	14,215	4,366	14,265	7,912	20,261	79,340	78,427
South East	12,988	4,397	15,357	4,827	13,514	8,864	19,738	79,685	82,638
South West	7,590	2,169	8,290	2,226	6,551	4,721	10,053	41,600	43,055
England	79,420	27,164	81,105	27,383	76,606	49,752	103,938	445,368	463,183
Wales	4,438	1,430	4,052	1,385	3,580	2,375	4,917	22,177	23,469
Scotland	9,141	2,941	7,371	2,414	7,288	5,184	9,386	43,726	45,634
Northern Ireland[4]	2,691	944	1,812	817	2,166	1,173	2,494	12,097	12,837

1 Based on the European System of Accounts 1995 (ESA95). See Notes and Definitions.
2 Expenditure by UK households and foreign residents in the United Kingdom.
3 Expenditure by UK consumers, including non-profit institutions serving households but excluding expenditure in the United Kingdom by foreign residents.
4 Domestic rates which are levied in Northern Ireland are treated as part of individual consumption expenditure. Council tax levied in Great Britain is treated as a deduction from income, and is therefore not part of individual consumption expenditure. These series are therefore not comparable with those for the regions of Great Britain.

Source: Office for National Statistics

13 Industry and Agriculture

Gross domestic product

The East Midlands had the highest percentage of GDP from industry and the lowest proportion from services in 1997.

(Map13.1)

In 1997 London had the highest percentage of GDP from services and the lowest proportion from industry.

(Map 13.1)

Businesses

Industry as a percentage of total local units ranged from 11.2 per cent in Inverclyde to 26.2 per cent in Leicester UA in 1999.

(Map 13.2)

In 1999 services as a percentage of total local units ranged from 39.7 per cent in the Orkney Islands to 87.6 per cent in Glasgow City.

(Map 13.2)

Almost 25 per cent of business sites in Northern Ireland were in the agriculture sector in 1999.

(Table 13.3)

Manufacturing

The level of gross value added at basic prices in manufacturing in 1997 ranged from £28,600 per person employed in Northern Ireland to £40,300 in the South East.

(Table 13.5)

Wages and salaries per person employed in manufacturing in 1997 were highest in London, at £19,800, and lowest in Northern Ireland, at £14,700.

(Table 13.5)

Exports and imports

In 1999 the percentage of direct exports to the EU ranged from 45 per cent in London to 80 per cent in the North East.

(Table 13.7)

The percentage of direct imports from the EU in 1999 ranged from just over 30 per cent in Scotland to over 64 per cent in the West Midlands.

(Table 13.7)

Investment

The greatest number of direct inward investment project successes in manufacturing in 1998-99 was in the Yorkshire and the Humber region, at 63; London had the highest number of non-manufacturing project successes at 104.

(Table 13.8)

Research and development

Expenditure on research and development in 1998 ranged from over 3 per cent of regional GDP in the East of England to under 1 per cent in the Yorkshire and the Humber region, Wales and Northern Ireland.

(Table 13.9)

Assisted areas

In Great Britain in 1998-99, 39 per cent of government expenditure on regional preferential assistance to industry was in Wales, with a further 32 per cent in Scotland.

(Table 13.10)

Businesses

Business registration rates in 1998 ranged from 20 per 10,000 resident adults in the North East to 70 in London.

(Table 13.12)

Business deregistration rates in 1998 ranged from 20 per 10,000 resident adults in the North East to 50 in London.

(Table 13.12)

Business survival rates three years after registration for businesses registered in 1995 were highest in Northern Ireland at 76 per cent and lowest in the North East at 57 per cent.

(Table 13.13)

Construction

In Great Britain in 1999 new work accounted for 49 per cent of the value of construction contractors' output in the South East (excluding Greater London) and 64 per cent in the North.

(Table 13.14)

Tourism

Almost a fifth of all tourist expenditure by UK residents was in the West Country in 1998.

(Table 13.15)

Overseas residents accounted for 54 per cent of the tourists and 86 per cent of the expenditure incurred in London in 1998.

(Table 13.15)

Agriculture

The highest proportion of GDP from agriculture in 1997 was in Northern Ireland.

(Map 13.16)

Agriculture as a percentage of total legal units was highest at 60.4 per cent in the Orkney Islands in 1999; in contrast Leicester UA had just 0.1 per cent.

(Map 13.17)

In 1999, the East of England had the highest percentage of arable land, at 74.5 per cent; Wales had the lowest percentage at 13.3 per cent.

(Table 13.18)

The North West had the largest proportion of dairy farms, at 22.2 per cent, in 1999.

(Table 13.19)

The East of England, at 490 thousand hectares, had the largest area under wheat in 1999; Scotland had the largest area under barley, at 339 thousand hectares.

(Table 13.20)

In 1999 Wales had more sheep and lambs on agricultural holdings, at 11.8 million, than any other region in the UK.

(Table 13.21)

13.1 Percentage of gross domestic product[1] derived from industry and services, 1997

Industry

Industry as a percentage
of total GDP

- 37.0 or over
- 32.0 to 36.9
- 27.0 to 31.9
- 22.0 to 26.9
- 21.9 or under

Services

Services as a percentage
of total GDP

- 75.0 or over
- 70.0 to 74.9
- 65.0 to 69.9
- 60.0 to 64.9
- 59.9 or under

1 At basic prices. See Notes and Definitions.

Source: Office for National Statistics

13.2 Industry and services[1] local units as a percentage of total local units, 1999[2]

Industry

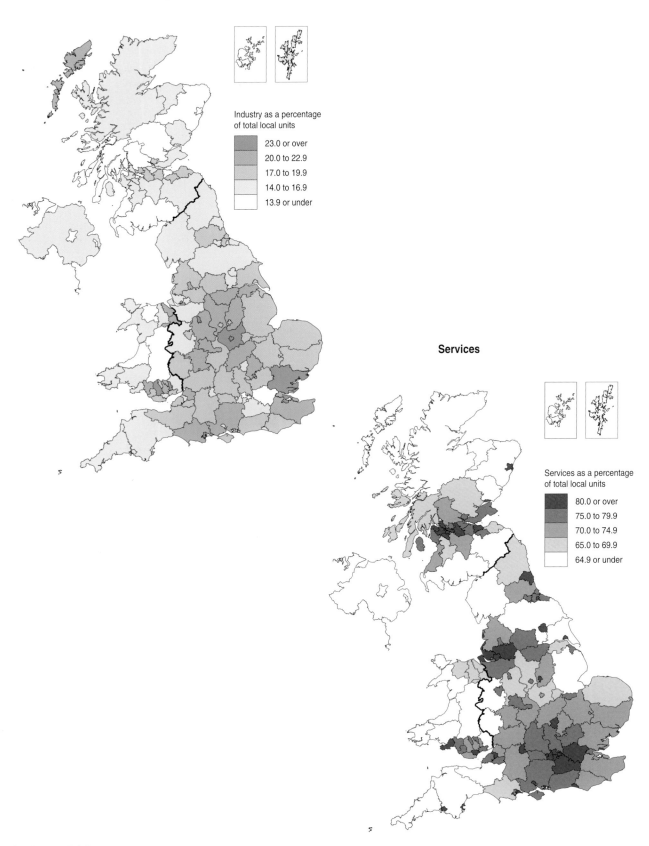

Industry as a percentage
of total local units

- 23.0 or over
- 20.0 to 22.9
- 17.0 to 19.9
- 14.0 to 16.9
- 13.9 or under

Services

Services as a percentage
of total local units

- 80.0 or over
- 75.0 to 79.9
- 70.0 to 74.9
- 65.0 to 69.9
- 64.9 or under

1 See Notes and Definitions.
2 Geographic boundaries relate to the sub-regions (Counties/UAs) in existence on 1 April 1998.

Source: Inter-Departmental Business Register, Office for National Statistics

13.3 Classification[1] of business sites[2], 1999[3]

Percentages and thousands

	Agriculture, hunting, forestry & fishing	Mining & quarrying, energy, water supply & manu-facturing	Con-struction	Distrib-ution, hotels & catering; repairs	Transport & com-munication	Financial intermed-iation, real estate renting & business activities	Education & health	Public admini-stration & other services	Total business sites (=100%) (thousands)
United Kingdom	7.2	8.4	8.9	29.6	4.3	25.1	6.5	10.0	2,508.0
North East	6.3	8.3	8.2	33.8	4.5	19.3	8.3	11.3	75.1
North West	5.5	8.9	8.3	33.4	4.5	22.9	7.1	9.3	253.7
Yorkshire and the Humber	7.4	9.6	8.8	33.3	4.8	19.9	6.8	9.5	187.9
East Midlands	8.1	11.4	9.6	30.5	4.8	20.4	6.4	8.8	169.5
West Midlands	6.9	11.6	9.2	31.0	4.3	21.6	6.6	8.6	208.6
East	6.6	8.9	11.2	27.4	4.8	26.5	5.6	8.9	241.1
London	0.3	6.8	5.8	26.7	3.9	38.7	5.5	12.3	378.4
South East	3.9	7.9	10.1	26.8	3.9	31.0	6.3	10.1	373.3
South West	11.7	7.8	10.0	29.5	3.8	21.6	6.6	8.9	230.7
England	5.6	8.7	9.0	29.5	4.3	26.7	6.4	9.9	2,118.3
Wales	16.5	7.2	9.5	30.6	4.2	15.5	6.3	10.1	115.1
Scotland	11.9	6.7	7.8	31.5	4.2	19.1	6.8	11.9	198.4
Northern Ireland	24.8	6.5	9.8	26.5	3.6	10.8	9.8	8.2	76.2

1 Based on SIC 1992. See Notes and Definitions.
2 Registered for VAT and/or PAYE, local unit basis eg an individual factory or shop. See Notes and Definitions.
3 At March.

Source: Inter-Departmental Business Register, Office for National Statistics

13.4 Manufacturing[1] industry business sites[2]: by employment size band[3], 1999[4]

Percentages and thousands

	Percentage of manufacturing local units with an employment sizeband[3] of								Total manu-facturing local units (=100%) (thousands)
	1-9	10-19	20-49	50-99	100-199	200-499	500-999	1,000 or over	
United Kingdom	73.7	11.2	7.8	3.3	2.2	1.4	0.3	0.1	205.0
North East	67.4	12.0	9.3	4.3	3.6	2.5	0.8	0.2	6.0
North West	69.5	12.7	8.9	3.8	2.9	1.7	0.4	0.2	22.1
Yorkshire and the Humber	69.7	12.2	9.1	4.0	2.7	1.7	0.5	0.1	17.4
East Midlands	70.5	12.2	8.8	3.9	2.7	1.5	0.3	0.2	18.8
West Midlands	69.7	12.7	9.2	3.8	2.5	1.5	0.3	0.1	23.9
East	76.7	10.2	7.0	2.9	1.7	1.1	0.3	0.1	20.9
London	82.2	9.4	5.0	1.6	0.9	0.6	0.1	0.1	25.2
South East	78.4	9.6	6.4	2.7	1.6	1.0	0.2	0.1	28.6
South West	77.1	10.2	6.3	2.7	1.9	1.3	0.3	0.1	17.5
England	74.3	11.1	7.5	3.2	2.1	1.3	0.3	0.1	180.3
Wales	71.0	11.0	8.3	3.9	3.1	2.0	0.6	0.3	7.9
Scotland	68.0	12.4	9.8	4.0	3.0	1.9	0.6	0.2	12.1
Northern Ireland	69.0	12.3	10.0	3.8	2.4	1.8	0.2	0.2	4.7

1 Based on SIC 1992 Section D. See Notes and Definitions.
2 Registered for VAT and/or PAYE, local unit basis eg individual factory. See Notes and Definitions.
3 Includes paid full and part-time employees and working proprietors.
4 At March.

Source: Inter-Departmental Business Register, Office for National Statistics

13.5 Turnover, expenditure and gross value added in manufacturing[1], 1997

£ million and £ per person employed

	Total turnover (£ million)	Purchases of goods and services (£ million)	Wages and salaries		Net capital expenditure		Gross value added at basic prices	
			£ million	£ per person employed	£ million	£ per person employed	£ million	£ per person employed
United Kingdom	451,877	286,580	73,475	17,564	20,091	4,673	144,519	33,615
North East	21,518	14,370	3,350	17,676	1,162	6,026	6,475	33,564
North West	59,537	36,465	9,412	17,611	2,759	5,041	18,758	34,270
Yorkshire and the Humber	41,208	27,285	7,143	17,017	1,732	4,030	13,139	30,570
East Midlands	36,968	21,548	6,760	16,073	1,662	3,850	13,637	31,595
West Midlands	50,843	32,847	9,351	16,759	2,484	4,338	17,582	30,709
East	39,641	26,088	6,619	18,447	1,606	4,340	12,597	34,045
London	32,541	20,623	5,656	19,833	1,299	4,315	11,381	37,812
South East	57,887	34,033	8,629	19,101	2,148	4,604	18,822	40,338
South West	31,667	20,328	5,444	17,661	1,429	4,507	10,520	33,178
England	371,810	233,587	62,365	17,685	16,281	4,486	122,910	33,870
Wales	29,391	18,796	3,737	17,309	1,427	6,475	7,106	32,242
Scotland	40,281	27,911	5,770	17,388	1,894	5,591	11,325	33,435
Northern Ireland	10,395	6,286	1,604	14,691	489	4,400	3,177	28,561

1 Based on SIC 1992 Section D. See Notes and Definitions.

Source: Annual Business Inquiry, Office for National Statistics

13.6 Gross value added in manufacturing[1]: by size of local unit, 1997

Percentages and £ million

	Percentage of gross value added by number employed[2]							Total (=100%) (£ million)
	1-19	20-49	50-99	100-199	200-499	500-999	1,000 or over	
United Kingdom	14.0	10.8	10.3	14.6	21.9	12.7	15.6	144,519
North East	9.4	8.0	8.5	14.0	27.8	19.3	13.0	6,475
North West	12.2	10.9	9.8	13.6	21.0	10.5	22.0	18,758
Yorkshire and the Humber	13.5	11.6	12.6	15.9	21.3	16.1	8.9	13,139
East Midlands	13.3	11.1	10.5	16.4	22.1	12.7	13.9	13,637
West Midlands	14.2	11.9	11.1	13.5	20.0	9.9	19.4	17,582
East	15.8	11.7	10.3	17.9	20.0	13.3	11.0	12,597
London	21.3	11.6	10.3	12.4	17.0	8.5	18.9	11,381
South East	14.3	10.3	9.7	13.5	26.7	13.6	11.9	18,822
South West	15.4	10.0	10.1	15.6	23.0	12.4	13.6	10,520
England	14.4	11.0	10.4	14.7	21.9	12.5	15.2	122,910
Wales	10.2	8.7	9.1	16.2	23.2	12.4	20.2	7,106
Scotland	11.3	10.0	10.2	13.3	22.3	16.2	16.6	11,325
Northern Ireland	16.0	11.7	10.8	14.1	18.0	11.0	18.5	3,177

1 Based on SIC 1992 Section D. See Notes and Definitions.
2 Average numbers employed during the year, including full and part-time employees and working proprietors.

Source: Annual Business Inquiry, Office for National Statistics

13.7 Export and import trade with EU and non-EU countries[1], 1999

£ million, Percentages and numbers

Exports

	£ million			Percentages		As a percentage of UK regional share of export trade			Average number of companies exporting[2]	
	Total	To the EU	Outside the EU	To the EU	Outside the EU	All export trade	To the EU	Outside the EU	To the EU[3]	Outside the EU[3]
United Kingdom	138,723	84,159	54,564	60.7	39.3	100.0	100.0	100.0	13,682	38,294
North East	5,668	4,554	1,114	80.3	19.7	4.1	5.4	2.0	360	818
North West	10,722	7,244	3,478	67.6	32.4	7.7	8.6	6.4	1,416	3,655
Yorkshire and the Humber	7,634	4,946	2,688	64.8	35.2	5.5	5.9	4.9	1,165	2,996
East Midlands	10,134	5,495	4,639	54.2	45.8	7.3	6.5	8.5	1,006	2,526
West Midlands	12,170	7,209	4,961	59.2	40.8	8.8	8.6	9.1	1,376	3,714
East	15,511	8,248	7,263	53.2	46.8	11.2	9.8	13.3	1,620	4,576
London	20,553	9,235	11,318	44.9	55.1	14.8	11.0	20.7	1,610	7,310
South East	23,865	14,311	9,554	60.0	40.0	17.2	17.0	17.5	2,333	6,970
South West	8,168	5,620	2,548	68.8	31.2	5.9	6.7	4.7	928	2,595
England	114,425	66,862	47,563	58.4	41.6	82.5	79.4	87.2	11,812	35,159
Wales	5,918	4,248	1,670	71.8	28.2	4.3	5.0	3.1	485	962
Scotland	15,649	11,260	4,389	72.0	28.0	11.3	13.4	8.0	789	1,870
Northern Ireland	2,731	1,789	942	65.5	34.5	2.0	2.1	1.7	596	303

Imports

	£ million			Percentages		As a percentage of UK regional share of import trade			Average number of companies importing[2]	
	Total	From the EU	From outside the EU	From the EU	From outside the EU	All import trade	From the EU	From outside the EU	From the EU[3]	From outside the EU[3]
United Kingdom	182,477	93,887	88,590	51.5	48.5	100.0	100.0	100.0	16,897	52,100
North East	3,856	1,998	1,858	51.8	48.2	2.1	2.1	2.1	373	993
North West	12,069	5,192	6,877	43.0	57.0	6.6	5.5	7.8	1,769	4,965
Yorkshire and the Humber	9,045	4,710	4,335	52.1	47.9	5.0	5.0	4.9	1,409	3,537
East Midlands	8,586	4,335	4,251	50.5	49.5	4.7	4.6	4.8	1,313	3,198
West Midlands	13,208	8,510	4,698	64.4	35.6	7.2	9.1	5.3	1,813	4,509
East	26,102	16,317	9,785	62.5	37.5	14.3	17.4	11.0	1,987	6,205
London	36,637	15,270	21,367	41.7	58.3	20.1	16.3	24.1	2,354	10,944
South East	45,072	27,690	17,382	61.4	38.6	24.7	29.5	19.6	2,995	9,767
South West	9,123	3,660	5,463	40.1	59.9	5.0	3.9	6.2	1,041	3,437
England	163,698	87,682	76,016	53.6	46.4	89.7	93.4	85.8	15,053	47,553
Wales	5,026	1,654	3,372	32.9	67.1	2.8	1.8	3.8	465	1,209
Scotland	11,025	3,366	7,659	30.5	69.5	6.0	3.6	8.6	793	2,572
Northern Ireland	2,728	1,185	1,543	43.4	56.6	1.5	1.3	1.7	587	766

1 See Notes and Definitions.
2 Over four quarters of 1999.
3 Companies who trade with both EU countries and countries outside the EU will appear more than once in the company count.

Source: HM Customs and Excise

13.8 Direct inward investment[1]: project successes[2]

Numbers

	Manufacturing					Non-manufacturing				
	1994-95	1995-96	1996-97	1997-98	1998-99	1994-95	1995-96[3]	1996-97	1997-98	1998-99
DTI Regions[4]										
United Kingdom	324	347	317	356	311	130	149	180	272	353
North East	31	49	36	35	28	10	13	10	12	7
North West	42	24	27	43	42	12	13	19	28	24
Yorkshire and the Humber	15	32	31	45	63	11	12	5	20	23
East Midlands	28	11	16	10	8	16	11	7	8	11
West Midlands	46	62	49	49	41	16	15	27	32	30
East	6	7	3	13	8	5	9	6	20	33
London	0	1	2	1	1	3	22	38	61	104
South East	17	25	15	21	23	20	25	31	36	51
South West	18	13	21	22	16	3	4	9	18	18
England	203	224	200	239	230	96	124	152	235	301
Wales	41	44	43	50	35	10	9	2	5	13
Scotland	63	57	53	48	26	18	15	23	27	28
Northern Ireland	17	22	21	19	20	6	1	3	5	11

1 See Notes and Definitions.
2 A project success is defined as a case where an overseas company specifies an interest and successfully completes investment in a UK company.
3 The UK figure includes one UK-wide project.
4 See map on page 240.

Source: Invest·UK, Department of Trade and Industry

13.9 Expenditure on research and development, 1998

£ million and percentages

	Expenditure within (£ million)			Expenditure as a percentage of regional GDP		
	Businesses	Government[1]	Higher education institutions	Businesses	Government[1]	Higher education institutions
United Kingdom	10,231	2,073	3,040	1.2	0.2	0.4
North East	178	3	105	0.6	-	0.4
North West	1,224	58	238	1.4	0.1	0.3
Yorkshire and the Humber	287	31	241	0.5	-	0.4
East Midlands	775	51	159	1.4	0.1	0.3
West Midlands	708	182	167	1.0	0.3	0.2
East	2,367	255	211	2.7	0.3	0.2
London	614	202	775	0.5	0.2	0.6
South East	2,542	698	460	1.9	0.5	0.3
South West	907	329	138	1.4	0.5	0.2
England	9,601	1,809	2,494	1.3	0.3	0.3
Wales	125	51	113	0.4	0.2	0.3
Scotland	424	200	375	0.6	0.3	0.5
Northern Ireland	81	12	57	0.4	0.1	0.3

1 Figures include estimates of NHS and local authorities' research and development.

Source: Office for National Statistics

13.10 Government expenditure on regional preferential assistance to industry

£ million

	1989-90	1990-91	1991-92	1992-93	1993-94	1994-95	1995-96	1996-97	1997-98	1998-99
Great Britain[1]	539.3	497.3	427.8	364.0	394.4	368.9	343.0	371.1	430.4	393.8
North East	117.0	85.0	63.8	48.3	52.7	38.4	46.4	24.3	38.1	22.3
North West	74.3	57.5	49.5	36.8	40.3	32.4	24.3	23.2	19.4	25.9
Yorkshire and the Humber	32.4	29.4	18.2	13.7	35.6	23.0	19.7	11.1	12.7	11.9
East Midlands	9.5	5.5	2.6	1.2	1.9	5.2	7.3	10.5	10.5	7.1
West Midlands	19.9	18.0	8.7	10.8	14.4	14.7	14.2	25.5	29.8	30.6
East	.	.	.	.	.	0.7	2.1	1.5	2.2	0.7
London	.	.	.	.	.	0.6	1.7	2.9	2.7	3.2
South East	.	.	.	.	.	0.9	4.2	4.1	5.4	3.3
South West	10.7	9.0	8.3	8.2	9.5	9.4	7.7	7.4	4.5	9.4
England[2]	263.8	204.4	151.1	119.0	154.4	125.3	127.6	110.5	125.3	114.4
Wales	131.7	133.7	153.9	140.6	118.8	109.2	98.0	132.4	172.6	153.9
Scotland	143.8	159.2	122.8	104.4	121.2	134.4	117.4	128.2	132.5	125.5
Northern Ireland	127.1	132.1	138.0	105.6	117.6	132.9	131.2	137.1	156.1	153.3

1 The system of assistance available in Northern Ireland is not comparable with that operating in Great Britain, and thus UK figures are not produced. See Notes and Definitions.
2 Payments for European Regional Incentives, General Consultancy Contracts and Regional Selective Assistance Payments to the European Commission are not included.

Source: Department of Trade and Industry, Department of Economic Development, Northern Ireland

13.11 Allocation of EU Structural Funds[1]

£ million at 1999 prices

	Objective 1[2]			Objective 2[2]			Objective 5b[2,3]			Objectives 1,2 and 5b		
	1998	1999	2000	1998	1999	2000	1998	1999	2000	1998	1999	2000
United Kingdom	346	375	596	644	663	485	117	118	.	1,107	1,156	1,081
North East	.	.	.	92	94	66	4	4	.	96	98	66
North West	120	130	126	112	115	85	4	4	.	236	249	211
Yorkshire and the Humber	.	.	111	94	97	50	7	7	.	101	104	161
East Midlands	.	.	.	25	27	36	9	9	.	34	36	36
West Midlands	.	.	.	114	117	86	6	7	.	120	124	86
East	.	.	.	.	.	15	9	9	.	9	9	15
London	.	.	.	25	27	23	.	.	.	25	27	23
South East	.	.	.	5	5	4	.	.	.	5	5	4
South West	.	.	47	9	9	18	31	32	.	40	41	65
England	120	130	284	476	491	383	70	71	.	666	692	667
Wales	.	.	176	55	56	14	26	26	.	81	82	190
Scotland	45	49	35	113	116	88	21	21	.	179	186	123
Northern Ireland	181	196	101	.	.	.	.	.	.	181	196	101

1 Only allocations resulting from the Commission's Single Programming Documents are shown. Allocations resulting from Community Initiatives, the value of which is about 8 per cent of the total Objective 1, 2 and 5b allocations, are not included because not all of these can be allocated to the Government Office Regions in the table.
2 See Notes and Definitions.
3 For the Structural Funds programme beginning in 2000, Objective 5b has been subsumed into Objective 2.

Source: Department of Trade and Industry

13.12 Business registrations and deregistrations[1]

Thousands and rates

	1997						1998					
	Regist-rations	Deregist-rations	Net change	End-year stock	Regist-ration rates[2]	Deregist-ration rates[2]	Regist-rations	Deregist-rations	Net change	End-year stock	Regist-ration rates[2]	Deregist-ration rates[2]
United Kingdom	182.6	164.5	18.1	1,621.3	39	35	186.3	155.9	30.3	1,651.6	40	33
North East	4.2	4.4	-0.2	41.8	20	21	4.2	4.0	0.2	42.0	20	20
North West	18.1	17.1	1.0	157.6	33	31	18.6	16.1	2.5	160.1	34	30
Yorkshire and the Humber	11.8	12.3	-0.4	117.2	30	31	11.8	11.4	0.5	117.7	30	28
East Midlands	11.7	11.2	0.5	110.0	36	34	11.9	10.7	1.2	111.2	36	32
West Midlands	13.6	13.9	-0.3	134.6	32	33	15.0	13.3	1.7	136.3	36	32
East	18.3	15.8	2.5	160.0	43	37	17.9	15.2	2.7	162.7	42	36
London	37.2	28.3	8.9	258.7	66	50	39.7	28.4	11.3	270.0	70	50
South East	30.0	25.7	4.3	246.1	47	41	29.9	23.0	6.9	253.0	47	36
South West	15.3	14.4	0.9	148.0	39	37	15.5	13.8	1.7	149.7	40	35
England	160.2	143.0	17.2	1,374.0	41	37	164.6	135.9	28.7	1,402.7	42	35
Wales	6.2	6.3	-0.1	75.3	27	27	6.1	6.2	-0.1	75.2	26	27
Scotland	12.3	11.5	0.7	118.3	30	28	11.8	10.9	0.9	119.2	29	26
Northern Ireland	3.9	3.6	0.2	53.8	31	29	3.7	2.8	0.9	54.6	30	23

1 Enterprises registered for VAT. See Notes and Definitions.
2 Registrations and deregistrations during the year per 10,000 of the resident adult population.

Source: Department of Trade and Industry

13.13 Business survival rates

Percentages

	The percentage of businesses surviving the stated number of months after year of registration								
	12 months				24 months			36 months	
	1994	1995	1996	1997	1994	1995	1996	1994	1995
United Kingdom	85.3	86.7	86.9	88.4	69.9	72.1	72.7	58.8	61.0
North East	82.1	83.9	85.6	88.2	64.8	68.9	71.2	54.1	57.0
North West	83.2	84.9	84.5	86.2	66.8	69.8	70.0	55.7	58.3
Yorkshire and the Humber	84.0	86.2	85.8	87.4	67.8	71.0	70.5	57.0	59.7
East Midlands	85.0	86.0	86.3	87.6	69.1	71.1	71.4	57.6	60.0
West Midlands	84.1	84.8	85.6	86.1	69.0	69.1	71.3	57.6	58.3
East	86.6	87.5	88.5	89.7	70.9	73.5	74.7	59.6	62.0
London	85.4	85.6	86.1	88.8	69.8	70.8	71.5	58.2	59.6
South East	85.9	88.2	88.5	89.5	71.1	73.7	74.7	60.2	62.8
South West	86.4	88.4	88.5	89.8	71.3	74.3	75.1	60.5	63.6
England	85.1	86.4	86.8	88.4	69.5	71.7	72.5	58.3	60.5
Wales	85.9	86.8	86.9	87.9	71.1	72.0	72.6	59.4	61.7
Scotland	85.7	87.6	85.8	87.2	70.2	72.9	72.1	59.1	61.7
Northern Ireland	91.5	93.0	93.0	92.7	82.3	84.6	83.3	74.5	76.2

Source: Department of Trade and Industry

13.14 Construction: value at current prices of contractors' output[1]

£ million and percentages

	Total work (£ million)						Of which new work (percentages)					
	1994	1995	1996	1997	1998	1999	1994	1995	1996	1997	1998	1999
Standard Statistical Regions												
Great Britain	45,870	48,942	51,969	55,191	59,027	62,416	54.0	53.9	53.4	53.7	54.8	56.3
North	2,139	2,186	2,506	2,601	2,887	2,883	58.5	56.9	61.1	59.4	60.6	63.7
North West (SSR)	4,576	5,072	5,412	5,577	5,785	5,984	55.0	55.4	54.0	52.4	52.6	54.7
Yorkshire and Humberside	3,703	3,862	4,324	4,576	4,843	5,235	53.3	50.2	51.7	51.4	52.9	54.0
East Midlands	3,301	3,446	3,827	4,252	4,382	4,425	56.5	55.2	56.6	58.3	58.2	58.8
West Midlands	4,039	4,157	4,399	4,813	5,248	5,404	52.7	51.8	50.1	50.9	54.2	56.8
East Anglia	1,977	2,064	2,021	2,137	2,273	2,334	52.7	52.9	51.0	53.3	52.4	54.9
South East (SSR)	15,598	16,690	17,965	19,094	20,956	23,017	51.6	52.0	51.8	52.3	53.9	54.9
Greater London	6,118	6,917	7,428	7,829	8,954	9,823	55.8	57.8	59.2	58.7	60.4	62.6
Rest of South East	9,481	9,774	10,537	11,265	12,002	13,194	48.8	47.8	46.6	47.9	49.0	49.1
South West	4,055	4,317	4,193	4,427	4,719	4,921	54.4	54.0	48.5	49.3	50.2	51.6
England	39,389	41,794	44,647	47,476	51,093	54,203	53.4	53.0	52.5	52.8	54.0	55.5
Wales	2,172	2,377	2,331	2,533	2,591	2,599	59.9	61.2	57.7	58.4	58.2	61.5
Scotland	4,310	4,771	4,991	5,182	5,343	5,614	57.1	58.5	59.6	60.1	60.4	62.4

1 Output of contractors, including estimates of unrecorded output by small firms and self-employed workers, classified to construction in SIC 1992. For new work, figures relate to the region in which the site is located; for repair and maintenance, figures are for the region in which the reporting unit is based.

Source: Department of the Environment, Transport and the Regions

13.15 Tourism, 1991 and 1998

Millions and £ million

	1991				1998			
	UK residents[1]		Overseas residents[2]		UK residents[1]		Overseas residents[2]	
	Number of tourists (millions)	Expenditure (£ million)	Number of tourists (millions)	Expenditure (£ million)	Number of tourists (millions)	Expenditure (£ million)	Number of tourists (millions)	Expenditure (£ million)
Tourist Board Regions[3]								
United Kingdom	94.4	10,470	17.1	7,305	122.3	14,030	25.7	12,671
Northumbria	3.4	255	0.3	95	4.2	340	0.5	139
Cumbria	2.7	330	0.2	35	2.9	380	0.3	49
North West	8.3	770	1.0	246	8.4	970	1.3	423
Yorkshire	7.4	680	0.9	183	9.2	940	1.0	307
East of England	9.9	925	1.4	372	13.0	1,290	1.8	628
Heart of England	11.8	950	1.7	388	16.8	1,160	2.2	668
London	6.6	720	9.2	3,924	11.6	1,055	13.5	6,736
Southern	9.4	920	1.7	473	10.9	1,245	2.2	829
South East England	6.4	600	2.0	552	10.5	825	2.5	885
West Country	12.9	1,765	1.4	306	16.6	2,670	1.7	529
England	76.0	7,925	15.1	6,595	101.9	10,880	21.9	11,204
Wales	8.7	900	0.7	133	9.8	1,100	0.8	176
Scotland	8.2	1,190	1.6	501	9.8	1,540	2.1	945
Northern Ireland	1.4	145	0.1	26	0.8	200	0.1	50

1 The United Kingdom figures include the value of tourism in the Channel Islands, the Isle of Man, and a small amount where the region was unknown.
2 The England figures include the value of tourism in the Channel Islands, the Isle of Man, and a small amount where the region was unknown. The United Kingdom figures also include an amount which cannot be allocated to an individual country. The Northern Ireland figures include the value of tourism created by visitors from the Republic of Ireland.
3 See map on page 240.

Source: National Tourist Boards; International Passenger Survey, Office for National Statistics

13.16 Percentage of gross domestic product[1] derived from agriculture, 1997

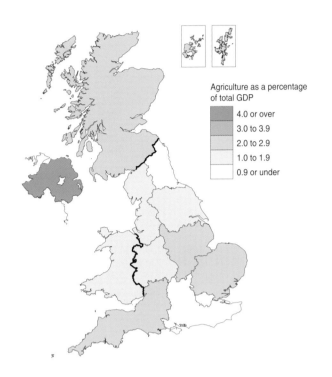

Agriculture as a percentage of total GDP

4.0 or over
3.0 to 3.9
2.0 to 2.9
1.0 to 1.9
0.9 or under

1 At basic prices. See Notes and Definitions.

Source: Office for National Statistics

13.17 Agricultural legal units as a percentage of total legal units[1,2], 1999

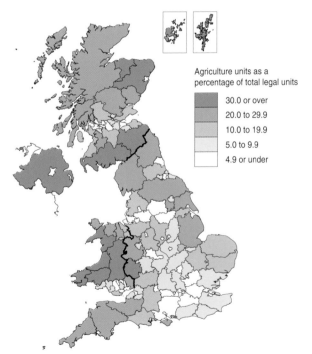

Agriculture units as a percentage of total legal units

30.0 or over
20.0 to 29.9
10.0 to 19.9
5.0 to 9.9
4.9 or under

1 The figures include only those enterprises that are registered for VAT. Some smaller holdings will therefore not be included. See Notes and Definitions.
2 Geographic boundaries relate to the sub-regions (Counties/UAs) in existence on 1 April 1998.

Source: Inter-Departmental Business Register, Office for National Statistics

13.18 Agricultural holdings[1]: by area of crops and grass, and by land use, June 1999

Percentages and numbers

	None[2]	Under 10 hectares	10-49.9 hectares	50 hectares or over	Total holdings (=100%) (numbers)	Arable land[3]	Grass five years old and over (including sole right rough grazing)	Set-aside land	Other land on agricultural holdings including woodland[4]	Total area on agricultural holdings (= 100%) (thousand hectares)
United Kingdom	4.9	26.4	36.8	31.9	239,538	33.9	55.1	3.3	7.7	17,490
North East	4.5	21.3	23.5	50.7	5,280	31.1	61.6	3.9	3.4	574
North West	4.7	27.2	36.7	31.3	17,609	22.1	73.9	1.0	3.0	883
Yorkshire and the Humber	4.5	26.6	31.9	37.0	16,204	52.4	38.8	5.5	3.3	1,083
East Midlands	3.1	24.2	34.2	38.5	15,715	65.4	23.9	7.2	3.5	1,220
West Midlands	4.0	29.3	36.2	30.5	19,161	49.3	42.0	4.3	4.4	942
East	5.4	27.9	27.7	39.0	17,064	74.5	11.0	8.3	6.2	1,460
South East (inc. London)	7.1	31.5	32.1	29.2	19,323	53.8	30.1	7.4	8.7	1,194
South West	3.7	30.0	37.1	29.3	36,864	39.9	51.5	3.5	5.1	1,795
England	4.6	28.2	33.8	33.4	147,220	50.8	38.9	5.4	4.9	9,151
Wales	5.0	21.0	41.1	32.9	28,018	13.3	82.5	0.4	3.8	1,493
Scotland	9.5	26.7	28.5	35.2	33,168	15.4	69.1	1.2	14.3	5,773
Northern Ireland	1.5	22.3	55.6	20.5	31,132	18.4	79.8	0.2	1.7	1,073

1 Includes estimates for minor holdings; figures for English regions exclude minor holdings hence their sum may be less than the England total. See Notes and Definitions.
2 These holdings consist only of rough grazing, woodland or other land.
3 Crops, bare fallow and all grass under five years old.
4 In Great Britain this includes farm roads, yards, buildings (except glasshouses), ponds and derelict land. In Northern Ireland it includes land under bog, water, roads, buildings etc and wasteland not used for agriculture.

Source: Ministry of Agriculture, Fisheries and Food; National Assembly for Wales; The Scottish Executive Rural Affairs Department; Department of Agriculture and Rural Development for Northern Ireland

13.19 Agricultural holdings by farm type, June 1999

Percentages and numbers

	Cereals	General cropping	Horti-culture	Pigs and poultry	Dairy	Cattle and sheep (LFA[1])	Cattle and sheep (Low-ground)	Mixed	Other	Total holdings (=100%) (numbers)
United Kingdom	10.7	5.8	3.9	2.9	11.2	22.1	17.1	6.3	20.1	239,538
North East	17.0	2.9	1.8	2.2	5.2	30.0	13.3	10.5	17.1	5,280
North West	3.0	3.9	4.6	3.2	22.2	16.7	20.1	3.8	22.7	17,609
Yorkshire and the Humber	17.1	11.1	3.3	5.3	9.2	13.1	14.1	8.9	18.0	16,204
East Midlands	24.5	13.5	5.0	3.4	7.8	4.9	16.5	7.9	16.5	15,715
West Midlands	8.9	7.2	5.1	2.9	12.8	6.3	26.5	9.1	21.2	19,161
East	30.7	21.6	8.4	6.0	1.4	0.0	8.3	5.9	17.9	17,064
London	11.8	4.0	21.6	5.0	3.5	0.0	11.8	6.0	36.3	399
South East	16.7	3.4	9.9	3.8	4.5	0.0	25.6	7.6	28.5	18,921
South West	7.3	1.7	4.4	3.0	16.0	5.5	29.2	7.2	25.7	36,864
England	14.2	7.6	5.6	3.7	11.1	7.2	21.2	7.3	22.1	147,220
Wales	1.0	0.6	1.2	1.5	13.0	46.4	12.7	2.4	21.1	28,018
Scotland	11.5	6.4	1.4	1.6	5.5	37.1	3.6	6.9	26.1	33,168
Northern Ireland	1.7	1.2	1.2	1.9	16.2	54.3	16.0	4.2	3.4	31,132

1 Less favoured areas. See Notes and Definitions.

Source: Ministry of Agriculture, Fisheries and Food; National Assembly for Wales; The Scottish Executive Rural Affairs Department; Department of Agriculture and Rural Development for Northern Ireland

13.20 Areas and estimated yields of selected crops[1], 1998 and 1999

Thousand hectares and tonnes per hectare

| | Area (thousand hectares) | | | | | | Estimated yields (tonnes per hectare) | | | | | |
| | Wheat | | Barley | | Rape (for oilseed)[2] | | Wheat | | Barley | | Rape (for oilseed)[2] | |
	1998	1999	1998	1999	1998	1999	1998	1999	1998	1999	1998	1999
United Kingdom	2,045	1,847	1,255	1,178	506	417	7.6	8.0	5.3	5.6	3.0	3.2
North East	76	64	46	44	27	22	6.5	7.4	5.2	5.6	..	..
North West	29	23	43	44	5	4	6.2	6.7	4.5	4.8	..	..
Yorkshire and the Humber	255	234	130	122	58	47	7.7	8.2	5.8	6.1	..	..
East Midlands	393	362	115	98	100	90	7.8	8.4	5.6	6.1	..	..
West Midlands	165	150	84	74	32	28	7.1	7.6	5.6	5.2	..	..
East	525	490	195	174	97	87	8.2	8.4	5.9	5.8	..	..
South East (inc. London)	265	243	107	94	79	61	7.3	7.9	5.5	5.8	..	..
South West	199	181	135	124	38	26	6.7	7.4	4.9	5.4	..	..
England	1,911	1,746	856	774	437	363	7.6	8.1	5.5	5.7	3.0	3.3
Wales	16	13	30	30	4	2	6.4	6.1	4.6	5.2	3.0	3.3
Scotland	111	84	334	339	65	51	7.4	7.8	4.9	5.4	2.8	3.2
Northern Ireland	7	3	35	36	1	0	6.9	6.8	4.4	4.7	2.8	3.0

1 Figures include minor holdings. Figures for English regions exclude minor holdings hence their sum may be less than the England total.
2 Excludes crops grown on set-aside scheme land.

Source: Ministry of Agriculture, Fisheries and Food; National Assembly for Wales; The Scottish Executive Rural Affairs Department; Department of Agriculture and Rural Development for Northern Ireland

13.21 Livestock on agricultural holdings[1], June 1999

Thousands

| | Cattle and calves | | | | | Poultry | | |
	Total herd[2]	Dairy cows	Beef cows	Sheep and lambs	Pigs	Total fowls[3]	Total laying flock[4]	Total Poultry
United Kingdom	11,423	2,440	1,924	44,656	7,284	149,867	38,841	165,242
North East	328	28	91	2,507	109	1,758	274	1,765
North West	1,095	362	106	3,900	274	8,461	3,291	8,541
Yorkshire and the Humber	622	136	96	2,633	1,839	10,952	2,482	11,553
East Midlands	576	127	79	1,664	624	17,591	5,736	18,193
West Midlands	859	240	104	2,945	402	17,555	4,077	17,579
East	256	43	50	456	1,565	24,523	3,763	25,891
South East (inc London)	570	134	85	1,770	542	14,857	5,721	14,951
South West	1,987	589	210	4,156	794	18,716	6,119	18,786
England	6,342	1,661	835	20,274	6,163	114,623	31,624	128,667
Wales	1,318	279	232	11,768	82	9,885	1,414	10,588
Scotland	2,044	214	525	9,705	549	10,829	2,881	10,938
Northern Ireland	1,719	286	332	2,909	490	14,530	2,921	15,048

1 Regional figures do not include minor holdings. Therefore they may not add up to the country and UK totals.
2 Includes bulls, in-calf heifers and fattening cattle and calves.
3 Excludes ducks, geese and turkeys
4 Excludes growing pullets (from day-old to point of lay).

Source: Ministry of Agriculture, Fisheries and Food; National Assembly for Wales; The Scottish Executive Rural Affairs Department; Department of Agriculture and Rural Development for Northern Ireland

14 Sub-regions of England

Government Office Regions, Counties and Unitary Authorities in England[1]

Unitary Authorities at 1 April 1998

1 Hartlepool
2 Middlesbrough
3 Redcar and Cleveland
4 Stockton-on-Tees
5 Darlington
6 Halton
7 Warrington
8 Blackburn with Darwen
9 Blackpool
10 East Riding of Yorkshire
11 City of Kingston upon Hull
12 North East Lincolnshire
13 North Lincolnshire
14 York
15 Derby
16 Leicester
17 Rutland
18 Nottingham
19 County of Herefordshire
20 Telford and Wrekin
21 Stoke-on-Trent
22 Luton
23 Peterborough
24 Southend-on-Sea
25 Thurrock
26 Bracknell Forest
27 Reading
28 Slough
29 West Berkshire (Newbury)
30 Windsor and Maidenhead
31 Wokingham
32 Milton Keynes
33 Brighton and Hove
34 Portsmouth
35 Southampton
36 Isle of Wight
37 Medway
38 Bath and North
    East Somerset
39 City of Bristol
40 North Somerset
41 South Gloucestershire
42 Plymouth
43 Torbay
44 Bournemouth
45 Poole
46 Swindon

1 Local government structure as at 1 April 1998. The unitary authorities are listed in the same order in which they are presented in Tables 14.1, 14.2, 14.4, 14.5 and 14.6.
  See Notes and Definitions.

14.1 Area and population: by local authority[1], 1998

	Area (sq km)	Persons per sq km	Population (thousands) Males	Population (thousands) Females	Population (thousands) Total	Total population percentage change 1981-1998	Total fertility rate (TFR)[2]	Standardised mortality ratio (UK=100) (SMR)[3]	Percentage of population aged Under 5	Percentage of population aged 5-15	Percentage of population aged 16 up to pension age[4]	Percentage of population aged Pension age[4] or over
United Kingdom	242,910	244	29,128	30,108	59,237	5.1	1.71	100	6.2	14.2	61.4	18.1
England	130,422	379	24,378	25,117	49,495	5.7	1.72	98	6.2	14.2	61.5	18.1
North East	8,592	301	1,268	1,321	2,590	-1.8	1.66	114	5.9	14.4	61.0	18.7
Darlington UA	197	515	50	52	101	2.9	1.74	110	6.0	14.5	60.4	19.1
Hartlepool UA	94	978	45	47	92	-3.1	1.81	113	6.4	15.3	60.0	18.3
Middlesbrough UA	54	2,687	71	74	145	-3.6	1.74	110	6.6	16.1	60.6	16.7
Redcar & Cleveland UA	245	565	68	70	138	-8.3	1.90	110	6.1	15.1	60.1	18.7
Stockton-on-Tees UA	204	887	89	91	181	4.1	1.71	115	6.2	15.6	61.9	16.4
Durham County	2,232	227	249	257	506	-1.2	1.67	114	5.7	14.1	61.5	18.7
Chester-le-Street	68	841	28	29	57	8.6	1.62	112	5.9	13.3	63.9	16.9
Derwentside	271	326	43	45	88	-0.2	1.67	128	5.8	13.9	61.0	19.2
Durham	187	483	44	46	90	2.7	1.47	104	4.9	13.6	64.6	16.8
Easington	145	645	46	48	93	-7.9	2.00	124	6.3	15.5	59.3	18.8
Sedgefield	217	411	44	45	89	-4.5	1.98	110	5.8	14.5	61.1	18.6
Teesdale	840	30	12	13	25	0.7	1.54	90	5.4	12.0	60.3	22.3
Wear Valley	505	125	30	33	63	-1.7	1.68	113	5.5	14.0	59.8	20.6
Northumberland	5,026	62	152	158	310	3.4	1.65	112	5.4	13.6	61.4	19.6
Alnwick	1,079	29	15	16	31	8.6	1.81	110	5.2	13.4	59.5	21.9
Berwick-upon-Tweed	972	27	13	14	27	1.4	1.36	90	4.6	11.7	58.5	25.2
Blyth Valley	70	1,153	40	41	81	3.8	1.64	130	5.9	13.9	64.3	15.9
Castle Morpeth	619	81	24	26	50	0.7	1.43	104	4.7	13.0	61.5	20.7
Tynedale	2,219	26	29	30	59	8.7	1.59	109	5.3	14.1	60.1	20.5
Wansbeck	67	927	31	32	62	-0.8	1.94	120	6.2	13.9	60.7	19.2
Tyne and Wear	540	2,066	545	571	1,116	-3.4	1.61	116	5.8	14.3	60.9	19.0
Gateshead	143	1,391	97	101	199	-6.7	1.62	123	5.8	13.6	61.2	19.5
Newcastle-upon-Tyne	112	2,465	136	140	276	-2.8	1.60	116	5.8	13.9	62.5	17.8
North Tyneside	84	2,310	93	101	194	-2.3	1.73	107	5.7	13.6	59.9	20.8
South Tyneside	64	2,415	75	79	155	-4.6	1.76	114	5.8	15.0	58.8	20.4
Sunderland	138	2,118	143	150	292	-1.7	1.65	120	5.9	15.2	61.1	17.8
Tees Valley	794	828	323	335	658	-1.7	1.75	111	6.3	15.4	60.7	17.6
Tees Valley less Darlington	597	932	273	283	556	-2.4	1.76	112	6.3	15.6	60.8	17.4
Former county of Durham	2,429	250	299	309	608	-0.6	1.68	113	5.8	14.2	61.3	18.7
North West	14,165	486	3,384	3,507	6,891	-0.7	1.73	109	6.1	14.8	60.9	18.2
Blackburn with Darwen UA	137	1,022	69	71	140	-1.8	2.24	122	7.9	16.9	59.6	15.5
Blackpool UA	35	4,301	74	77	151	1.0	1.69	111	5.5	12.7	59.9	21.9
Halton UA	74	1,645	60	62	122	-1.2	0.95	129	6.3	16.5	62.2	15.0
Warrington UA	176	1,081	94	96	190	12.0	1.77	110	6.3	14.4	62.7	16.6
Cheshire County	2,081	323	331	341	672	5.2	1.67	97	5.8	14.1	61.3	18.9
Chester	448	265	58	60	119	1.6	2.36	91	5.8	14.0	60.1	20.1
Congleton	211	419	44	45	88	10.5	1.56	94	5.5	13.8	62.3	18.4
Crewe and Nantwich	430	266	57	57	114	16.1	1.52	112	6.1	14.5	62.3	17.1
Ellesmere Port and Neston	87	919	39	41	80	-3.1	1.73	97	6.0	15.0	60.5	18.5
Macclesfield	525	292	75	78	153	2.2	1.64	96	5.5	13.2	61.1	20.2
Vale Royal	380	310	58	60	118	5.6	1.65	97	6.0	14.4	61.4	18.2
Cumbria	6,824	72	242	250	493	2.4	1.61	103	5.6	13.5	60.3	20.6
Allerdale	1,258	76	47	49	96	0.4	1.65	102	5.6	13.3	60.3	20.8
Barrow-in-Furness	78	902	35	35	70	-4.2	1.82	120	6.4	14.5	61.1	18.1
Carlisle	1,040	99	50	53	103	2.0	1.60	106	5.6	13.8	60.1	20.5
Copeland	738	95	35	35	70	-3.9	1.60	118	5.6	14.5	61.7	18.3
Eden	2,156	23	25	25	50	15.8	1.61	95	5.5	12.7	60.9	20.9
South Lakeland	1,554	66	50	53	103	8.8	1.45	90	5.0	12.5	58.7	23.8
Greater Manchester (Met. County)	1,286	2,004	1,269	1,308	2,577	-1.6	1.76	113	6.4	15.2	61.4	17.0
Bolton	140	1,911	132	135	267	2.0	1.85	110	6.5	15.1	61.5	16.9
Bury	99	1,846	91	92	183	3.5	1.83	112	6.0	14.7	62.5	16.8
Manchester	116	3,705	212	218	430	-7.1	1.83	127	7.1	17.0	60.6	15.2
Oldham	141	1,555	107	112	219	-1.0	2.05	117	6.9	15.7	60.9	16.5
Rochdale	160	1,301	102	106	208	0.0	1.97	116	6.9	15.7	61.1	16.3
Salford	97	2,329	112	114	226	-8.6	1.82	113	6.4	14.7	60.4	18.5
Stockport	126	2,324	140	152	293	0.8	1.59	95	5.7	14.2	61.2	19.0
Tameside	103	2,140	109	111	220	0.9	1.85	121	6.2	15.2	61.6	17.0
Trafford	106	2,078	108	112	220	-0.6	1.69	101	5.9	14.5	61.2	18.4
Wigan	199	1,560	154	156	310	0.1	1.65	115	6.0	14.3	63.3	16.4

Regional Trends 35, © Crown copyright 2000

14.1 *(continued)*

	Area (sq km)	Persons per sq km	Population (thousands)			Total population percentage change 1981-1998	Total fertility rate (TFR)[2]	Standardised mortality ratio (UK=100) (SMR)[3]	Percentage of population aged			
			Males	Females	Total				Under 5	5-15	16 up to pension age[4]	Pension age[4] or over
Lancashire County	2,897	392	559	577	1,136	3.9	1.74	103	5.9	14.7	60.5	19.0
Burnley	111	820	45	46	91	-1.9	1.95	117	6.7	16.3	59.5	17.5
Chorley	203	482	48	50	98	6.2	1.65	101	5.7	14.1	63.8	16.5
Fylde	166	456	37	39	76	9.6	1.56	90	4.7	11.9	57.5	25.8
Hyndburn	73	1,088	39	40	79	0.0	2.08	119	7.1	15.6	59.9	17.4
Lancaster	576	237	67	70	137	9.1	1.72	101	5.4	14.2	60.2	20.2
Pendle	169	493	41	42	83	-3.6	2.13	116	6.7	16.0	59.9	17.5
Preston	142	951	68	67	135	6.8	1.78	108	6.6	15.2	61.5	16.7
Ribble Valley	584	91	27	26	53	-1.5	1.71	94	4.9	13.6	61.4	20.0
Rossendale	138	466	32	33	64	-1.0	1.83	104	6.2	15.0	61.7	17.1
South Ribble	113	917	50	53	104	7.0	1.61	97	5.6	14.8	62.2	17.5
West Lancashire	338	329	55	56	111	3.5	1.81	97	5.8	15.2	61.1	17.8
Wyre	284	370	50	55	105	5.8	1.76	101	5.0	13.5	56.5	24.9
Merseyside (Met. County)	655	2,152	684	725	1,409	-7.4	1.65	112	6.0	14.9	60.4	18.8
Knowsley	97	1,593	75	79	155	-11.2	1.81	130	7.0	17.0	59.8	16.3
Liverpool	113	4,084	226	235	461	-10.7	1.66	124	6.2	15.1	61.6	17.0
St Helens	133	1,342	138	150	288	-4.2	1.71	114	5.8	14.1	62.2	17.8
Sefton	153	1,881	88	90	178	-6.2	1.60	98	5.4	14.2	58.6	21.9
Wirral	159	2,058	157	170	327	-4.0	1.68	104	5.8	14.7	59.4	20.2
Former county of Cheshire	2,331	422	485	499	984	5.6	1.72	103	5.9	14.4	61.7	17.9
Former county of Lancashire	3,070	465	702	725	1,427	3.0	1.78	106	6.0	14.7	60.3	19.0
Yorkshire and the Humber	15,411	327	2,489	2,553	5,043	2.5	1.75	103	6.2	14.5	61.1	18.3
East Riding of Yorkshire UA	2,415	130	153	160	313	15.2	1.65	102	5.2	13.7	60.6	20.5
City of Kingston upon Hull UA	71	3,687	130	131	262	-4.4	1.75	112	6.6	15.7	60.4	17.4
North East Lincolnshire UA	192	814	76	80	156	-3.1	1.91	103	6.2	15.6	59.1	19.1
North Lincolnshire UA	833	183	75	77	152	0.9	1.80	107	5.8	14.3	60.6	19.3
York UA	271	654	87	90	177	7.2	1.57	93	5.4	13.3	61.6	19.6
North Yorkshire County	8,038	70	276	290	565	10.5	1.70	93	5.6	13.5	59.9	21.0
Craven	1,179	44	25	27	52	8.9	1.85	94	5.5	14.1	57.0	23.5
Hambleton	1,311	66	42	44	86	14.4	1.58	89	5.7	13.7	61.2	19.4
Harrogate	1,305	115	72	78	150	9.5	1.51	95	5.7	13.4	61.3	19.6
Richmondshire	1,319	37	25	24	49	13.5	1.64	89	6.1	13.2	63.3	17.5
Ryedale	1,506	32	24	24	48	11.0	1.97	91	5.1	12.2	57.4	25.2
Scarborough	817	132	52	56	108	5.7	1.74	93	5.2	13.3	57.1	24.4
Selby	601	119	35	36	71	14.3	2.16	100	6.2	14.3	61.2	18.3
South Yorkshire (Met. County)	1,559	837	647	657	1,304	-1.0	1.72	107	6.1	14.2	61.4	18.4
Barnsley	328	695	112	116	228	1.0	1.74	114	6.2	14.3	61.1	18.4
Doncaster	581	500	144	147	290	-0.1	1.92	116	6.3	15.0	60.5	18.3
Rotherham	283	899	126	128	254	0.7	1.82	105	6.2	14.8	61.4	17.7
Sheffield	367	1,447	265	266	531	-3.0	1.67	102	6.0	13.4	62.0	18.7
West Yorkshire (Met. County)	2,034	1,039	1,045	1,068	2,113	2.2	1.81	104	6.5	14.8	61.5	17.2
Bradford	366	1,320	239	245	483	4.0	2.12	107	7.2	16.1	60.0	16.6
Calderdale	363	531	94	98	193	0.0	1.90	106	6.5	14.5	61.3	17.6
Kirklees	410	953	193	198	391	3.6	1.90	105	6.7	14.9	61.5	17.0
Leeds	562	1,294	361	367	727	1.3	1.62	97	6.1	14.2	62.2	17.6
Wakefield	333	957	158	161	319	1.5	1.75	110	6.1	14.7	62.2	17.0
The Humber	3,511	252	435	448	883	3.0	1.73	106	5.9	14.7	60.3	19.1
Former county of North Yorkshire	8,309	89	362	380	742	9.7	1.66	93	5.6	13.4	60.3	20.6
East Midlands	15,627	267	2,063	2,106	4,169	8.2	1.69	100	6.0	14.2	61.4	18.4
Derby UA	78	3,023	117	119	236	8.5	1.74	100	6.4	14.6	61.1	17.9
Leicester UA	73	4,031	147	147	294	4.0	1.82	105	7.1	16.0	61.1	15.8
Nottingham UA	75	3,824	142	145	287	3.1	1.54	108	6.6	15.3	61.3	16.7
Rutland UA	394	91	17	19	36	8.2	1.11	75	5.2	14.9	61.8	18.0
Derbyshire County	2,551	288	364	370	734	5.4	1.69	103	5.8	13.5	61.8	19.0
Amber Valley	265	440	58	58	117	6.6	1.75	97	5.4	12.8	62.2	19.6
Bolsover	160	445	35	36	71	0.2	1.76	112	6.3	13.0	60.5	20.2
Chesterfield	66	1,519	49	51	100	2.3	1.63	119	5.7	13.3	61.8	19.2
Derbyshire Dales	795	89	35	36	71	4.2	1.70	90	5.1	12.9	60.4	21.6
Erewash	109	989	54	54	108	4.1	1.79	97	6.3	14.0	61.3	18.4
High Peak	540	165	44	45	89	7.8	1.70	101	6.1	14.8	62.4	16.7
North East Derbyshire	277	355	49	49	98	2.2	1.69	103	5.5	13.1	61.5	19.9
South Derbyshire	338	239	40	40	81	17.8	1.49	108	5.8	13.7	64.1	16.4

14.1 *(continued)*

	Area (sq km)	Persons per sq km	Population (thousands)			Total population percentage change 1981-1998	Total fertility rate (TFR)[2]	Standardised mortality ratio (UK=100) (SMR)[3]	Percentage of population aged			
			Males	Females	Total				Under 5	5-15	16 up to pension age[4]	Pension age[4] or over
Leicestershire County	2,084	287	298	300	599	10.3	1.70	91	5.8	13.9	62.5	17.9
Blaby	130	665	43	43	86	11.9	1.83	90	6.4	13.8	63.3	16.5
Charnwood	279	557	77	78	155	10.9	1.64	87	5.3	13.9	63.0	17.7
Harborough	593	127	37	38	75	22.3	1.59	83	5.9	14.6	62.0	17.4
Hinckley and Bosworth	297	329	49	49	98	11.0	1.82	88	5.7	13.2	62.6	18.4
Melton	481	96	23	23	46	6.3	1.90	97	6.4	13.6	61.8	18.2
North West Leicestershire	279	305	43	42	85	7.4	1.74	104	5.6	14.0	62.4	17.9
Oadby and Wigston	24	2,197	26	27	53	-0.6	1.87	90	5.6	14.1	60.5	19.8
Lincolnshire	5,921	105	304	319	623	12.7	1.70	98	5.5	13.5	59.2	21.9
Boston	362	151	27	28	55	4.1	1.76	95	5.3	12.9	59.4	22.4
East Lindsey	1,760	71	60	65	125	18.3	1.73	100	5.1	12.6	56.1	26.3
Lincoln	36	2,299	41	42	83	8.2	1.80	110	6.3	13.9	61.3	18.5
North Kesteven	922	95	43	45	88	9.2	1.71	94	5.7	13.3	58.8	22.1
South Holland	742	99	36	37	73	17.9	1.69	94	4.9	12.3	58.8	24.1
South Kesteven	943	130	59	63	123	24.7	1.56	94	5.9	14.6	61.2	18.3
West Lindsey	1,156	67	38	39	77	-0.7	1.72	103	5.3	14.4	59.2	21.0
Northamptonshire	2,367	260	305	310	616	15.6	1.74	100	6.4	15.0	62.4	16.3
Corby	80	642	25	26	51	-2.3	2.14	102	6.3	16.2	60.8	16.7
Daventry	666	102	34	34	68	16.9	1.75	105	6.3	14.3	63.3	16.0
East Northamptonshire	510	144	37	37	73	17.7	1.68	102	5.9	14.9	62.0	17.2
Kettering	233	353	41	42	82	15.2	1.67	103	6.5	14.2	62.1	17.2
Northampton	81	2,418	97	99	196	23.3	1.74	103	6.5	15.3	63.0	15.2
South Northamptonshire	634	122	39	39	77	20.0	1.68	95	6.1	15.2	62.7	16.0
Wellingborough	163	415	33	34	68	4.4	1.91	91	6.6	14.6	61.1	17.7
Nottinghamshire County	2,085	357	369	376	745	4.0	1.75	101	5.7	13.8	61.7	18.8
Ashfield	110	981	53	54	108	1.1	1.80	110	6.1	13.7	61.6	18.7
Bassetlaw	637	166	52	53	106	2.7	2.02	111	5.9	13.8	61.5	18.8
Broxtowe	81	1,355	54	55	110	5.1	1.64	97	5.4	13.6	62.2	18.8
Gedling	120	927	55	57	111	2.8	1.71	92	5.6	13.4	61.8	19.2
Mansfield	77	1,289	49	50	99	-0.7	1.88	106	5.8	14.9	61.0	18.2
Newark and Sherwood	651	161	52	53	105	4.2	1.71	103	5.7	14.0	61.3	19.1
Rushcliffe	409	259	52	54	106	13.9	1.59	88	5.3	13.4	62.4	18.9
Former county of Derbyshire	2,629	369	481	489	970	6.1	1.70	102	5.9	13.7	61.6	18.7
Former county of Leicestershire	2,551	364	462	467	929	8.2	1.71	94	6.1	14.9	62.0	17.2
Former county of Nottinghamshire	2,160	478	510	521	1,032	3.8	1.66	102	6.0	14.2	61.6	18.2
West Midlands	13,004	410	2,639	2,694	5,333	2.8	1.80	101	6.4	14.6	60.9	18.1
County of Herefordshire UA	2,162	78	82	86	168	12.4	1.79	92	5.8	13.9	58.4	21.8
Stoke-on-Trent UA	93	2,705	125	127	252	-0.3	1.61	109	6.1	14.4	61.3	18.2
Telford and Wrekin UA	290	517	74	75	150	19.4	1.88	106	7.1	15.4	63.3	14.2
Shropshire County	3,197	88	140	141	280	9.8	1.70	94	5.5	13.5	60.3	20.7
Bridgnorth	633	83	27	25	52	3.5	1.64	96	5.2	12.8	63.0	19.1
North Shropshire	679	80	27	27	55	6.5	1.80	96	5.7	13.2	59.0	22.2
Oswestry	256	135	17	18	35	9.8	1.65	109	5.5	13.5	60.1	21.0
Shrewsbury and Atcham	602	162	48	49	97	11.0	1.63	89	5.8	14.3	60.5	19.5
South Shropshire	1,027	40	21	21	42	21.4	1.88	86	5.1	13.3	58.4	23.2
Staffordshire County	2,623	309	404	406	810	5.6	1.69	101	5.7	14.2	62.5	17.7
Cannock Chase	79	1,158	46	45	92	7.5	1.81	109	6.4	14.7	64.0	15.0
East Staffordshire	390	265	52	52	103	7.3	1.70	105	6.2	14.4	60.9	18.5
Lichfield	329	286	47	47	94	5.8	1.69	102	5.6	13.1	63.8	17.5
Newcastle-under-Lyme	211	584	61	63	123	2.3	1.83	98	5.4	14.3	60.6	19.7
South Staffordshire	408	251	51	52	102	5.2	1.73	94	5.0	13.9	62.6	18.4
Stafford	599	211	63	64	127	8.0	1.54	94	5.5	13.9	62.2	18.4
Staffordshire Moorlands	576	164	47	47	94	-1.5	1.67	109	4.9	13.3	62.5	19.2
Tamworth	31	2,383	37	37	74	13.2	1.79	112	6.7	16.2	64.9	12.2
Warwickshire	1,979	256	251	256	507	6.2	1.66	98	5.8	13.8	62.1	18.3
North Warwickshire	285	217	31	31	62	3.2	1.81	99	6.1	14.4	62.5	17.0
Nuneaton and Bedworth	79	1,499	59	59	118	4.0	1.85	107	6.2	15.1	62.1	16.6
Rugby	356	247	44	44	88	0.3	1.76	99	6.1	14.5	61.2	18.1
Stratford-on-Avon	977	117	56	59	115	13.9	1.46	94	5.3	12.2	62.3	20.2
Warwick	282	439	61	63	124	7.6	1.65	92	5.5	13.1	62.3	19.1

14.1 *(continued)*

	Area (sq km)	Persons per sq km	Population (thousands)			Total population percentage change 1981-1998	Total fertility rate (TFR)[2]	Standardised mortality ratio (UK=100) (SMR)[3]	Percentage of population aged			
			Males	Females	Total				Under 5	5-15	16 up to pension age[4]	Pension age[4] or over
West Midlands (Met. County)	899	2,923	1,298	1,330	2,628	-1.7	1.90	103	6.7	15.3	60.2	17.8
Birmingham	265	3,824	501	512	1,013	-0.7	2.01	104	7.3	16.0	60.0	16.7
Coventry	97	3,137	150	154	304	-4.7	1.82	101	6.5	15.5	60.2	17.8
Dudley	98	3,178	155	157	311	3.5	1.76	101	6.1	13.8	61.3	18.8
Sandwell	86	3,378	143	148	291	-6.2	1.97	118	6.7	15.0	59.8	18.4
Solihull	179	1,149	101	105	206	3.7	1.58	83	5.8	14.4	61.3	18.5
Walsall	106	2,464	130	132	261	-2.4	2.15	97	6.5	15.0	60.0	18.5
Wolverhampton	69	3,502	119	123	242	-5.8	1.94	105	6.6	15.2	59.2	19.0
Worcestershire County	1,761	306	265	273	538	10.5	1.71	97	5.9	13.8	61.5	18.8
Bromsgrove	220	387	42	44	85	-3.4	1.71	112	5.5	13.2	61.0	20.2
Malvern Hills	595	125	36	38	74	8.8	1.70	94	4.9	14.1	58.1	22.9
Redditch	54	1,421	38	39	77	13.4	1.96	89	6.8	15.1	63.5	14.5
Worcester	33	2,887	47	48	95	23.5	1.70	86	6.7	13.8	63.0	16.5
Wychavon	664	167	54	57	111	17.4	1.58	93	5.6	13.3	60.7	20.3
Wyre Forest	195	494	48	48	96	5.0	1.68	103	5.7	13.8	62.4	18.1
Herefordshire and Worcestershire	3,923	180	347	359	706	11.0	1.73	95	5.9	13.9	60.8	19.5
Former county of Shropshire	3,488	123	214	216	430	13.0	1.77	97	6.1	14.2	61.3	18.4
Former county of Staffordshire	2,715	391	529	533	1,062	4.1	1.67	103	5.8	14.2	62.2	17.9
East	19,120	281	2,654	2,723	5,377	10.8	1.70	93	6.2	13.9	61.4	18.5
Luton UA	43	4,262	92	91	183	11.2	2.05	101	7.8	16.6	62.1	13.5
Peterborough UA	344	453	78	78	156	16.6	1.93	107	7.0	15.3	62.0	15.7
Southend-on-Sea UA	42	4,191	85	91	176	11.7	1.56	97	6.0	13.2	60.2	20.6
Thurrock UA	164	823	67	68	135	5.9	1.95	102	7.3	14.7	63.4	14.7
Bedfordshire County	1,192	313	186	188	373	8.2	1.70	94	6.4	14.2	63.4	16.0
Bedford	477	294	70	70	140	5.1	1.90	92	6.3	13.8	62.6	17.2
Mid Bedfordshire	503	245	61	62	123	17.8	1.53	92	6.6	14.2	64.4	14.8
South Bedfordshire	213	517	55	55	110	2.6	1.64	99	6.3	14.7	63.2	15.7
Cambridgeshire County	3,056	184	280	284	564	23.8	1.48	90	6.0	13.5	64.0	16.5
Cambridge	41	2,943	60	60	121	19.5	1.09	80	5.2	11.3	68.5	15.0
East Cambridgeshire	655	111	36	36	73	35.1	1.55	91	5.9	13.9	62.4	17.8
Fenland	546	149	40	41	81	22.3	2.02	106	6.3	13.0	59.1	21.6
Huntingdonshire	912	172	77	80	156	24.9	1.57	94	6.6	14.9	64.2	14.3
South Cambridgeshire	902	147	66	67	132	21.7	1.64	82	5.8	14.1	63.4	16.6
Essex County	3,469	373	636	659	1,295	8.1	1.70	92	6.0	13.7	61.3	19.0
Basildon	110	1,495	81	83	164	7.8	2.06	92	6.9	14.7	60.9	17.5
Braintree	612	212	64	66	130	15.4	1.70	102	6.0	13.8	63.4	16.8
Brentwood	149	479	35	37	71	-1.4	1.33	96	5.5	13.0	61.8	19.7
Castle Point	45	1,877	42	43	84	-2.7	1.65	101	5.4	13.4	62.7	18.5
Chelmsford	342	452	76	79	155	10.7	1.50	84	5.8	14.1	63.3	16.9
Colchester	334	468	77	80	156	13.1	1.58	92	6.1	13.6	63.5	16.8
Epping Forest	340	352	59	61	120	2.5	1.79	93	6.2	12.9	61.6	19.3
Harlow	30	2,528	37	39	76	-4.7	1.86	87	6.8	14.0	61.0	18.1
Maldon	360	157	28	28	56	16.7	1.88	92	5.9	14.2	61.6	18.3
Rochford	169	457	38	39	77	4.9	1.71	81	6.0	13.5	60.4	20.1
Tendring	337	400	64	71	135	17.6	1.76	96	5.2	12.3	53.2	29.2
Uttlesford	641	108	34	35	69	10.4	1.71	86	6.0	14.4	62.5	17.1
Hertfordshire	1,639	631	512	522	1,034	6.9	1.74	92	6.5	14.2	62.1	17.2
Broxbourne	52	1,583	41	42	82	3.1	1.66	92	6.3	13.9	62.7	17.1
Dacorum	212	647	68	70	137	4.8	1.80	89	6.6	14.6	61.6	17.3
East Hertfordshire	477	266	64	63	127	16.0	1.77	90	6.5	13.4	64.6	15.5
Hertsmere	98	993	48	50	97	9.8	1.56	96	6.4	14.6	60.4	18.5
North Hertfordshire	375	311	58	59	117	7.5	1.79	99	6.4	13.9	61.4	18.2
St Albans	161	816	65	66	131	4.8	1.75	81	6.5	13.1	63.3	17.1
Stevenage	26	3,018	39	39	78	5.1	1.80	109	7.2	16.8	60.8	15.3
Three Rivers	89	972	43	44	87	6.7	1.76	75	5.9	13.4	61.5	19.2
Watford	21	3,891	41	41	82	9.5	1.66	119	7.5	14.5	64.3	13.8
Welwyn Hatfield	127	748	47	48	95	0.9	1.92	88	6.4	14.8	59.5	19.3

14.1 *(continued)*

	Area (sq km)	Persons per sq km	Population (thousands)			Total population percentage change 1981-1998	Total fertility rate (TFR)[2]	Standardised mortality ratio (UK=100) (SMR)[3]	Percentage of population aged			
			Males	Females	Total				Under 5	5-15	16 up to pension age[4]	Pension age[4] or over
Norfolk	5,372	147	388	403	790	12.4	1.65	91	5.6	13.0	59.2	22.3
Breckland	1,305	90	58	60	118	21.5	1.64	92	5.7	13.1	59.1	22.0
Broadland	552	214	58	60	118	20.4	1.60	93	5.5	13.0	60.6	21.0
Great Yarmouth	174	513	44	46	89	9.7	1.91	103	6.0	13.9	57.9	22.2
Kings Lynn and West Norfolk	1,429	93	66	68	134	9.4	1.86	93	5.7	13.0	57.6	23.6
North Norfolk	965	104	49	51	100	20.2	1.47	88	4.9	12.1	55.7	27.3
Norwich	39	3,180	61	63	124	-1.6	1.49	86	5.7	13.2	62.2	18.9
South Norfolk	908	119	53	55	108	13.1	1.70	88	5.3	12.6	60.4	21.7
Suffolk	3,798	177	331	340	671	11.6	1.71	92	6.0	14.4	59.6	20.0
Babergh	595	134	39	40	80	7.4	1.92	90	5.2	13.7	60.4	20.7
Forest Heath	374	185	35	34	69	31.4	1.50	104	7.7	18.8	60.4	13.1
Ipswich	39	2,921	56	58	114	-5.2	1.89	97	6.6	14.2	59.7	19.4
Mid Suffolk	871	94	41	41	82	15.8	1.97	83	5.8	13.6	59.9	20.7
St Edmundsbury	657	148	48	49	97	11.5	1.54	97	6.0	12.9	62.8	18.3
Suffolk Coastal	892	136	60	61	121	25.1	1.44	85	5.4	14.9	58.7	21.0
Waveney	370	292	52	56	108	8.1	1.89	93	5.7	13.4	56.3	24.6
Former county of Bedfordshire	1,236	451	278	279	557	9.1	1.82	96	6.9	15.0	63.0	15.2
Former county of Cambridgeshire	3,400	212	358	362	720	22.1	1.57	94	6.2	13.9	63.5	16.3
Former county of Essex	3,675	437	788	818	1,606	8.3	1.70	94	6.1	13.7	61.3	18.8
London[5]	1,580	4,549	3,548	3,639	7,187	5.6	1.77	95	7.0	13.7	64.4	14.9
Inner London	322	8,573	1,361	1,400	2,761	8.3	1.75	98	7.4	13.3	66.1	13.2
Inner London - West	110	9,160	494	513	1,008	8.8	..	91	6.1	10.4	69.7	13.8
Camden	22	8,573	93	96	189	5.3	1.71	95	6.4	11.3	67.8	14.5
City of London	3	1,737	3	2	5	-3.6	2.21	87	5.2	8.1	64.9	21.8
Hammersmith and Fulham	16	9,842	75	83	157	4.1	1.39	101	6.6	10.8	69.9	12.8
Kensington and Chelsea	12	14,161	84	86	170	21.3	1.23	79	5.6	10.3	70.8	13.3
Wandsworth	35	7,588	129	136	266	1.4	1.46	99	6.8	10.5	69.1	13.6
Westminster	22	10,038	110	110	221	17.3	1.09	79	5.1	9.4	71.0	14.5
Inner London - East	212	8,269	867	886	1,753	7.9	..	102	8.1	15.0	64.0	12.8
Hackney	20	9,734	96	98	195	5.2	2.38	97	8.6	15.2	64.0	12.2
Haringey	30	7,386	110	111	222	6.9	1.84	92	7.6	13.4	66.8	12.2
Islington	15	11,931	87	92	179	7.8	1.55	98	6.8	13.3	66.5	13.4
Lambeth	27	9,981	133	137	269	6.6	1.66	99	7.5	14.0	66.2	12.2
Lewisham	35	6,965	118	126	244	3.1	1.83	109	7.6	14.5	63.5	14.4
Newham	36	6,425	116	115	231	8.7	2.69	112	9.6	18.4	60.3	11.7
Southwark	29	7,999	115	117	232	6.2	2.10	103	8.4	14.6	63.3	13.7
Tower Hamlets	20	9,063	91	90	181	24.8	2.30	101	8.9	16.9	61.0	13.1
Outer London	1,258	3,519	2,187	2,240	4,427	4.0	1.78	94	6.9	13.9	63.4	15.9
Outer London - East and North East	438	3,505	755	780	1,535	0.0	..	96	7.0	14.5	61.7	16.8
Barking and Dagenham	34	4,575	76	79	156	2.6	2.15	105	7.8	15.9	58.2	18.1
Bexley	61	3,571	107	111	218	0.3	1.81	96	6.4	14.5	61.3	17.8
Enfield	81	3,270	131	134	265	1.5	2.05	89	7.4	14.1	62.1	16.4
Greenwich	48	4,481	105	110	215	-0.2	1.96	103	7.2	15.6	61.9	15.3
Havering	118	1,935	112	116	228	-5.7	1.70	96	5.6	13.6	61.4	19.5
Redbridge	56	4,141	115	117	232	1.2	1.85	93	6.7	14.5	62.4	16.4
Waltham Forest	40	5,535	109	112	221	1.8	2.01	98	7.8	14.7	63.2	14.3
Outer London - South	358	3,196	562	582	1,144	4.8	..	93	6.5	13.4	63.6	16.4
Bromley	152	1,955	145	153	297	-0.4	1.71	88	6.1	12.7	61.8	19.5
Croydon	87	3,888	167	171	338	5.3	1.74	94	6.7	14.4	64.2	14.7
Kingston upon Thames	38	3,877	73	74	147	9.7	1.67	97	6.2	12.8	65.3	15.7
Merton	38	4,851	91	93	184	10.0	1.67	91	7.0	13.0	65.2	14.8
Sutton	43	4,118	86	91	177	4.1	1.62	101	6.7	13.8	62.7	16.8
Outer London - West and North West	462	3,783	870	878	1,748	7.2	..	91	6.9	13.5	64.8	14.8
Barnet	89	3,725	163	168	332	12.3	1.54	91	6.5	13.4	64.4	15.7
Brent	44	5,754	127	126	253	-0.4	1.84	103	7.4	14.1	65.5	13.0
Ealing	55	5,493	152	150	302	7.1	1.76	100	7.1	13.6	65.8	13.4
Harrow	51	4,143	104	107	211	6.2	1.72	83	6.5	14.2	63.5	15.8
Hillingdon	110	2,283	125	126	251	7.7	1.76	89	6.7	13.7	63.6	16.0
Hounslow	58	3,648	107	105	212	3.8	1.97	92	7.4	13.7	64.8	14.1
Richmond-upon-Thames	55	3,395	91	95	187	15.4	1.58	79	6.2	11.8	65.9	16.1

14.1 *(continued)*

	Area (sq km)	Persons per sq km	Population (thousands)			Total population percentage change 1981-1998	Total fertility rate (TFR)[2]	Standardised mortality ratio (UK=100) (SMR)[3]	Percentage of population aged			
			Males	Females	Total				Under 5	5-15	16 up to pension age[4]	Pension age[4] or over
South East	19,096	419	3,934	4,070	8,004	10.3	1.70	91	6.1	13.9	61.4	18.5
Bracknell Forest UA	109	1,015	56	54	111	30.7	1.75	98	7.3	15.0	65.2	12.5
Brighton and Hove UA	82	3,120	125	131	256	7.8	1.40	98	5.3	11.7	63.9	19.1
Isle of Wight UA	380	334	62	65	127	7.6	1.80	91	5.1	12.8	56.3	25.8
Medway UA	192	1,263	120	123	243	1.0	1.71	105	6.8	15.0	63.5	14.6
Milton Keynes UA	309	658	101	102	203	61.3	1.74	102	7.1	15.9	65.4	11.6
Portsmouth UA	40	4,749	97	93	190	-0.8	1.72	103	6.1	13.4	63.2	17.4
Reading UA	40	3,696	75	73	148	7.6	1.66	88	6.8	13.5	64.9	14.9
Slough UA	27	4,136	55	57	112	14.4	1.78	108	7.7	14.9	64.2	13.2
Southampton UA	50	4,321	110	106	216	3.0	1.65	97	6.0	13.7	62.9	17.4
West Berkshire UA (Newbury)	704	205	72	73	144	17.6	1.71	89	6.1	15.0	64.1	14.8
Windsor and Maidenhead UA	198	710	70	70	140	3.7	1.60	94	5.9	13.4	63.2	17.4
Wokingham UA	179	811	74	72	145	24.2	1.80	95	6.6	14.8	64.6	14.0
Buckinghamshire County	1,568	305	238	241	479	7.7	1.71	91	6.3	14.4	63.1	16.3
Aylesbury Vale	903	175	79	79	158	17.7	1.79	96	6.7	14.7	64.1	14.5
Chiltern	196	478	46	48	94	3.4	1.66	85	5.5	14.3	61.4	18.7
South Buckinghamshire	145	440	32	32	64	2.2	1.66	96	5.9	14.3	60.9	18.9
Wycombe	325	503	82	82	163	3.9	1.66	87	6.5	14.2	63.8	15.5
East Sussex County	1,713	287	231	260	491	14.8	1.79	85	5.6	13.5	55.2	25.8
Eastbourne	44	2,056	42	49	90	16.7	1.57	85	5.3	11.9	55.1	27.8
Hastings	30	2,712	38	43	81	7.5	2.08	97	6.2	14.5	57.3	21.9
Lewes	292	294	41	45	86	8.7	2.31	84	5.7	13.9	54.0	26.5
Rother	511	179	42	49	91	19.4	1.59	80	5.0	12.9	50.9	31.3
Wealden	836	170	68	74	142	19.3	1.69	84	5.7	14.1	57.5	22.7
Hampshire County	3,689	336	607	631	1,238	13.8	1.68	91	6.0	14.3	61.7	18.0
Basingstoke and Deane	634	233	73	75	148	12.1	1.92	97	7.1	14.8	63.8	14.4
East Hampshire	515	216	54	57	111	21.5	1.75	93	6.0	15.4	61.3	17.3
Eastleigh	80	1,406	56	57	112	21.1	1.71	89	6.3	14.5	63.1	16.1
Fareham	74	1,421	51	54	105	18.2	1.50	89	5.2	13.8	60.8	20.1
Gosport	25	3,069	36	41	77	-1.4	1.51	103	6.7	15.2	60.7	17.4
Hart	215	401	44	42	86	23.3	1.61	86	5.9	13.2	66.9	14.0
Havant	55	2,150	57	61	118	2.2	1.90	88	5.5	14.8	58.2	21.5
New Forest	753	227	83	88	171	17.6	1.57	89	5.3	12.9	57.6	24.2
Rushmoor	39	2,237	43	44	87	0.2	1.81	100	7.3	14.3	65.1	13.2
Test Valley	637	175	55	57	111	18.9	1.48	97	5.6	14.8	63.1	16.5
Winchester	661	166	54	56	110	18.4	1.72	83	5.6	13.8	61.1	19.5
Kent County	3,543	376	652	680	1,332	7.1	1.81	94	6.1	14.1	60.2	19.7
Ashford	581	172	49	51	100	14.8	1.79	92	6.3	14.6	60.8	18.3
Canterbury	309	451	68	71	139	14.1	1.55	87	5.1	13.2	59.1	22.6
Dartford	73	1,170	43	43	85	5.2	1.78	109	6.5	13.6	63.6	16.4
Dover	315	345	53	56	109	5.0	1.84	90	6.0	14.3	58.4	21.3
Gravesham	99	927	45	47	92	-3.9	2.01	99	6.6	14.8	60.3	18.3
Maidstone	393	359	69	72	141	7.9	1.70	90	6.0	14.1	62.6	17.4
Sevenoaks	368	304	55	57	112	1.9	1.83	79	6.2	14.1	60.5	19.2
Shepway	357	280	48	52	100	15.9	1.70	96	5.9	13.3	58.6	22.2
Swale	373	318	60	59	119	7.8	1.92	96	6.3	14.6	61.7	17.4
Thanet	103	1,231	60	66	127	4.0	1.90	103	5.7	13.7	55.5	25.1
Tonbridge and Malling	240	442	53	53	106	8.4	1.88	93	6.7	14.3	62.0	17.0
Tunbridge Wells	332	309	50	53	103	4.3	2.03	95	6.2	14.3	60.0	19.5
Oxfordshire	2,606	237	307	309	617	13.8	1.60	91	6.2	14.4	63.4	16.0
Cherwell	589	229	66	68	135	23.3	1.63	93	6.8	15.7	63.0	14.5
Oxford	46	3,132	73	72	144	10.5	1.31	85	5.4	12.9	67.3	14.4
South Oxfordshire	679	188	63	65	127	8.8	1.84	97	6.8	15.0	61.3	17.0
Vale of White Horse	579	195	58	56	113	9.4	1.90	85	5.9	14.5	62.1	17.5
West Oxfordshire	714	136	48	49	97	19.3	1.63	94	6.4	14.0	62.1	17.4

14.1 *(continued)*

	Area (sq km)	Persons per sq km	Population (thousands)			Total population percentage change 1981-1998	Total fertility rate (TFR)[2]	Standardised mortality ratio (UK=100) (SMR)[3]	Percentage of population aged			
			Males	Females	Total				Under 5	5-15	16 up to pension age[4]	Pension age[4] or over
Surrey	1,677	632	522	539	1,061	4.5	1.73	87	6.2	13.5	61.9	18.5
Elmbridge	97	1,333	63	67	129	14.9	1.59	84	6.4	13.6	62.8	17.3
Epsom and Ewell	34	2,068	35	36	70	1.4	1.69	78	5.9	13.2	61.6	19.3
Guildford	271	467	63	64	127	1.4	1.69	79	5.8	13.2	62.7	18.3
Mole Valley	258	305	38	40	79	1.6	1.94	83	5.8	12.5	59.4	22.3
Reigate and Banstead	129	932	59	61	120	2.7	1.85	97	6.4	13.3	61.8	18.5
Runnymede	78	979	38	38	76	5.0	1.57	86	5.9	12.8	62.0	19.3
Spelthorne	57	1,554	44	44	89	-4.5	1.88	83	6.1	12.5	61.6	19.7
Surrey Heath	95	883	42	42	84	10.3	1.65	86	6.5	14.7	64.1	14.7
Tandridge	250	317	38	41	79	1.7	1.72	88	6.0	14.3	60.5	19.2
Waverley	345	334	56	59	115	2.9	1.85	91	6.1	14.0	59.7	20.2
Woking	64	1,434	46	46	92	12.0	1.85	98	6.7	13.8	64.0	15.6
West Sussex	1,988	378	361	390	752	12.8	1.68	86	5.7	13.2	58.3	22.8
Adur	42	1,393	28	30	59	-0.1	1.83	90	5.8	13.2	56.2	24.9
Arun	221	637	67	73	141	18.8	1.62	86	4.9	11.8	54.4	28.8
Chichester	786	137	49	58	107	8.6	1.46	82	5.4	13.1	56.2	26.3
Crawley	44	2,196	47	49	97	17.8	1.66	90	7.1	13.9	63.1	15.9
Horsham	530	230	60	62	122	21.3	1.66	81	6.0	14.8	60.5	18.7
Mid Sussex	333	382	63	64	127	10.1	1.85	88	5.8	13.8	62.2	18.2
Worthing	32	3,115	47	53	100	7.8	1.72	90	5.4	12.2	55.7	26.8
Former county of Berkshire	1,259	637	402	398	800	15.2	1.71	94	6.7	14.4	64.3	14.6
Former county of Buckinghamshire	1,877	363	339	343	682	19.6	1.73	93	6.5	14.8	63.8	14.9
Former county of East Sussex	1,795	416	356	391	747	12.3	1.60	89	5.5	12.9	58.2	23.5
Former county of Hampshire	3,779	435	813	830	1,644	10.4	1.67	93	6.0	14.1	62.0	17.9
Former county of Kent	3,735	422	772	803	1,575	6.1	1.79	95	6.2	14.2	60.7	18.9
South West	23,829	206	2,398	2,504	4,901	11.9	1.68	89	5.7	13.6	59.7	21.1
Bath and North East Somerset UA	351	477	82	86	167	3.6	1.61	81	5.3	12.7	60.9	21.1
Bournemouth UA	46	3,531	77	85	162	13.3	1.50	91	5.2	11.5	58.6	24.7
City of Bristol UA	110	3,657	200	202	402	0.3	1.62	93	6.2	13.7	62.9	17.2
North Somerset UA	373	506	92	96	189	15.9	1.72	89	5.3	13.3	59.9	21.5
Plymouth UA	80	3,162	124	129	253	-0.1	1.54	98	5.6	14.2	61.9	18.3
Poole UA	65	2,176	69	73	141	17.6	1.57	87	5.7	13.5	59.0	21.8
South Gloucestershire UA	497	485	121	120	241	18.6	1.74	85	6.8	13.9	63.1	16.3
Swindon UA	230	781	90	90	180	18.5	1.78	99	6.6	14.5	63.0	15.9
Torbay UA	63	1,952	58	65	123	8.7	1.66	92	5.2	12.7	55.8	26.2
Cornwall and the Isles of Scilly	3,559	138	239	252	490	15.0	1.78	88	5.2	13.6	58.2	23.0
Caradon	664	122	40	41	81	20.0	1.71	86	5.1	14.0	59.5	21.4
Carrick	461	185	41	44	85	12.7	1.77	82	5.0	13.5	56.6	24.9
Kerrier	473	191	44	47	91	8.3	1.79	87	5.6	13.8	58.7	21.9
North Cornwall	1,190	68	39	42	81	24.5	1.49	88	5.4	13.6	58.0	23.0
Penwith	304	195	29	30	59	9.7	2.30	94	4.9	12.2	57.4	25.5
Restormel	452	202	45	46	91	16.2	1.82	93	5.3	13.9	58.8	22.0
Isles of Scilly	15	133	1	1	2	1.2	7.96	80	5.8	17.7	54.0	22.5
Devon County	6,562	106	335	357	692	15.6	1.71	88	5.3	13.3	57.8	23.6
East Devon	814	153	59	66	124	15.5	1.80	81	4.7	11.8	53.4	30.1
Exeter	47	2,361	55	56	111	10.3	1.88	95	5.6	13.3	62.8	18.3
Mid Devon	915	74	32	35	67	14.8	1.82	89	5.8	14.1	58.8	21.4
North Devon	1,086	80	43	44	87	11.3	1.81	93	5.6	13.1	58.3	23.0
South Hams	887	90	39	42	80	20.3	1.69	86	5.3	13.8	58.2	22.7
Teignbridge	674	177	58	62	119	24.6	1.72	89	5.4	13.1	57.2	24.4
Torridge	979	57	27	28	55	14.0	1.77	92	5.4	14.9	56.8	22.9
West Devon	1,160	41	23	25	47	11.1	1.89	84	5.0	13.7	58.0	23.4
Dorset County	2,542	152	188	199	387	15.7	1.59	81	5.0	12.8	56.9	25.3
Christchurch	50	877	21	23	44	14.6	1.71	79	4.8	11.0	51.6	32.6
East Dorset	354	234	40	43	83	19.8	1.49	74	4.5	11.7	56.6	27.2
North Dorset	609	99	30	31	60	23.4	1.39	75	5.1	14.3	58.8	21.8
Purbeck	404	114	23	24	46	14.0	1.49	84	5.2	13.1	59.0	22.7
West Dorset	1,082	84	44	47	91	13.8	1.79	84	4.8	13.0	55.1	27.1
Weymouth and Portland	42	1,500	31	32	63	8.6	1.69	94	5.7	13.9	60.3	20.1

Regional Trends 35, © Crown copyright 2000

14.1 (continued)

	Area (sq km)	Persons per sq km	Population (thousands)			Total population percentage change 1981-1998	Total fertility rate (TFR)[2]	Standardised mortality ratio (UK=100) (SMR)[3]	Percentage of population aged			
			Males	Females	Total				Under 5	5-15	16 up to pension age[4]	Pension age[4] or over
Gloucestershire	2,653	210	275	282	557	10.1	1.77	91	5.9	14.0	60.3	19.8
Cheltenham	47	2,252	52	54	106	3.0	1.68	87	5.5	13.0	61.2	20.3
Cotswold	1,165	71	40	42	83	17.6	1.54	84	5.5	13.8	59.3	21.4
Forest of Dean	526	148	39	39	78	6.3	1.79	99	5.7	13.2	61.2	19.9
Gloucester	41	2,620	53	54	107	7.2	2.02	96	7.2	15.3	60.4	17.1
Stroud	461	236	54	55	109	13.4	1.80	93	5.6	14.3	60.0	20.1
Tewkesbury	414	180	37	38	74	17.3	1.88	91	6.0	14.1	59.4	20.4
Somerset	3,452	142	238	251	489	13.6	1.77	90	5.7	14.1	58.2	22.0
Mendip	739	134	48	51	99	10.4	1.85	98	6.1	15.8	58.7	19.4
Sedgemoor	564	185	51	53	104	15.9	1.76	94	5.7	13.3	59.2	21.8
South Somerset	959	160	75	79	154	15.4	1.75	85	5.8	13.9	58.5	21.7
Taunton Deane	462	216	48	52	100	12.9	1.68	90	5.4	14.2	58.4	22.1
West Somerset	727	45	15	17	32	10.1	2.07	86	5.0	11.5	51.7	31.8
Wiltshire County	3,246	131	209	216	426	14.0	1.76	90	6.3	14.2	61.0	18.5
Kennet	957	83	39	40	79	21.0	1.63	85	6.6	14.7	61.3	17.5
North Wiltshire	768	161	62	62	124	18.0	1.79	91	6.8	14.1	62.4	16.6
Salisbury	1,004	113	54	59	114	11.0	1.63	91	5.7	14.1	60.2	20.0
West Wiltshire	517	210	54	55	109	8.4	2.01	91	6.1	14.1	60.1	19.7
Bristol/Bath area	1,333	751	495	504	999	7.6	1.65	88	6.0	13.5	62.0	18.5
Former county of Devon	6,703	159	518	551	1,068	10.6	1.65	90	5.4	13.4	58.6	22.6
Former county of Dorset	2,653	261	334	357	691	15.5	1.55	85	5.2	12.7	57.7	24.4
Former county of Wiltshire	3,476	174	299	306	606	15.3	1.77	92	6.4	14.3	61.6	17.7

1 Local government structure as at 1 April 1998. See Notes and Definitions.
2 The total fertility rate (TFR) is the average number of children which would be born to a woman if the current pattern of fertility persisted throughout her child-bearing years. Previously called total period fertility rate (TPFR).
3 Adjusted for the age structure of the population. See Notes and Definitions to the Population chapter.
4 Pension age is 65 for males and 60 for females.
5 London is presented by NUTS levels 1, 2, 3 and 4. See Notes and Definitions.

Source: Office for National Statistics

14.2 Vital[1,2] and social statistics: by sub-region[3]

	Live births[2] per 1,000 population		Deaths[2] per 1,000 population		Perinatal mortality rate[4]	Infant mortality rate[5]	Percentage of live births under 2.5 kg	Percentage of live births outside marriage	Children looked after by LAs per 1,000 population aged under 18
	1991	1998	1991	1998	1997-1999	1997-1999	1998	1998	1998[6]
United Kingdom	13.7	12.1	11.2	10.6	8.3	5.8	..	37.6	..
England	13.7	12.2	11.1	10.5	8.3	5.7	7.5	37.5	4.9
North East	13.4	11.1	12.2	11.8	8.2	5.4	7.4	47.9	6.5
Darlington UA	14.3	11.5	13.9	12.3	10.7	5.4	5.8	46.8	5.8
Hartlepool UA	14.9	11.9	11.5	11.1	9.1	6.4	7.2	57.7	5.0
Middlesbrough UA	15.1	12.4	11.1	10.4	8.6	5.4	8.7	52.5	4.7
Redcar & Cleveland UA	14.0	11.3	10.9	11.1	6.8	4.1	8.0	53.5	5.0
Stockton-on-Tees UA	14.9	11.6	10.4	10.0	8.4	5.3	7.2	44.8	4.0
Durham County	12.8	11.0	12.1	11.8	7.7	5.8	7.2	45.5	4.3
Northumberland	11.8	10.3	12.6	12.3	8.0	6.3	6.7	39.0	4.4
Tyne and Wear	13.4	11.0	12.7	12.2	8.1	5.2	7.6	49.6	6.8
Tees Valley	..	11.7	..	10.8	8.6	5.2	7.5	51.3	4.8
Tees Valley less Darlington	14.7	11.8	10.9	10.6	8.2	5.2	7.8	51.1	4.6
Former county of Durham	13.0	11.1	12.4	11.9	8.2	5.8	7.0	45.7	4.5
North West	14.2	11.8	12.0	11.4	8.8	6.5	8.0	44.1	5.9
Blackburn with Darwen UA	..	15.7	..	11.1	7.3	8.4	11.6	37.3	7.4
Blackpool UA	..	10.8	..	14.3	10.8	7.4	8.3	52.0	7.4
Halton UA	..	13.2	..	10.4	5.9	7.3	6.4	54.0	4.4
Warrington UA	..	12.6	..	10.4	5.8	5.0	6.6	36.3	3.3
Cheshire County	..	10.9	..	10.5	7.2	5.3	6.1	33.2	3.2
Cumbria	12.7	10.3	12.7	12.2	8.4	5.7	6.7	38.6	4.9
Greater Manchester (Met. County)	14.9	12.4	11.8	11.2	9.1	6.5	8.7	45.3	6.2
Lancashire County	..	11.3	..	11.5	9.1	6.9	7.7	40.3	5.7
Merseyside (Met. County)	14.0	11.4	12.4	11.9	9.4	7.2	8.2	52.1	6.9
Former county of Cheshire	13.5	11.5	10.7	10.5	6.8	5.3	6.2	36.8	3.4
Former county of Lancashire	13.9	11.7	12.7	11.8	8.9	7.0	8.3	41.0	6.0
Yorkshire and the Humber	13.8	11.8	11.5	11.0	8.6	6.5	8.0	40.5	5.5
East Riding of Yorkshire UA	11.8	9.7	12.0	11.9	6.6	5.5	6.3	33.5	3.3
City of Kingston upon Hull UA	15.6	12.5	10.9	11.3	9.3	7.3	8.4	57.5	10.9
North East Lincolnshire UA	14.5	11.9	10.9	11.3	9.1	5.2	8.3	52.2	6.8
North Lincolnshire UA	13.0	11.2	11.2	11.6	7.6	6.6	8.1	41.0	5.8
York UA	12.3	10.8	11.6	10.8	8.0	5.3	5.9	36.8	3.4
North Yorkshire County	11.8	10.6	12.4	11.6	6.9	4.7	6.0	29.0	2.7
South Yorkshire (Met. County)	13.7	11.7	11.7	11.3	9.1	6.6	8.0	44.9	5.5
West Yorkshire (Met. County)	14.5	12.6	11.2	10.4	8.8	7.1	8.6	38.7	5.8
The Humber	13.7	11.2	11.3	11.5	8.1	6.2	7.7	46.3	6.8
Former county of North Yorkshire	11.8	10.7	12.2	11.4	7.2	4.9	6.0	30.9	2.8
East Midlands	13.4	11.6	10.9	10.6	7.8	5.8	7.6	38.8	4.2
Derby UA	15.4	12.8	11.8	10.4	9.0	6.9	8.5	41.6	8.3
Leicester UA	16.5	13.7	11.3	9.9	11.1	7.8	9.9	40.0	6.0
Nottingham UA	..	12.4	..	10.5	7.6	6.9	9.2	55.3	8.8
Rutland UA	9.8	8.7	8.7	7.5	7.2	5.1	7.7	25.1	0.8
Derbyshire County	12.8	11.1	11.6	11.2	7.3	5.0	6.7	38.6	3.8
Leicestershire County	12.5	11.3	9.4	9.5	8.2	5.1	6.9	31.0	2.0
Lincolnshire	11.7	10.7	11.8	12.0	8.0	6.4	7.3	37.1	3.9
Northamptonshire	14.0	12.2	10.3	9.8	7.4	5.8	7.5	38.3	4.0
Nottinghamshire County	..	11.1	..	10.9	6.8	5.0	7.2	39.2	3.0
Former county of Derbyshire	13.4	11.5	11.6	11.0	7.7	5.5	7.2	39.4	3.8
Former county of Leicestershire	13.7	12.0	10.0	9.5	9.2	6.1	8.0	34.1	2.0
Former county of Nottinghamshire	13.7	11.5	10.8	10.8	6.7	5.4	7.8	44.0	3.0
West Midlands	14.1	12.2	10.8	10.4	9.6	6.8	8.4	38.3	4.9
County of Herefordshire UA	..	10.8	..	11.5	6.9	5.2	6.7	35.2	5.6
Stoke-on-Trent UA	14.3	11.5	12.1	11.2	10.5	8.1	9.0	48.0	6.0
Telford and Wrekin UA	..	13.4	..	8.7	9.6	6.8	8.1	44.7	3.8
Shropshire County	..	10.5	..	11.2	6.2	4.9	6.3	32.2	3.0
Staffordshire County	12.9	11.0	10.3	10.2	9.2	5.4	7.3	36.0	3.1
Warwickshire	12.5	11.2	10.5	10.2	8.0	5.5	6.7	33.5	3.3
West Midlands (Met. County)	15.2	13.2	11.0	10.4	10.4	7.8	9.6	40.0	5.9
Worcestershire County	..	11.3	..	10.5	9.5	5.5	6.6	34.1	4.6
Herefordshire and Worcestershire	12.6	11.2	10.7	10.7	8.9	5.4	6.6	34.3	4.8
Former county of Shropshire	13.3	11.5	10.4	10.3	7.0	5.4	7.0	37.3	3.3
Former county of Staffordshire	13.2	11.2	10.7	10.4	9.5	6.0	7.8	38.9	3.8

14.2 *(continued)*

	Live births[2] per 1,000 population		Deaths[2] per 1,000 population		Perinatal mortality rate[4]	Infant mortality rate[5]	Percentage of live births under 2.5 kg	Percentage of live births outside marriage	Children looked after by LAs per 1,000 population aged under 18
	1991	1998	1991	1998	1997-1999	1997-1999	1998	1998	1998[6]
East	13.3	12.0	10.3	10.0	7.3	4.8	6.8	33.0	4.0
Luton UA	18.5	15.7	8.8	7.8	11.0	7.7	8.3	30.5	5.3
Peterborough UA	..	13.9	..	9.5	9.5	5.9	9.5	38.7	9.2
Southend-on-Sea UA	..	11.9	..	12.5	6.6	4.2	6.2	44.5	6.8
Thurrock UA	..	14.9	..	8.4	6.6	4.8	6.8	42.1	5.3
Bedfordshire County	13.6	12.5	9.1	8.7	6.7	4.6	7.2	30.4	3.6
Cambridgeshire County	..	11.3	..	9.0	7.3	5.5	6.0	28.6	3.7
Essex County	..	11.6	..	10.2	7.0	4.2	6.3	33.5	3.7
Hertfordshire	13.6	12.8	9.5	9.2	6.9	4.2	6.5	29.0	3.1
Norfolk	11.6	10.5	12.1	11.9	7.2	5.6	7.0	36.5	4.0
Suffolk	12.8	11.4	11.2	10.8	7.3	4.4	7.0	34.5	3.5
Former county of Bedfordshire	15.2	13.5	9.0	8.4	8.4	5.8	7.6	30.4	4.3
Former county of Cambridgeshire	13.6	11.9	9.2	9.1	7.5	5.5	6.9	31.1	5.0
Former county of Essex	13.3	11.9	10.6	10.3	7.0	4.4	6.4	35.6	4.2
London	15.4	14.7	10.0	8.7	9.0	5.9	7.9	34.7	5.8
South East	13.0	11.9	10.8	10.2	7.0	4.7	6.6	32.5	3.9
Bracknell Forest UA	..	14.6	..	7.5	5.5	2.6	6.7	26.6	2.5
Brighton and Hove UA	12.1	11.2	13.9	12.2	7.3	5.7	7.0	45.3	7.1
Isle of Wight UA	10.6	9.6	15.5	14.2	7.5	4.5	7.3	44.5	5.7
Medway UA	..	13.0	..	8.8	7.4	5.4	7.3	41.1	5.3
Milton Keynes UA	17.1	14.2	7.2	7.1	8.6	6.7	7.2	39.9	5.9
Portsmouth UA	..	11.8	..	8.0	8.6	7.2	6.9	43.8	5.5
Reading UA	14.2	13.8	12.0	11.1	8.0	6.1	8.8	39.0	4.3
Slough UA	..	16.0	..	8.1	8.6	6.3	10.8	31.9	4.5
Southampton UA	..	11.9	..	8.2	7.4	5.3	7.8	45.9	7.6
West Berkshire UA (Newbury)	13.8	12.6	11.1	10..0	6.1	4.5	6.2	28.5	3.8
Windsor and Maidenhead UA	..	11.6	..	9.2	9.0	4.7	6.7	25.5	2.4
Wokingham UA	..	12.1	..	7.6	6.6	4.3	6.3	20.7	2.7
Buckinghamshire County	13.3	12.0	8.8	8.6	6.0	4.1	6.4	23.2	2.1
East Sussex County	11.2	10.4	15.5	14.0	7.8	5.4	5.9	37.9	4.2
Hampshire County	12.7	11.7	9.5	9.7	6.8	4.7	6.4	30.0	2.9
Kent County	..	11.9	..	11.0	6.8	4.8	6.6	37.9	4.7
Oxfordshire	13.7	12.1	8.8	8.7	6.7	4.5	5.9	28.9	3.3
Surrey	12.3	12.2	10.4	9.8	6.1	3.8	6.2	24.3	2.7
West Sussex	11.7	11.2	13.5	12.1	7.2	4.1	6.6	30.6	4.2
Former county of Berkshire	14.3	13.3	8.5	8.1	7.4	4.8	7.6	29.0	3.4
Former county of Buckinghamshire	14.3	12.7	8.4	8.2	6.9	5.0	6.6	28.7	3.3
Former county of East Sussex	11.5	10.7	15.0	13.4	7.6	5.5	6.3	40.6	5.1
Former county of Hampshire	13.1	11.7	10.0	9.9	7.1	5.1	6.6	33.7	3.8
Former county of Kent	13.4	12.1	11.2	10.7	6.9	4.8	6.7	38.4	4.8
South West	12.2	11.0	11.9	11.3	7.9	5.1	6.5	36.1	4.4
Bath and North East Somerset UA	11.1	10.5	11.4	10.4	7.2	3.6	5.4	34.1	4.0
Bournemouth UA	11.1	10.4	16.0	14.7	8.7	5.6	7.6	42.4	6.0
City of Bristol UA	13.9	12.3	11.2	9.9	8.6	5.6	7.1	45.0	6.3
North Somerset UA	11.0	10.3	12.2	11.8	7.4	6.4	6.4	32.4	3.5
Plymouth UA	..	11.0	..	10.5	7.8	5.6	7.0	43.3	6.4
Poole UA	12.2	10.5	12.7	11.6	9.5	6.0	7.0	33.1	3.6
South Gloucestershire UA	13.9	12.5	8.4	7.9	6.2	4.4	5.5	27.8	3.0
Swindon UA	15.1	13.3	9.5	8.8	7.6	5.1	7.0	39.5	4.6
Torbay UA	..	10.2	..	15.7	8.3	5.4	7.1	44.1	8.9
Cornwall and the Isles of Scilly	11.5	10.3	12.9	11.8	8.7	5.3	6.2	41.6	5.3
Devon County	..	10.0	..	12.6	7.4	5.6	6.5	35.3	4.5
Dorset County	10.6	9.4	13.0	12.5	7.8	3.9	6.8	31.2	3.2
Gloucestershire	12.6	11.6	10.8	10.6	8.3	5.5	6.5	33.7	3.9
Somerset	12.1	10.9	11.9	11.8	8.2	4.6	6.3	36.2	3.4
Wiltshire County	12.9	12.1	10.5	9.9	8.2	4.7	6.3	27.7	2.6
Bristol/Bath area	12.9	11.7	10.8	9.8	7.6	5.1	6.3	36.8	4.6
Former county of Devon	11.8	10.3	12.9	12.5	7.3	5.4	6.7	38.3	5.5
Former county of Dorset	11.1	9.9	13.7	12.8	8.4	4.8	7.0	34.4	3.9
Former county of Wiltshire	13.5	12.5	10.2	9.6	8.0	4.8	6.5	31.4	3.2

1 Births and deaths are based on the usual area of residence of the mother/deceased. See Notes and Definitions to the Population chapter for details of the inclusion/ exclusion of births to non-resident mothers and deaths of non-resident persons.
2 Births data are on the basis of year of occurrence in England and Wales and year of registration in Scotland and Northern Ireland. All deaths data relate to year of registration.
3 Counties and Unitary Authorities in existence from 1 April 1998. See Notes and Definitions.
4 Still births and deaths of infants under 1 week of age per 1,000 live and still births.
5 Death of infants under 1 year of age per 1,000 live births.
6 At 31 March. Under 18 mid-1998 population estimates used.

Source: Office for National Statistics; Department of Health

14.3　Education and training: by sub-region[1]

	Day nursery places per 1,000 population aged under 5 years[2] March 1999	Children under 5 in education[3] (percentages) Jan. 2000	Pupil/teacher ratio[4] 1999/00 (numbers)		Pupils and students participating in post-compulsory education[5] (percentages) 1997/98	Percentage of pupils in last year of compulsory schooling[6,7] 1998/99 with		Average A/AS level points score[7,8] 1998/99
			Primary schools	Secondary schools		No graded results	5 or more A*-Cs at GCSE	
United Kingdom	..	64	22.7	16.6	78	5.4	47.3	17.8
England	81.3	63	23.3	17.2	76	5.5	45.7	17.9
North East	51.7	86	23.1	17.3	68	7.0	39.4	17.8
Darlington UA	66.5	89	24.4	17.4	66	5.7	42.9	16.8
Hartlepool UA	59.3	96	24.7	17.9	69	6.2	38.8	19.5
Middlesbrough UA	63.5	93	24.0	16.4	65	9.5	31.0	16.1
Redcar & Cleveland UA	63.3	103	23.7	17.2	64	6.0	44.6	-
Stockton-on-Tees UA	69.2	95	23.3	17.2	73	5.9	41.3	19.7
Durham County	45.2	81	22.9	17.4	68	7.1	38.2	18.0
Northumberland	27.8	80	23.5	19.0	72	6.0	46.3	18.3
Tyne and Wear	52.5	..	..	..	66	7.3	37.7	17.2
North West	101.9	71	23.4	16.8	74	5.9	44.2	19.4
Blackburn with Darwen UA	56.2	69	23.3	16.5	..	7.1	35.9	18.6
Blackpool UA	109.1	51	24.7	17.3	..	7.3	36.5	17.5
Halton UA	72.4	61	22.0	16.6	..	5.7	37.3	12.3
Warrington UA	85.5	60	23.8	17.4	..	3.2	49.7	17.8
Cheshire County	112.3	57	23.3	17.1	..	3.4	54.5	20.4
Cumbria	77.4	68	22.5	17.0	77	5.0	47.6	19.3
Greater Manchester (Met. County)	90.9	..	..	..	70	6.0	42.5	19.8
Lancashire County	133.1	59	24.1	17.0	..	4.6	48.0	21.6
Merseyside (Met. County)	112.2	80	23.0	16.6	74	8.3	39.6	18.1
Yorkshire and the Humber	81.6	73	23.7	17.5	74	7.0	40.5	18.5
East Riding of Yorkshire UA	100.1	63	24.9	17.9	79	4.1	48.7	18.0
City of Kingston upon Hull UA	62.5	75	24.4	18.9	59	10.0	23.4	15.8
North East Lincolnshire UA	34.1	75	23.9	18.5	68	7.8	35.5	15.8
North Lincolnshire UA	46.8	66	24.2	18.7	76	3.7	41.6	17.0
York UA	96.3	69	23.3	16.9	87	5.0	51.5	21.8
North Yorkshire County	95.0	58	23.4	16.6	85	3.4	56.8	21.1
South Yorkshire (Met. County)	87.4	..	..	..	68	8.1	37.1	18.4
West Yorkshire (Met. County)	79.9	..	..	..	74	7.7	38.6	17.4
East Midlands	84.4	63	24.0	17.5	74	5.4	45.3	17.8
Derby UA	127.9	77	24.2	16.9	65	6.0	40.9	16.2
Leicester UA	77.5	73	22.7	16.4	80	8.0	37.1	14.3
Nottingham UA	114.2	81	22.8	16.7	..	10.4	28.7	15.5
Rutland UA	151.7	52	21.2	17.0	96	1.0	58.9	-
Derbyshire County	56.7	73	25.2	17.8	72	4.1	49.1	19.2
Leicestershire County	89.0	32	23.4	17.9	82	4.0	48.3	17.6
Lincolnshire	79.4	58	24.7	17.1	73	4.9	50.3	21.5
Northamptonshire	90.6	59	23.6	17.5	75	5.2	46.7	16.2
Nottinghamshire County	78.3	69	24.5	17.9	..	5.6	43.4	15.9
West Midlands	91.5	70	23.5	17.1	74	5.5	43.5	17.9
County of Herefordshire UA	127.3	49	22.8	17.0	..	5.0	52.0	17.3
Stoke-on-Trent UA	82.5	78	24.7	17.6	58	5.9	35.8	17.7
Telford and Wrekin UA	115.8	53	24.2	17.0	..	6.2	45.3	22.7
Shropshire County	110.1	43	23.4	17.0	..	2.7	55.4	15.8
Staffordshire County	93.7	65	24.3	17.9	..	4.6	48.1	17.9
Warwickshire	96.0	58	23.7	17.2	78	4.8	47.1	18.4
West Midlands (Met. County)	80.3	..	..	..	72	6.2	39.5	17.6
Worcestershire County	123.0	56	23.9	18.5	..	5.6	47.9	18.0

14.3 *(continued)*

| | Day nursery places per 1,000 population aged under 5 years[2] March 1999 | Children under 5 in education[3] (percent-ages) Jan. 2000 | Pupil/teacher ratio[4] 1999/00 (numbers) | | Pupils and students participating in post-compulsory education[5] (percentages) 1997/98 | Percentage of pupils in last year of compulsory schooling[6,7] 1998/99 with | | Average A/AS level points score[7,8] 1998/99 |
			Primary schools	Secondary schools		No graded results	5 or more A*-Cs at GCSE	
East	68.3	54	23.1	17.2	79	4.7	49.9	17.7
Luton UA	45.8	60	22.4	17.3	72	3.2	35.8	10.2
Peterborough UA	89.6	43	23.2	17.1	..	7.4	41.1	17.4
Southend-on-Sea UA	376.6	51	23.4	16.9	..	5.9	54.1	22.4
Thurrock UA	37.0	46	22.8	18.5	..	7.8	35.9	-
Bedfordshire County	63.4	61	23.5	18.7	90	5.1	47.2	16.9
Cambridgeshire County	83.8	47	24.6	18.7	..	5.8	52.7	16.3
Essex County	66.0	44	23.2	17.2	..	4.3	49.3	19.0
Hertfordshire	56.6	72	23.1	16.6	91	4.2	54.9	17.8
Norfolk	38.5	54	22.7	16.6	74	5.1	47.7	17.0
Suffolk	44.0	48	22.1	16.9	76	3.3	53.8	16.3
London	76.0	69	22.3	16.5	79	5.2	43.6	16.4
Inner London	92.9	73	21.5	15.9	74	5.7	34.1	15.4
Outer London	64.4	66	22.8	16.6	82	4.9	48.1	16.6
South East	75.6	47	23.2	17.3	80	4.8	51.2	17.8
Bracknell Forest UA	82.0	33	24.2	16.9	..	4.2	45.4	14.9
Brighton and Hove UA	139.3	59	22.6	16.5	78	5.9	42.1	14.6
Isle of Wight UA	53.3	42	22.1	18.1	80	6.0	44.6	16.9
Medway UA	97.2	43	23.6	17.4	..	5.2	45.4	16.9
Milton Keynes UA	87.2	45	23.7	17.1	62	7.2	39.5	15.9
Portsmouth UA	85.3	50	23.1	17.8	78	10.5	32.2	14.3
Reading UA	121.0	54	23.1	16.2	..	11.3	42.9	20.2
Slough UA	48.1	73	22.0	16.4	..	3.3	48.1	18.3
Southampton UA	84.8	47	22.7	16.7	75	6.1	40.5	19.0
West Berkshire UA (Newbury)	110.5	47	22.4	16.5	..	3.4	56.8	17.9
Windsor and Maidenhead UA	57.2	53	22.6	17.3	..	3.9	57.9	17.1
Wokingham UA	85.9	37	24.1	16.9	..	4.5	58.2	17.9
Buckinghamshire County	35.3	48	22.8	18.4	83	4.9	61.0	20.8
East Sussex County	91.6	49	22.4	17.1	80	4.4	49.6	15.8
Hampshire County	48.1	43	23.6	17.4	76	3.5	53.1	18.1
Kent County	59.0	44	23.7	17.3	..	4.9	50.9	18.5
Oxfordshire	73.4	39	23.3	18.1	81	5.4	47.9	16.1
Surrey	84.3	58	23.1	17.3	82	4.8	56.6	17.6
West Sussex	57.1	44	22.7	17.2	81	3.8	53.7	16.3
South West	85.7	49	23.7	17.7	79	4.3	50.2	18.5
Bath and North East Somerset UA	107.6	54	22.7	16.8	93	3.9	54.7	16.0
Bournemouth UA	84.6	45	24.7	18.2	92	6.1	48.8	21.8
City of Bristol UA	78.1	73	23.4	15.7	73	10.7	31.0	14.1
North Somerset UA	70.5	47	24.2	17.3	78	3.3	51.7	17.1
Plymouth UA	118.7	53	23.8	17.5	..	4.3	47.8	18.8
Poole UA	79.1	35	25.3	17.1	82	4.0	55.9	20.9
South Gloucestershire UA	75.8	47	24.0	17.0	82	3.7	47.6	17.3
Swindon UA	94.6	47	24.2	17.8	67	4.9	45.2	16.9
Torbay UA	80.8	66	24.5	17.0	..	5.5	47.4	22.4
Cornwall and the Isles of Scilly	52.4	54	23.5	18.2	81	3.3	51.7	17.0
Devon County	122.9	50	23.5	17.8	..	4.0	48.4	17.4
Dorset County	95.7	43	23.2	18.4	79	3.6	53.5	18.3
Gloucestershire	110.6	40	23.4	17.4	80	3.0	56.1	20.6
Somerset	61.7	46	23.7	18.6	77	3.1	52.7	17.2
Wiltshire County	47.0	43	23.5	17.9	76	4.9	52.3	18.4

1 Local government structure as at 1 April 1998. See Notes and Definitions.
2 Local authority provided and registered day nurseries only. A small number of places provided by facilities exempt from registration are excluded.
3 Figures relate to all pupils as a percentage of the three and four year old population. As some pupils are aged two, this can lead to participation rates greater than 100 per cent.
4 Public sector schools only.
5 Pupils and students aged 16 in education as a percentage of the 16 year old population (ages measured at the beginning of the academic year).
6 Pupils in their last year of compulsory schooling as a percentage of the school population of the same age.
7 Figures relate to maintained schools only; hence they are not directly comparable with those in Tables 4.4, 16.3 and 17.3 which are for all schools.
8 Figure for United Kingdom relates to England and Wales average.

Source: Department for Education and Employment

14.4 Housing and households: by local authority[1]

	Housing completions[2] 1999 (numbers)		Stock of dwellings 1991[3] (thousands)	Households 1998					Local authority tenants: average weekly unrebated rent per dwelling (£) April 1999[5]	Council Tax (£)[6] April 1999
	Private enterprise	Social registered landlords & councils		All house-holds (thousands)	Average household size (number of people)	Lone parents[4] as a percentage of all households	One-person households as a percentage of all households			
England	122,335	17,605	19,709	20,540	2.37	5.8	29.4	43.80	798	
North East	6,513	725	1,074	1,087	2.35	6.8	30.4	36.80	895	
Darlington UA	386	48	..	42	2.36	5.9	30.9	37.20	688	
Hartlepool UA	505	103	..	38	2.42	7.3	28.8	37.60	978	
Middlesbrough UA	241	75	..	58	2.45	10.5	29.4	43.20	806	
Redcar & Cleveland UA	180	73	..	58	2.38	6.6	28.8	40.80	1,033	
Stockton-on-Tees UA	993	53	..	74	2.42	6.6	28.7	39.30	853	
Durham County	1,333	76	..	211	2.37	5.9	28.5	37.70	896	
Chester-le-Street	20	9	22	24	2.34	4.8	27.2	35.00	830	
Derwentside	229	28	37	37	2.36	6.8	29.2	39.20	900	
Durham	308	0	34	36	2.44	5.5	28.2	39.20	848	
Easington	196	24	41	39	2.39	6.3	29.3	38.30	951	
Sedgefield	353	15	37	37	2.36	6.3	27.4	36.40	967	
Teesdale	68	0	11	11	2.33	2.9	30.1	36.80	845	
Wear Valley	159	0	27	27	2.32	6.6	28.9	37.20	869	
Northumberland	973	96	129	128	2.37	4.6	28.5	34.30	898	
Alnwick	110	20	14	13	2.36	4.9	27.2	37.50	901	
Berwick-upon-Tweed	..	..	13	12	2.21	3.3	34.6	37.00	893	
Blyth Valley	..	..	33	34	2.36	5.2	28.2	31.80	883	
Castle Morpeth	117	6	20	20	2.41	3.7	27.4	38.90	908	
Tynedale	139	40	24	24	2.40	4.4	28.7	40.00	903	
Wansbeck	172	0	26	26	2.40	5.1	27.7	30.50	902	
Tyne and Wear	1,902	200	474	478	2.31	7.3	32.5	35.60	906	
Gateshead	294	26	86	87	2.27	6.5	32.5	35.50	962	
Newcastle-upon-Tyne	251	83	120	119	2.28	8.0	35.9	36.70	978	
North Tyneside	765	16	84	86	2.24	6.3	32.6	32.20	905	
South Tyneside	..	..	66	67	2.29	8.0	33.1	31.70	884	
Sunderland	527	55	118	120	2.42	7.5	28.7	38.80	815	
Tees Valley	2,305	352	..	..	..	..	..	..	873	
Tees Valley less Darlington	1,919	304	222	550	2.42	7.7	28.9	..	908	
Former county of Durham	1,719	125	249	253	2.37	5.9	28.9	..	860	
North West	15,718	2,437	2,222	2,842	2.39	7.0	29.8	40.60	901	
Blackburn with Darwen UA	339	29	..	54	2.54	8.0	28.3	48.00	915	
Blackpool UA	88	26	..	64	2.24	5.5	33.0	35.10	683	
Halton UA	..	..	..	49	2.49	10.1	26.2	34.50	679	
Warrington UA	486	29	..	79	2.37	6.9	29.2	37.80	730	
Cheshire County	2,299	157	..	277	2.40	4.1	27.1	37.30	865	
Chester	268	6	49	50	2.33	4.9	30.6	38.40	866	
Congleton	419	0	34	36	2.45	2.8	25.7	.	860	
Crewe and Nantwich	484	2	43	47	2.43	4.0	26.8	38.10	865	
Ellesmere Port and Neston	173	0	32	33	2.43	5.4	25.7	29.70	876	
Macclesfield	253	65	63	65	2.33	3.8	27.5	40.30	858	
Vale Royal	..	..	45	47	2.46	3.8	25.1	39.90	872	
Cumbria	1,173	133	210	209	2.32	4.2	30.3	39.20	868	
Allerdale	136	45	41	40	2.34	3.9	31.4	.	856	
Barrow-in-Furness	114	13	31	30	2.33	5.4	29.5	43.40	893	
Carlisle	316	24	43	44	2.31	4.5	30.3	38.00	883	
Copeland	163	34	29	29	2.39	5.5	28.6	38.30	865	
Eden	122	0	20	21	2.35	3.1	27.6	.	860	
South Lakeland	322	17	46	45	2.25	3.0	32.1	39.20	856	
Greater Manchester (Met. County)	5,172	994	1,051	1,067	2.39	7.6	30.6	40.60	885	
Bolton	540	205	106	110	2.40	6.2	29.6	35.40	889	
Bury	252	54	72	75	2.40	5.5	28.7	39.70	811	
Manchester	1,013	368	184	178	2.38	13.2	35.9	47.60	987	
Oldham	475	23	89	90	2.41	7.4	29.7	35.50	929	
Rochdale	509	58	84	85	2.43	7.8	29.8	37.50	871	
Salford	451	68	98	96	2.33	8.6	33.9	40.80	980	
Stockport	289	60	118	124	2.35	5.8	29.4	34.80	917	
Tameside	229	96	91	91	2.40	6.2	28.4	41.40	899	
Trafford	236	62	87	91	2.40	6.3	30.0	40.00	689	
Wigan	..	..	123	128	2.41	5.4	27.3	35.80	835	

14.4 *(continued)*

| | Housing completions[2] 1999 (numbers) | | Stock of dwellings 1991[3] (thousands) | Households 1998 | | | | Local authority tenants: average weekly unrebated rent per dwelling (£) April 1999[5] | Council Tax (£)[6] April 1999 |
	Private enterprise	Social registered landlords & councils		All house-holds (thousands)	Average household size (number of people)	Lone parents[4] as a percentage of all households	One-person households as a percentage of all households		
Lancashire County	3,110	235	..	467	2.39	6.0	29.0	38.80	914
Burnley	204	8	38	37	2.40	8.5	29.0	41.60	948
Chorley	520	19	38	40	2.43	5.6	25.1	32.50	897
Fylde	222	42	31	33	2.21	3.5	32.4	35.50	885
Hyndburn	197	0	33	32	2.41	7.1	29.0	40.70	940
Lancaster	216	109	53	56	2.36	6.4	30.8	39.00	884
Pendle	157	18	36	34	2.41	6.4	30.8	37.80	947
Preston	282	8	52	55	2.41	7.2	32.8	41.10	959
Ribble Valley	..	..	21	21	2.47	3.5	27.7	35.90	898
Rossendale	112	11	27	26	2.40	6.1	27.8	39.40	944
South Ribble	417	10	40	42	2.45	4.9	24.7	.	890
West Lancashire	354	0	42	45	2.45	6.8	26.2	37.80	911
Wyre	285	5	43	45	2.29	4.4	30.0	.	895
Merseyside (Met. County)	2,845	825	575	576	2.41	8.9	30.3	42.90	1,020
Knowsley	..	..	57	60	2.55	14.8	25.4	44.90	970
Liverpool	895	343	194	188	2.42	9.6	32.8	41.10	1,172
St Helens	426	76	71	72	2.44	5.6	24.8	43.20	920
Sefton	384	80	116	118	2.37	7.2	30.8	41.60	967
Wirral	503	233	137	138	2.34	8.4	31.3	46.20	977
Former county of Cheshire	2,991	195	386	405	2.40	5.4	27.4	..	822
Former county of Lancashire	3,537	290	574	585	2.39	6.1	29.4	..	891
Yorkshire and the Humber	12,163	1,415	2,025	2,098	2.37	5.8	29.4	..	810
East Riding of Yorkshire UA	1,178	33	..	129	2.38	3.8	26.6	36.20	857
City of Kingston upon Hull UA	251	32	..	109	2.37	7.9	32.9	35.20	794
North East Lincolnshire UA	412	13	..	65	2.38	7.3	27.3	36.70	865
North Lincolnshire UA	541	0	..	62	2.41	5.4	25.6	34.70	997
York UA	662	220	..	75	2.33	4.5	29.8	41.50	695
North Yorkshire County	1,443	93	..	234	2.35	4.2	28.6	42.10	754
Craven	134	0	22	22	2.33	4.6	30.3	42.70	760
Hambleton	205	36	32	34	2.45	3.3	26.0	.	695
Harrogate	488	19	60	62	2.33	4.4	29.2	45.30	765
Richmondshire	124	0	19	19	2.44	4.0	27.2	41.40	772
Ryedale	92	13	29	20	2.35	2.7	27.9	.	779
Scarborough	124	19	49	47	2.22	5.1	32.1	41.60	757
Selby	276	3	36	29	2.43	3.9	24.6	39.20	772
South Yorkshire (Met. County)	2,520	330	528	546	2.36	5.8	29.0	32.80	828
Barnsley	601	5	91	93	2.42	5.8	26.1	32.50	801
Doncaster	946	53	116	120	2.40	6.0	26.9	33.30	761
Rotherham	570	177	101	105	2.41	5.8	26.9	30.20	809
Sheffield	403	95	221	228	2.31	5.7	32.4	33.90	886
West Yorkshire (Met. County)	5,156	624	840	878	2.38	6.4	30.2	35.80	804
Bradford	958	274	183	192	2.49	7.2	29.6	37.50	808
Calderdale	454	87	81	82	2.33	5.9	30.1	35.90	876
Kirklees	..	..	155	162	2.39	6.1	30.1	40.20	870
Leeds	1,772	145	293	310	2.32	6.7	31.6	34.20	768
Wakefield	1,104	42	128	132	2.39	4.8	28.1	34.50	756
The Humber	2,382	148	356	365	2.38	5.9	28.4	..	867
Former county of North Yorkshire	2,105	313	301	309	2.34	4.2	28.9	..	741
East Midlands	14,986	1,159	1,638	1,718	2.40	5.1	27.3	38.10	834
Derby UA	593	26	..	97	2.39	6.6	29.7	37.00	787
Leicester UA	394	272	..	114	2.54	8.5	30.8	42.90	792
Nottingham UA	..	..	..	119	2.37	10.0	31.7	35.80	886
Rutland UA	..	..	..	14	2.46	6.0	23.7	45.20	945
Derbyshire County	2,729	179	..	307	2.37	4.1	27.1	35.20	873
Amber Valley	244	15	47	49	2.35	3.6	26.8	39.60	873
Bolsover	344	1	29	29	2.43	3.9	24.9	32.00	905
Chesterfield	184	37	43	43	2.28	4.3	30.8	33.20	855
Derbyshire Dales	193	15	29	29	2.37	3.0	27.0	36.80	878
Erewash	318	45	44	45	2.35	5.0	27.8	34.80	857
High Peak	699	46	35	37	2.39	4.7	27.5	41.70	866
North East Derbyshire	209	20	40	41	2.35	4.1	26.1	32.20	905
South Derbyshire	538	0	29	32	2.48	3.8	24.5	37.40	855

14.4 (continued)

	Housing completions[2] 1999 (numbers)		Stock of dwellings 1991[3] (thousands)	Households 1998				Local authority tenants: average weekly unrebated rent per dwelling (£) April 1999[5]	Council Tax (£)[6] April 1999
	Private enterprise	Social registered landlords & councils		All households (thousands)	Average household size (number of people)	Lone parents[4] as a percentage of all households	One-person households as a percentage of all households		
Leicestershire County	2,236	72	..	243	2.44	3.7	25.3	37.50	803
Blaby	380	14	33	35	2.48	3.5	22.0	35.50	814
Charnwood	332	40	57	63	2.44	4.1	27.1	35.60	804
Harborough	620	0	27	30	2.46	3.0	25.8	44.50	809
Hinckley and Bosworth	245	0	39	41	2.38	3.4	26.2	40.20	770
Melton	142	0	19	19	2.40	3.5	25.4	35.70	799
North West Leicestershire	402	13	32	35	2.44	4.0	25.7	37.20	828
Oadby and Wigston	..	..	21	21	2.51	4.0	22.5	35.30	802
Lincolnshire	3,547	222	250	262	2.33	4.3	27.1	37.40	787
Boston	161	64	23	23	2.33	3.9	28.0	37.70	789
East Lindsey	..	..	52	53	2.29	3.8	27.5	.	774
Lincoln	240	30	36	37	2.23	6.2	33.8	34.60	801
North Kesteven	943	0	33	36	2.36	3.9	23.4	36.30	798
South Holland	556	0	29	31	2.33	2.9	25.1	40.90	795
South Kesteven	620	19	46	51	2.40	4.9	25.9	39.30	769
West Lindsey	487	26	31	32	2.41	3.9	25.9	37.40	806
Northamptonshire	2,612	155	236	249	2.44	5.2	25.9	40.50	765
Corby	76	0	21	20	2.51	9.5	22.2	38.10	768
Daventry	..	..	25	27	2.47	3.4	23.4	39.10	720
East Northamptonshire	734	6	28	30	2.44	3.7	24.9	38.10	772
Kettering	322	49	32	34	2.42	4.5	26.5	38.80	774
Northampton	402	36	75	80	2.42	5.8	28.4	43.50	790
South Northamptonshire	335	19	28	31	2.47	3.5	23.7	46.30	797
Wellingborough	222	0	28	28	2.39	6.5	26.8	36.40	682
Nottinghamshire County	2,427	176	..	311	2.37	4.4	26.9	38.80	925
Ashfield	464	20	44	45	2.39	4.6	26.6	37.70	930
Bassetlaw	..	..	43	44	2.37	4.1	26.0	41.00	919
Broxtowe	88	0	45	47	2.32	4.8	27.6	34.60	918
Gedling	232	24	45	47	2.34	3.5	27.8	35.10	909
Mansfield	..	..	42	41	2.39	5.8	26.6	42.30	935
Newark and Sherwood	424	46	42	43	2.40	4.6	26.4	37.60	970
Rushcliffe	534	21	40	44	2.36	3.8	27.6	40.00	900
Former county of Derbyshire	3,322	205	387	404	2.37	4.7	27.7	..	854
Former county of Leicestershire	2,769	351	351	371	2.47	5.3	26.9	..	806
Former county of Nottinghamshire	2,736	226	414	430	2.37	6.0	28.3	..	916
West Midlands	12,525	2,566	2,083	2,156	2.44	5.9	27.9	39.80	812
County of Herefordshire UA	571	149	..	68	2.42	4.2	26.4	38.70	729
Stoke-on-Trent UA	526	202	..	103	2.43	5.8	28.3	39.60	751
Telford and Wrekin UA	954	162	..	59	2.51	6.4	24.2	.	757
Shropshire County	867	43	..	115	2.39	3.8	27.9	38.70	791
Bridgnorth	..	..	20	21	2.44	4.1	25.3	40.20	783
North Shropshire	248	6	21	22	2.43	2.4	26.6	35.30	811
Oswestry	102	27	14	15	2.32	3.9	30.3	38.80	805
Shrewsbury and Atcham	316	2	38	40	2.37	4.8	30.0	39.80	776
South Shropshire	162	0	17	17	2.39	2.9	25.6	.	801
Staffordshire County	2,569	340	..	325	2.46	4.1	24.7	38.10	760
Cannock Chase	383	44	34	36	2.52	4.9	23.1	42.90	783
East Staffordshire	..	..	40	42	2.46	3.8	26.0	36.30	786
Lichfield	291	72	35	38	2.46	3.2	22.7	.	766
Newcastle-under-Lyme	294	66	49	50	2.42	4.4	28.6	33.50	770
South Staffordshire	173	51	40	42	2.42	3.7	22.9	.	688
Stafford	295	0	47	50	2.48	3.9	25.1	37.30	763
Staffordshire Moorlands	274	0	38	38	2.44	3.4	24.3	38.00	789
Tamworth	423	75	26	29	2.54	6.4	22.4	42.80	736
Warwickshire	2,016	306	197	209	2.40	4.5	26.9	40.30	823
North Warwickshire	78	0	24	25	2.46	4.3	26.0	37.20	870
Nuneaton and Bedworth	419	90	46	48	2.47	5.2	24.9	39.90	851
Rugby	350	60	34	36	2.39	5.1	26.9	40.00	839
Stratford-on-Avon	561	124	44	48	2.36	2.8	27.6	.	790
Warwick	608	32	49	52	2.36	5.0	28.4	42.70	804

14.4 *(continued)*

| | Housing completions[2] 1999 (numbers) | | Stock of dwellings 1991[3] (thousands) | Households 1998 | | | | Local authority tenants: average weekly unrebated rent per dwelling (£) April 1999[5] | Council Tax (£)[6] April 1999 |
	Private enterprise	Social registered landlords & councils		All households (thousands)	Average household size (number of people)	Lone parents[4] as a percentage of all households	One-person households as a percentage of all households		
West Midlands (Met. County)	3,375	1,182	1,036	1,056	2.46	7.2	29.8	40.10	865
Birmingham	1,041	683	392	404	2.48	9.0	32.2	41.40	893
Coventry	483	69	123	125	2.40	8.1	31.1	37.70	968
Dudley	372	167	123	127	2.43	4.3	26.7	40.00	779
Sandwell	562	30	120	118	2.45	6.0	29.7	44.60	856
Solihull	318	38	79	84	2.44	5.2	25.7	41.70	742
Walsall	409	59	101	104	2.50	5.4	26.3	33.10	803
Wolverhampton	152	116	99	96	2.50	7.7	29.5	37.80	955
Worcestershire County	1,647	182	..	221	2.40	4.6	25.9	40.10	736
Bromsgrove	158	7	36	35	2.41	3.1	22.7	37.20	721
Malvern Hills	143	75	36	30	2.41	3.5	27.0	.	728
Redditch	300	61	30	31	2.48	6.6	24.2	41.80	759
Worcester	410	6	34	40	2.33	6.0	28.4	37.60	725
Wychavon	447	25	41	46	2.39	4.1	25.6	.	728
Wyre Forest	189	6	38	40	2.40	4.4	27.2	41.60	757
Herefordshire and Worcestershire	2,218	331	274	289	2.41	4.5	26.0	..	734
Former county of Shropshire	1,821	205	165	174	2.43	4.7	26.6	..	781
Former county of Staffordshire	3,095	542	411	427	2.45	4.5	25.5	..	758
East	16,472	1,894	2,098	2,224	2.38	4.7	27.6	45.60	768
Luton UA	..	..	..	73	2.50	6.7	27.3	50.30	707
Peterborough UA	532	34	..	65	2.38	6.9	29.2	47.30	754
Southend-on-Sea UA	85	68	..	77	2.25	6.0	32.1	47.60	677
Thurrock UA	773	53	..	54	2.50	6.1	23.8	46.70	672
Bedfordshire County	1,011	215	..	154	2.40	4.5	27.0	46.90	877
Bedford	405	99	54	59	2.36	5.1	29.3	.	864
Mid Bedfordshire	..	..	44	50	2.42	3.2	25.9	44.50	859
South Bedfordshire	255	0	43	45	2.41	5.1	25.2	48.20	912
Cambridgeshire County	2,308	109	..	232	2.39	4.4	27.1	47.40	723
Cambridge	..	..	41	52	2.26	6.7	35.4	43.00	737
East Cambridgeshire	..	..	25	30	2.45	2.5	24.8	.	703
Fenland	459	9	32	34	2.35	3.4	27.1	43.50	727
Huntingdonshire	..	..	58	63	2.44	5.0	23.7	53.80	737
South Cambridgeshire	680	32	47	53	2.46	3.0	24.3	49.00	703
Essex County	4,437	608	..	535	2.39	4.4	27.4	47.30	786
Basildon	622	328	64	68	2.40	5.6	27.2	48.50	806
Braintree	569	28	49	54	2.39	4.6	26.9	44.30	778
Brentwood	103	2	29	29	2.40	3.6	27.6	50.60	762
Castle Point	92	24	34	34	2.44	3.4	23.8	54.90	797
Chelmsford	447	61	61	64	2.40	3.9	26.8	50.10	781
Colchester	585	32	59	64	2.41	6.0	27.2	45.20	767
Epping Forest	348	27	48	50	2.37	4.0	27.3	48.80	784
Harlow	408	0	30	31	2.40	6.3	29.4	46.60	872
Maldon	327	34	22	23	2.45	3.2	26.2	.	764
Rochford	369	15	29	31	2.45	3.6	25.3	45.20	790
Tendring	327	15	59	59	2.22	3.6	32.8	43.60	769
Uttlesford	240	42	26	28	2.47	3.5	24.7	48.00	783
Hertfordshire	2,703	277	394	421	2.42	4.5	27.6	49.40	763
Broxbourne	409	55	32	33	2.49	3.6	23.8	55.40	723
Dacorum	92	0	54	57	2.37	5.2	28.9	45.90	743
East Hertfordshire	397	22	47	51	2.48	3.0	25.5	49.90	745
Hertsmere	262	25	35	39	2.46	4.7	26.3	.	763
North Hertfordshire	..	..	46	49	2.36	4.0	29.4	55.10	765
St Albans	..	..	50	53	2.42	4.0	27.9	49.80	770
Stevenage	..	..	30	32	2.47	6.9	25.8	47.80	762
Three Rivers	..	..	31	35	2.41	3.8	28.2	50.90	767
Watford	104	5	30	34	2.40	6.3	30.2	57.00	832
Welwyn Hatfield	121	17	39	39	2.43	4.7	29.4	42.60	781

14.4 (continued)

	Housing completions[2] 1999 (numbers)		Stock of dwellings 1991[3] (thousands)	All house-holds (thousands)	Households 1998			Local authority tenants: average weekly unrebated rent per dwelling (£) April 1999[5]	Council Tax (£)[6] April 1999
	Private enterprise	Social registered landlords & councils			Average household size (number of people)	Lone parents[4] as a percentage of all households	One-person households as a percentage of all households		
Norfolk	2,369	293	328	336	2.31	4.6	28.2	36.50	779
Breckland	396	72	46	49	2.35	4.2	26.3	.	750
Broadland	607	44	44	48	2.41	3.2	23.3		771
Great Yarmouth	215	47	38	38	2.31	5.3	29.8	33.60	775
Kings Lynn and West Norfolk	304	0	58	43	2.26	3.8	29.7	35.00	775
North Norfolk	235	28	45	57	2.16	7.6	35.1	40.40	779
Norwich	124	84	55	45	2.35	3.8	25.1	35.90	818
South Norfolk	488	18	43	56	2.35	3.9	27.5	41.60	788
Suffolk	2,096	177	269	278	2.38	4.2	27.5	41.60	761
Babergh	211	3	33	33	2.37	3.7	25.6	45.50	765
Forest Heath	102	45	22	26	2.60	5.7	25.6	40.70	730
Ipswich	165	44	50	48	2.33	6.2	30.6	39.30	829
Mid Suffolk	336	16	33	33	2.42	2.8	25.3	40.00	760
St Edmundsbury	429	31	37	40	2.36	3.6	25.5	45.00	745
Suffolk Coastal	526	10	47	50	2.38	3.1	28.3	.	753
Waveney	327	22	48	46	2.30	4.5	29.2	40.00	730
Former county of Bedfordshire	1,169	275	209	227	2.43	5.2	27.1	..	829
Former county of Cambridgeshire	2,840	143	267	297	2.39	4.9	27.6	..	729
Former county of Essex	5,295	729	632	666	2.38	4.8	27.7	..	767
London	9,565	2,885	2,916	3,061	2.32	7.8	33.7	..	731
Inner London	4,533	1,322	..	1,239	2.18	10.2	38.7	63.80	684
Inner London - West	1,629	434	..	478	2.04	7.5	43.2	..	575
Camden	147	123	85	86	2.09	7.9	44.0	59.80	897
City of London	169	21	3	3	1.75	4.2	52.8	58.60	536
Hammersmith and Fulham	84	31	74	76	2.03	8.9	41.2	58.00	827
Kensington and Chelsea	304	22	79	85	1.94	6.8	48.7	68.20	581
Wandsworth	476	90	115	121	2.15	8.2	35.5	65.60	373
Westminster	449	147	101	107	1.96	5.9	48.1	74.00	350
Inner London - East	2,904	888	..	761	2.28	12.0	35.8	56.10	769
Hackney	..	..	80	84	2.28	12.8	36.9	56.60	790
Haringey	38	75	89	98	2.24	10.5	35.7	58.90	898
Islington	425	27	77	81	2.15	11.5	38.7	60.00	912
Lambeth	..	..	114	125	2.13	13.7	37.2	57.70	642
Lewisham	217	76	104	107	2.25	11.3	33.2	52.20	728
Newham	..	..	85	87	2.63	11.1	30.3	49.40	704
Southwark	..	..	104	104	2.21	12.5	38.0	54.70	809
Tower Hamlets	..	..	70	74	2.41	11.8	37.0	58.70	674
Outer London	5,032	1,563	..	1,822	2.40	6.1	30.4	..	760
Outer London - East and North East	2,122	750	..	627	2.43	6.4	29.5	54.60	781
Barking and Dagenham	..	..	60	62	2.49	7.3	30.6	49.70	738
Bexley	228	0	89	90	2.42	5.0	26.5	.	750
Enfield	88	87	106	108	2.44	6.2	29.0	55.60	733
Greenwich	..	..	89	89	2.38	9.9	32.0	56.30	883
Havering	290	55	92	93	2.43	4.6	26.4	46.80	790
Redbridge	476	43	92	92	2.50	4.4	29.4	67.70	750
Waltham Forest	82	429	91	94	2.35	7.9	33.2	59.50	840
Outer London - South	925	518	..	481	2.35	5.6	30.4	59.80	741
Bromley	..	..	124	126	2.32	4.6	30.5	.	670
Croydon	159	262	131	140	2.39	7.1	29.6	64.20	758
Kingston upon Thames	..	..	57	61	2.37	4.3	32.2	64.00	794
Merton	151	61	73	78	2.35	5.8	30.1	54.80	787
Sutton	109	99	72	75	2.33	5.2	30.7	54.20	749
Outer London - West and North West	1,985	295	..	714	2.42	6.2	31.1	61.50	757
Barnet	672	36	121	133	2.45	5.6	31.0	55.40	761
Brent	..	..	99	103	2.43	9.8	31.3	72.60	678
Ealing	135	124	113	124	2.40	7.2	32.4	60.90	703
Harrow	95	3	79	83	2.52	4.7	26.9	66.80	788
Hillingdon	..	..	95	102	2.43	5.3	28.9	66.70	764
Hounslow	..	..	83	86	2.44	6.1	30.5	52.90	795
Richmond-upon-Thames	416	36	73	83	2.23	4.2	36.6	61.10	834

14.4 *(continued)*

	Housing completions[2] 1999 (numbers)			Households 1998				Local authority tenants: average weekly	
	Private enterprise	Social registered landlords & councils	Stock of dwellings 1991[3] (thousands)	All house- holds (thousands)	Average household size (number of people)	Lone parents[4] as a percentage of all households	One-person households as a percentage of all households	unrebated rent per dwelling (£) April 1999[5]	Council Tax (£)[6] April 1999
South East	20,004	2,958	3,106	3,302	2.38	4.6	28.2	50.30	764
Bracknell Forest UA	175	3	..	44	2.47	5.1	24.8	52.80	711
Brighton and Hove UA	434	116	..	117	2.13	5.7	37.7	46.10	698
Isle of Wight UA	356	42	..	54	2.27	4.4	30.5	.	793
Medway UA	412	78	..	96	2.49	5.4	24.6	47.10	670
Milton Keynes UA	1,200	239	..	83	2.42	7.0	26.6	39.60	744
Portsmouth UA	123	89	..	78	2.38	7.1	31.4	48.70	682
Reading UA	192	48	..	63	2.31	6.3	31.7	60.90	868
Slough UA	446	0	..	45	2.48	7.5	28.3	53.20	702
Southampton UA	285	39	..	91	2.35	6.5	32.0	42.20	709
West Berkshire UA (Newbury)	..	..	..	57	2.48	3.8	24.6	.	827
Windsor and Maidenhead UA	280	126	..	58	2.40	3.2	28.0	.	744
Wokingham UA	521	25	..	56	2.55	2.9	22.3	50.60	825
Buckinghamshire County	1,091	130	..	193	2.45	3.6	25.4	51.90	766
Aylesbury Vale	667	51	58	63	2.45	3.3	25.3	49.00	762
Chiltern	112	0	35	38	2.43	3.8	26.0	.	774
South Buckinghamshire	157	8	25	26	2.44	4.0	24.4	.	757
Wycombe	..	..	62	66	2.46	3.7	25.7	55.40	769
East Sussex County	1,090	170	..	214	2.24	4.6	32.3	46.80	804
Eastbourne	..	..	40	41	2.11	5.8	36.3	46.00	789
Hastings	91	50	37	36	2.19	6.6	35.9	.	813
Lewes	82	0	39	38	2.23	4.0	31.8	50.90	811
Rother	161	8	39	40	2.24	3.7	32.0	.	775
Wealden	..	..	55	59	2.36	3.4	27.9	43.60	822
Hampshire County	3,663	399	..	506	2.40	4.4	25.5	53.80	779
Basingstoke and Deane	599	101	56	60	2.45	4.5	24.2	.	759
East Hampshire	..	..	41	45	2.43	4.1	25.1	.	800
Eastleigh	267	24	43	46	2.43	4.4	23.6	.	794
Fareham	525	48	40	43	2.38	3.7	24.0	46.00	763
Gosport	44	0	31	32	2.38	7.7	26.3	49.70	786
Hart	147	85	30	35	2.44	3.4	23.0	.	779
Havant	168	9	48	49	2.39	5.2	26.4	.	775
New Forest	406	60	69	73	2.30	3.7	28.1	53.00	799
Rushmoor	196	7	31	35	2.42	5.1	25.4	.	779
Test Valley	712	8	40	45	2.43	4.2	25.0	53.10	756
Winchester	353	15	39	44	2.46	3.1	27.9	62.30	774
Kent County	3,337	430	..	550	2.38	4.6	28.3	50.30	765
Ashford	..	..	38	41	2.40	4.8	27.4	51.90	746
Canterbury	369	15	53	58	2.35	4.6	30.2	49.60	756
Dartford	250	46	32	35	2.41	4.0	26.9	50.50	766
Dover	187	0	44	45	2.33	5.1	30.8	55.60	768
Gravesham	91	36	37	37	2.44	5.7	26.2	48.60	731
Maidstone	392	87	54	57	2.43	4.1	25.7	50.70	797
Sevenoaks	74	12	43	45	2.45	3.6	26.6	.	775
Shepway	436	14	42	43	2.26	5.8	31.7	45.70	780
Swale	418	112	47	48	2.45	4.8	24.9	.	744
Thanet	140	3	56	55	2.24	6.0	32.5	47.30	776
Tonbridge and Malling	252	40	40	42	2.48	3.8	24.6	.	771
Tunbridge Wells	209	21	41	43	2.35	3.3	30.7	.	758
Oxfordshire	2,093	408	219	244	2.48	4.7	26.5	49.20	771
Cherwell	657	31	47	53	2.50	5.1	24.5	47.00	759
Oxford	539	10	46	57	2.45	7.0	32.3	49.90	826
South Oxfordshire	..	..	47	50	2.52	3.8	25.0	.	786
Vale of White Horse	230	52	43	45	2.49	2.8	24.4	.	749
West Oxfordshire	..	..	36	39	2.45	4.1	24.9	50.00	727

14.4 (continued)

	Housing completions[2] 1999 (numbers)		Stock of dwellings 1991[3] (thousands)	All house-holds (thousands)	Average household size (number of people)	Lone parents[4] as a percentage of all households	One-person households as a percentage of all households	Local authority tenants: average weekly unrebated rent per dwelling (£) April 1999[5]	Council Tax (£)[6] April 1999
	Private enterprise	Social registered landlords & councils							
Surrey	2,111	305	412	433	2.40	3.6	27.7	56.70	764
Elmbridge	438	29	48	53	2.42	4.4	27.8	57.80	780
Epsom and Ewell	140	11	26	28	2.41	3.6	29.1	.	749
Guildford	..	..	50	51	2.42	4.0	27.9	58.00	766
Mole Valley	94	2	32	33	2.32	2.7	29.2	48.30	745
Reigate and Banstead	189	79	48	49	2.38	3.3	28.1	57.00	774
Runnymede	245	14	29	32	2.34	3.1	29.3	61.80	689
Spelthorne	127	45	38	38	2.33	3.2	27.9	.	761
Surrey Heath	217	13	30	33	2.50	3.9	22.9	.	759
Tandridge	243	50	30	32	2.44	3.0	27.2	47.90	772
Waverley	257	0	46	47	2.42	3.9	28.0	57.40	782
Woking	176	58	35	38	2.39	4.3	26.9	62.30	791
West Sussex	1,737	306	304	321	2.29	4.0	30.4	53.10	759
Adur	67	77	25	25	2.32	4.1	30.0	51.70	800
Arun	375	60	60	63	2.17	3.5	32.4	56.40	766
Chichester	..	..	45	46	2.27	4.0	30.9	50.40	749
Crawley	..	..	35	39	2.45	6.2	26.0	50.30	755
Horsham	536	105	45	50	2.38	3.0	27.7	59.90	744
Mid Sussex	150	20	49	52	2.38	3.8	27.8	.	763
Worthing	76	5	44	45	2.14	4.2	37.4	.	757
Former county of Berkshire	2,072	206	290	323	2.44	4.7	26.7	..	785
Former county of Buckinghamshire	2,291	369	250	276	2.44	4.6	25.8	..	761
Former county of East Sussex	1,524	287	321	331	2.20	5.0	34.2	..	771
Former county of Hampshire	4,071	527	629	674	2.39	5.0	27.1	..	762
Former county of Kent	3,749	508	924	646	2.40	4.8	27.7	..	752
South West	14,389	1,566	1,973	2,052	2.34	4.7	28.8	43.70	782
Bath and North East Somerset UA	337	8	..	70	2.33	4.4	29.2	41.20	795
Bournemouth UA	212	0	..	72	2.17	5.4	34.5	45.90	738
City of Bristol UA	197	60	..	172	2.30	6.8	32.4	40.70	992
North Somerset UA	..	..	..	78	2.35	4.0	27.4	51.30	740
Plymouth UA	..	..	..	104	2.36	6.7	28.8	39.50	702
Poole UA	316	16	..	60	2.33	4.2	29.3	45.00	703
South Gloucestershire UA	..	..	..	97	2.45	4.1	24.8	44.40	782
Swindon UA	..	..	..	74	2.40	5.3	25.7	38.80	699
Torbay UA	315	65	..	53	2.25	6.2	31.7	45.40	752
Cornwall and the Isles of Scilly	1,802	120	207	205	2.34	4.8	28.0	43.90	748
Caradon	415	0	34	34	2.37	4.5	25.5	41.50	745
Carrick	207	39	37	36	2.29	4.2	29.6	45.20	756
Kerrier	..	..	37	37	2.38	5.5	26.8	.	757
North Cornwall	297	12	34	34	2.34	4.6	27.7	42.00	759
Penwith	154	8	28	26	2.26	5.3	29.9	.	738
Restormel	..	..	36	38	2.36	4.9	28.8	46.90	740
Isles of Scilly	2	4	1	1	2.56	5.9	20.9	45.70	532
Devon County	2,710	311	..	292	2.31	4.0	29.7	42.80	776
East Devon	495	12	54	55	2.21	3.8	31.5	39.00	761
Exeter	157	75	41	47	2.33	5.2	33.0	37.70	749
Mid Devon	..	..	27	37	2.30	4.4	29.5	41.20	794
North Devon	281	76	37	34	2.32	3.5	28.2	51.70	791
South Hams	302	13	37	50	2.30	3.6	28.7	.	777
Teignbridge	584	98	47	28	2.36	3.5	27.6	47.20	783
Torridge	425	12	23	22	2.43	3.6	27.0	45.70	773
West Devon	181	0	19	20	2.35	3.8	28.4	.	802
Dorset County	1,226	168	..	166	2.28	3.8	29.8	47.20	829
Christchurch	126	0	20	20	2.16	4.3	33.1	.	816
East Dorset	274	28	33	36	2.29	3.0	26.0	.	844
North Dorset	224	61	23	25	2.38	3.9	28.7	.	809
Purbeck	76	36	19	19	2.34	2.8	28.4	50.80	826
West Dorset	..	..	39	40	2.25	3.0	32.1	.	831
Weymouth and Portland	176	19	26	27	2.26	6.5	30.7	45.20	834

14.4 *(continued)*

	Housing completions[2] 1999 (numbers)		Stock of dwellings 1991[3] (thousands)	Households 1998					Local authority tenants: average weekly unrebated rent per dwelling (£) April 1999[5]	Council Tax (£)[6] April 1999
	Private enterprise	Social registered landlords & councils		All house-holds (thousands)	Average household size (number of people)	Lone parents[4] as a percentage of all households	One-person households as a percentage of all households		
Gloucestershire	1,440	224	222	233	2.36	4.5	28.5	47.40	771
Cheltenham	138	29	45	47	2.22	5.4	34.9	52.20	762
Cotswold	173	64	33	35	2.37	3.3	28.0	.	760
Forest of Dean	167	13	31	31	2.44	3.3	25.2	42.00	787
Gloucester	365	85	42	44	2.38	6.6	28.2	46.90	761
Stroud	326	33	43	45	2.40	4.0	26.7	47.10	818
Tewkesbury	271	0	29	31	2.37	3.4	26.2		726
Somerset	1,558	242	195	203	3.51	4.1	28.5	42.90	770
Mendip	249	29	40	40	2.45	4.2	28.7	43.50	779
Sedgemoor	..	..	41	44	2.35	4.2	27.2	45.80	761
South Somerset	388	58	60	64	2.38	3.6	27.7	.	781
Taunton Deane	568	91	40	42	2.32	5.1	30.2	40.60	752
West Somerset	92	8	15	14	2.20	2.9	30.0	.	777
Wiltshire County	1,317	204	..	171	2.43	4.3	25.2	52.90	781
Kennet	..	..	28	30	2.53	4.2	23.7	.	769
North Wiltshire	396	0	45	49	2.46	4.7	23.3	.	790
Salisbury	246	26	43	46	2.40	4.3	26.1	54.20	771
West Wiltshire	560	99	44	45	2.36	3.8	27.3	51.10	791
Bristol/Bath area	2,753	827	391	418	2.35	5.2	29.2	..	852
Former county of Devon	3,266	519	440	449	2.31	4.9	29.7	..	759
Former county of Dorset	1,754	390	287	297	2.26	4.3	30.8	..	783
Former county of Wiltshire	1,816	246	231	246	2.42	4.6	25.3	..	759

1 The table reflects the local government structure at 1 April 1998. For some new areas data are not available. See Notes and Definitions.
2 District figures do not always add to county totals. See Notes and Definitions.
3 The figures for housing stock at local authority level shown in this table are derived using different methods from the regional stock figures shown in Table 6.1. This has led to small discrepancies between the two sets of figures. The figures in Table 6.1 provide the definitive regional estimates.
4 Lone parents with dependent children only.
5 Some local authorities have no housing stock following large scale voluntary transfers to social registered landlords.
6 See Notes and Definitions.

Source: Department of the Environment, Transport and the Regions; Department of Social Security

14.5 Labour market statistics[1]: by sub-region

	Total in employment[2] 1998-1999[3] (thousands)	Employment rate 1998-1999[3] (percentages)[3]	ILO unemployment rate 1998-1999[3] (percentages)[4]	Average gross weekly full-time earnings[5], April 1999 (£)						
				Males			Females			All persons total
				Total	10 per cent earned		Total	10 per cent earned		
					Less than	More than		Less than	More than	
United Kingdom	26,508	73.5	6.3	440.7	210.0	709.9	325.6	168.8	518.9	398.7
England	22,404	74.2	6.0	448.1	212.3	721.2	330.6	170.9	529.2	405.4
North East	1,056	67.1	8.4	384.6	199.8	595.8	289.8	156.0	466.4	349.6
Darlington UA	46	74.3	..	371.1	209.9	548.2	..	..	..	339.0
Hartlepool UA	35	67.2	..	364.7	202.4	567.2	..	..	..	332.2
Middlesbrough UA	53	60.9	14.5	390.4	198.8	678.9	306.4	168.0	476.1	362.3
Redcar & Cleveland UA	57	66.0	..	446.6	206.7	641.1	..	..	..	403.1
Stockton-on-Tees UA	80	69.9	8.2	413.7	190.6	662.2	260.4	157.4	394.8	361.4
Durham County	211	67.1	8.1	382.1	201.2	576.9	283.3	151.8	445.2	343.4
Northumberland	131	73.3	..	343.4	183.9	518.2	276.2	142.1	448.8	314.9
Tyne and Wear	441	65.2	9.5	384.3	199.8	598.0	299.5	158.3	473.7	352.6
Tees Valley	272	67.4	..	..	..	..	..	..	..	..
Tees Valley less Darlington	225	66.2	..	..	..	..	..	..	..	..
Former county of Durham	257	68.2	7.6	..	..	..	..	..	..	..
North West	2,933	70.3	6.5	415.1	205.0	666.1	299.4	166.0	474.9	372.6
Blackburn with Darwen UA	50	62.0	..	350.5	197.6	516.5	..	..	..	332.9
Blackpool UA	59	75.3	..	..	..	..	288.0	165.1	473.7	319.1
Halton UA	50	64.0	11.7	415.1	231.6	657.8	306.4	159.0	505.8	379.5
Warrington UA	97	80.0	..	405.4	215.0	658.4	291.0	152.5	472.7	365.0
Cheshire County	316	74.5	5.1	449.1	215.3	722.8	304.5	170.8	496.8	396.5
Cumbria	223	72.6	6.1	405.8	200.0	667.2	278.4	160.4	450.3	361.1
Greater Manchester (Met. County)	1,126	70.4	6.2	420.0	205.0	678.6	304.9	169.2	482.3	377.0
Lancashire County	500	75.7	4.3	400.3	200.0	633.4	286.8	153.9	453.5	363.6
Merseyside (Met. County)	513	62.2	11.0	416.7	209.9	641.2	304.0	168.4	487.1	371.8
Former county of Cheshire	462	74.2	5.3	..	..	..	..	..	..	..
Former county of Lancashire	609	74.4	4.3	..	..	..	..	..	..	..
Yorkshire and the Humber	2,230	72.9	7.0	395.8	203.1	614.1	297.9	163.0	472.8	361.0
East Riding of Yorkshire UA	139	75.9	5.0	380.7	191.0	576.9	275.1	144.0	445.2	351.3
City of Kingston upon Hull UA	99	64.5	13.8	373.2	190.5	596.0	292.9	161.3	477.9	344.3
North East Lincolnshire UA	61	66.3	9.7	394.8	208.7	601.1	..	..	..	359.1
North Lincolnshire UA	71	72.6	9.0	..	..	..	269.3	152.6	444.0	..
York UA	81	76.5	..	421.4	202.4	665.5	293.6	172.7	439.2	382.0
North Yorkshire County	270	80.0	..	392.7	192.5	629.6	267.4	152.1	445.2	347.2
South Yorkshire (Met. County)	550	68.8	9.1	380.9	203.8	586.2	297.8	158.7	472.7	350.6
West Yorkshire (Met. County)	959	74.2	6.2	402.4	208.1	625.0	309.3	168.6	491.4	368.2
The Humber	..	70.3	..	..	..	..	..	..	..	..
Former county of North Yorkshire	352	79.1	3.5	..	..	..	..	..	..	..
East Midlands	1,925	75.7	5.2	398.3	205.2	610.6	286.7	157.5	460.7	361.2
Derby UA	102	72.3	6.7	451.9	217.4	659.3	313.2	160.4	500.9	414.8
Leicester UA	112	67.2	8.6	391.6	180.0	616.4	293.4	164.2	460.7	351.7
Nottingham UA	119	65.7	7.9	398.0	201.5	671.8	294.6	168.0	436.5	361.1
Rutland UA	13	76.2	..	..	..	..	..	..	..	..
Derbyshire County	336	74.9	5.1	392.0	202.5	556.9	275.4	155.8	449.2	357.6
Leicestershire County	314	83.0	3.4	406.2	218.8	627.9	289.0	157.3	479.9	369.3
Lincolnshire	284	75.2	4.9	367.3	206.6	576.9	270.1	146.0	449.8	335.5
Northamptonshire	300	80.2	4.1	416.0	221.5	625.7	295.2	175.2	430.1	376.3
Nottinghamshire County	347	75.1	5.9	383.3	191.9	602.0	277.5	149.7	470.2	348.7
Former county of Derbyshire	438	74.3	5.5	..	..	..	..	..	..	..
Former county of Leicestershire	438	78.0	4.7	..	..	..	..	..	..	..
Former county of Nottinghamshire	465	72.4	6.4	..	..	..	..	..	..	..
West Midlands	2,398	74.2	6.6	414.6	210.6	640.0	301.0	166.5	483.2	375.6
County of Herefordshire UA	70	77.7	..	340.0	203.1	540.9	286.4	153.0	473.4	320.0
Stoke-on-Trent UA	100	68.0	7.4	367.1	187.2	568.1	284.5	164.8	471.8	334.9
Telford and Wrekin UA	71	75.5	..	377.9	200.8	568.1	255.8	157.7	400.6	334.7
Shropshire County	133	75.3	..	371.8	197.1	576.9	277.9	160.9	445.2	339.3
Staffordshire County	392	78.7	4.9	405.3	210.0	614.4	298.5	160.0	513.1	369.9
Warwickshire	257	80.1	3.7	448.7	215.0	673.7	303.8	177.2	471.7	401.6
West Midlands (Met. County)	1,104	70.3	8.6	427.3	218.8	671.8	312.1	169.9	498.7	387.9
Worcestershire County	269	81.1	4.2	395.5	205.5	582.8	282.6	157.5	471.8	356.6
Herefordshire and Worcestershire	341	80.3	4.5	..	..	..	..	..	..	..
Former county of Shropshire	204	75.3	5.2	..	..	..	..	..	..	..
Former county of Staffordshire	492	76.3	5.4	..	..	..	..	..	..	..

14.5   (continued)

	Total in employment[2] 1998-1999[3] (thousands)	Employment rate 1998-1999[3] (percentages)	ILO unemployment rate 1998-1999[3] (percentages)[4]	Average gross weekly full-time earnings[5], April 1999 (£)						
				Males			Females			All persons total
					10 per cent earned			10 per cent earned		
				Total	Less than	More than	Total	Less than	More than	
East	2,542	77.7	4.6	436.0	215.6	701.0	323.9	172.5	514.6	396.6
Luton UA	88	77.5	..	493.8	243.8	838.6	318.4	162.5	498.5	431.9
Peterborough UA	67	74.9	..	419.7	218.8	670.2	299.1	174.9	487.3	378.6
Southend-on-Sea UA	74	70.5	8.1	..	..	..	328.3	172.7	500.9	370.8
Thurrock UA	58	72.8	..	..	..	..	..	..	..	..
Bedfordshire County	188	79.5	5.2	439.6	232.7	678.1	332.2	167.5	540.4	401.9
Cambridgeshire County	285	79.4	..	440.6	224.7	702.5	336.6	179.3	528.4	404.9
Essex County	604	76.7	4.3	435.5	213.1	718.9	323.0	171.2	500.9	396.8
Hertfordshire	516	81.5	2.8	486.8	230.3	785.0	368.2	192.7	587.3	443.6
Norfolk	339	75.0	6.6	394.2	205.6	635.3	287.4	161.8	445.2	356.7
Suffolk	323	77.5	5.5	382.4	201.3	585.4	281.4	161.3	471.0	348.9
Former county of Bedfordshire	276	78.9	4.7	..	..	..	..	..	..	..
Former county of Cambridgeshire	353	78.5	3.7	..	..	..	..	..	..	..
Former county of Essex	735	75.7	5.0	..	..	..	..	..	..	..
London	3,223	70.4	8.0	584.4	242.3	1,008.6	422.8	212.3	665.6	520.0
South East	3,845	79.1	4.3	471.2	221.8	770.3	341.0	182.2	539.9	423.2
Bracknell Forest UA	60	83.9	..	624.6	260.2	1,150.6	..	..	..	540.7
Brighton and Hove UA	119	75.5	5.7	383.7	193.3	652.6	313.2	175.4	494.8	352.4
Isle of Wight UA	50	72.8	..	350.2	180.0	500.9	..	..	..	323.1
Medway UA	113	76.1	7.0	429.2	235.6	630.3	298.5	156.2	472.7	386.9
Milton Keynes UA	98	76.8	6.9	461.5	238.2	729.4	335.4	182.0	533.8	412.4
Portsmouth UA	82	73.2	7.6	447.9	221.8	737.6	309.7	183.1	484.6	396.0
Reading UA	76	78.0	..	548.6	265.9	921.3	387.3	206.3	600.5	487.2
Slough UA	47	73.7	..	..	..	..	389.1	207.4	648.8	481.4
Southampton UA	97	70.8	8.9	431.4	220.7	671.8	326.4	182.6	484.1	395.7
West Berkshire UA (Newbury)	76	85.0	..	512.8	253.1	836.0	365.7	216.6	575.8	463.6
Windsor and Maidenhead UA	73	85.0	..	..	..	..	..	..	..	513.9
Wokingham UA	83	85.0	..	..	..	..	..	..	..	498.7
Buckinghamshire County	242	80.6	4.5	522.7	225.0	840.0	369.4	202.1	575.8	466.8
East Sussex County	204	76.6	5.5	380.0	199.5	576.9	318.2	168.4	508.1	355.0
Hampshire County	621	80.4	3.6	467.0	220.4	758.8	326.9	180.0	500.9	417.5
Kent County	610	76.6	5.5	418.9	208.0	647.0	317.1	171.3	499.0	381.4
Oxfordshire	316	80.5	3.3	453.4	227.5	691.0	335.5	190.0	500.9	410.1
Surrey	529	82.5	2.8	534.2	230.3	898.4	377.3	203.5	607.2	476.1
West Sussex	349	80.9	3.4	458.4	219.5	753.3	331.2	175.1	518.4	410.5
Former county of Berkshire	415	82.0	2.7	..	..	..	..	..	..	..
Former county of Buckinghamshire	339	79.5	5.2	..	..	..	..	..	..	..
Former county of East Sussex	322	76.1	5.6	..	..	..	..	..	..	..
Former county of Hampshire	802	78.3	4.7	..	..	..	..	..	..	..
Former county of Kent	723	76.5	5.8	..	..	..	..	..	..	..
South West	2,253	78.1	4.6	402.9	200.0	641.9	297.8	161.5	472.7	364.9
Bath and North East Somerset UA	78	77.2	..	..	..	..	..	..	..	409.2
Bournemouth UA	66	75.8	..	391.9	211.5	576.9	..	..	..	347.4
City of Bristol UA	174	71.9	8.2	447.8	220.8	716.9	321.5	174.9	500.0	399.9
North Somerset UA	85	77.0	..	405.8	211.1	643.0	288.0	166.2	434.3	365.0
Plymouth UA	110	70.0	6.9	385.5	190.2	605.9	287.5	153.6	445.2	348.0
Poole UA	67	82.0	..	437.9	214.7	725.6	283.3	169.0	445.2	384.2
South Gloucestershire UA	132	83.7	..	433.5	231.5	681.2	293.7	163.2	441.8	395.0
Swindon UA	98	85.2	..	472.0	245.0	724.8	311.8	188.1	471.8	420.2
Torbay UA	45	73.0	..	..	..	..	..	..	..	316.9
Cornwall and the Isles of Scilly	208	72.3	5.8	322.5	174.1	493.3	258.0	149.9	413.6	297.1
Devon County	324	80.3	4.3	352.8	182.0	545.1	280.2	153.4	460.6	328.0
Dorset County	176	81.1	..	376.0	191.9	582.9	303.4	161.2	483.7	350.4
Gloucestershire	266	80.4	..	427.6	223.1	662.2	301.5	172.3	491.5	383.3
Somerset	223	78.9	..	388.2	201.8	626.8	302.3	159.7	500.9	358.8
Wiltshire County	202	80.5	..	397.7	200.4	639.8	289.7	156.5	458.7	357.0
Bristol/Bath area	469	76.7	..	..	..	..	..	..	..	..
Former county of Devon	479	76.9	5.5	..	..	..	..	..	..	..
Former county of Dorset	308	80.1	3.8	..	..	..	..	..	..	..
Former county of Wiltshire	300	82.1	3.3	..	..	..	..	..	..	..

1 Local government structure as at 1 April 1998. See Notes and Definitions to the Labour market chapter. In some cases sample sizes are too small to provide reliable estimates.
2 Includes those on government-supported employment and training schemes and unpaid family workers.
3 For those of working age. Data are from the Labour Force Survey and relate to the period March 1998 to February 1999.
4 As a percentage of the economically active.
5 Earnings estimates have been derived from the New Earnings Survey and relate to full-time employees whose pay for the survey pay-period was not affected by absence.

Source: Office for National Statistics

14.6 Labour market[1] and economic statistics:by local authority[2]

| | Economically active 1998-1999[3] (percentages) | Claimant count[4] March 2000 | | | Income Support beneficiaries[6] Feb.2000 (percentages) | Businesses registered for VAT 1998 | | Stock of businesses end 1998 (thousands) |
		Total (thousands)	Of which females (percentages)	Of which long-term claimants[5] (percentages)		Registration rates[7] (percentages)	Deregistration rates[7] (percentages)	
United Kingdom	78.4	1,194.3	23.5	22.1	10	11	10	1,651.6
England	79.0	958.8	23.8	22.1	9	12	10	1,402.7
North East	73.3	79.6	20.0	22.6	12	10	10	42.0
Darlington UA	78.3	2.9	20.7	21.4	10	12	9	2.1
Hartlepool UA	75.8	3.7	16.7	27.4	14	11	10	1.2
Middlesbrough UA	71.2	6.6	18.0	22.0	15	10	10	1.9
Redcar & Cleveland UA	70.2	4.9	18.5	22.8	12	8	10	1.8
Stockton-on-Tees UA	76.1	6.6	19.9	22.3	10	12	13	2.7
Durham County	73.1	11.2	22.3	18.8	11	9	9	8.7
Chester-le-Street	70.0	1.1	22.3	20.1	8	13	11	0.8
Derwentside	74.9	2.0	21.3	18.9	13	10	8	1.4
Durham	79.1	1.7	26.1	17.4	7	11	9	1.4
Easington	65.0	2.1	20.5	18.7	15	10	10	1.1
Sedgefield	72.2	2.1	22.4	14.9	11	9	10	1.5
Teesdale	83.1	0.4	22.2	21.1	6	6	6	1.1
Wear Valley	72.8	1.8	21.9	23.4	14	8	8	1.4
Northumberland	76.5	7.5	23.2	20.1	8	8	7	7.6
Alnwick	71.3	0.7	28.7	18.8	7	8	7	1.1
Berwick-upon-Tweed	82.0	0.7	30.8	11.2	8	4	7	1.0
Blyth Valley	77.3	2.3	21.8	16.7	10	10	9	1.0
Castle Morpeth	75.5	0.9	23.4	20.2	6	9	8	1.3
Tynedale	78.0	0.9	23.3	19.6	6	8	6	2.3
Wansbeck	74.2	2.1	20.2	27.4	10	9	9	0.7
Tyne and Wear	72.0	36.2	19.4	24.0	14	11	11	16.1
Gateshead	76.4	5.2	19.4	22.0	14	10	9	3.3
Newcastle-upon-Tyne	68.9	9.1	18.8	26.6	14	11	15	4.6
North Tyneside	77.8	6.0	20.6	19.5	13	11	10	2.7
South Tyneside	72.5	6.6	17.7	23.2	14	10	7	1.9
Sunderland	67.8	9.3	20.4	26.1	14	11	10	3.6
Tees Valley	..	24.6	18.7	23.0	12	11	10	9.7
Tees Valley less Darlington	..	21.7	18.5	23.2	13	11	11	7.6
Former county of Durham	73.8	14.1	22.0	19.4	11	10	9	10.8
North West	75.2	151.3	21.7	19.6	12	12	10	160.1
Blackburn with Darwen UA	66.3	3.5	20.4	14.7	14	11	11	2.9
Blackpool UA	77.6	3.9	20.2	9.3	14	11	12	3.0
Halton UA	72.5	3.6	21.7	19.4	14	12	9	1.9
Warrington UA	81.8	3.1	25.2	15.6	8	14	9	4.7
Cheshire County	78.5	8.4	23.0	13.9	7	11	9	20.7
Chester	76.6	1.4	20.7	17.9	8	10	11	3.7
Congleton	80.7	0.9	26.9	12.4	5	10	8	2.9
Crewe and Nantwich	79.0	1.7	27.1	14.0	7	10	8	2.9
Ellesmere Port and Neston	73.4	1.3	20.6	12.8	8	11	9	1.4
Macclesfield	80.3	1.4	19.8	10.8	6	10	9	6.5
Vale Royal	79.3	1.8	23.3	14.8	7	12	8	3.4
Cumbria	77.3	9.5	23.0	18.5	8	7	7	16.0
Allerdale	72.9	2.2	20.9	21.0	8	6	7	3.1
Barrow-in-Furness	72.1	1.8	18.1	18.3	13	9	8	1.0
Carlisle	83.1	2.1	26.5	17.1	8	8	7	2.9
Copeland	78.3	2.0	22.0	23.1	9	10	8	1.5
Eden	80.7	0.4	27.8	10.0	5	6	7	3.0
South Lakeland	76.4	0.9	29.4	10.0	4	8	7	4.5
Greater Manchester (Met. County)	75.0	53.4	21.2	18.6	12	13	12	57.8
Bolton	76.5	5.4	20.8	14.9	11	12	11	5.7
Bury	82.2	2.3	23.4	7.9	10	14	12	4.2
Manchester	63.5	15.3	20.0	24.8	20	15	19	9.5
Oldham	77.4	4.5	22.3	16.1	12	11	10	4.5
Rochdale	75.3	4.6	21.2	17.5	13	11	11	4.3
Salford	71.1	4.4	19.6	18.0	15	15	11	4.6
Stockport	83.5	3.7	20.1	18.4	8	11	10	7.9
Tameside	79.2	3.9	22.3	16.8	11	13	10	4.7
Trafford	75.6	3.2	22.3	18.8	9	14	11	6.3
Wigan	74.3	6.0	23.0	14.7	10	12	9	6.1

14.6 *(continued)*

	Economically active 1998-1999[3] (percentages)	Claimant count[4] March 2000			Income Support beneficiaries[6] Feb.2000 (percentages)	Businesses registered for VAT 1998		Stock of businesses end 1998 (thousands)
		Total (thousands)	Of which females (percentages)	Of which long-term claimants[5] (percentages)		Registration rates[7] (percentages)	Deregistration rates[7] (percentages)	
Lancashire County	79.1	18.0	22.7	13.6	10	10	9	29.3
Burnley	71.5	1.5	23.5	9.1	12	11	9	1.8
Chorley	84.8	1.3	23.8	11.5	7	11	8	2.7
Fylde	82.0	0.6	24.4	8.1	7	14	19	1.7
Hyndburn	74.6	1.2	25.3	8.0	12	11	9	1.9
Lancaster	78.9	2.9	21.3	17.7	10	8	8	3.2
Pendle	75.2	1.6	26.2	9.6	12	9	9	2.1
Preston	74.2	2.9	18.4	19.5	13	12	11	3.3
Ribble Valley	81.3	0.3	27.4	8.0	5	9	7	2.1
Rossendale	87.1	0.9	27.0	11.2	11	12	9	2.0
South Ribble	87.8	1.1	23.3	11.7	6	12	9	2.6
West Lancashire	77.3	2.3	22.4	15.0	10	8	8	3.0
Wyre	78.3	1.5	22.7	12.9	8	10	8	3.0
Merseyside (Met. County)	69.9	47.9	21.4	25.8	16	15	10	23.6
Knowsley	64.9	6.0	21.7	26.0	21	12	9	1.7
Liverpool	65.5	19.9	21.3	29.0	20	13	12	7.4
St Helens	76.4	4.7	21.5	21.8	12	10	9	2.9
Sefton	74.5	7.5	20.6	24.8	12	23	8	6.6
Wirral	71.4	9.7	22.0	21.8	13	12	10	5.1
Former county of Cheshire	78.4	15.1	23.2	15.6	8	11	9	27.3
Former county of Lancashire	77.7	25.3	22.0	13.1	10	11	10	35.3
Yorkshire and the Humber	78.4	118.6	22.4	20.1	10	10	10	117.7
East Riding of Yorkshire UA	79.9	5.6	26.9	18.7	7	8	8	9.2
City of Kingston upon Hull UA	74.9	10.3	21.6	23.3	16	11	12	4.2
North East Lincolnshire UA	73.5	5.8	21.6	18.1	12	9	10	3.2
North Lincolnshire UA	79.8	3.0	25.1	14.7	9	8	8	4.0
York UA	81.3	2.8	24.6	14.8	6	17	15	2.4
North Yorkshire County	82.3	7.4	27.4	16.8	6	7	7	24.2
Craven	92.0	0.5	26.0	13.0	7	6	6	2.7
Hambleton	82.4	0.9	30.1	19.0	5	6	7	3.9
Harrogate	83.9	1.3	27.1	11.1	5	9	7	6.0
Richmondshire	83.5	0.4	34.5	12.7	3	6	8	2.0
Ryedale	84.0	0.5	31.0	20.8	5	4	4	3.5
Scarborough	75.4	2.7	24.8	18.6	11	7	8	3.1
Selby	80.3	1.1	28.2	18.4	5	8	7	3.0
South Yorkshire (Met. County)	75.7	35.9	21.7	21.3	11	11	11	22.7
Barnsley	71.8	5.8	22.4	17.6	12	11	10	3.8
Doncaster	77.1	7.7	23.0	18.5	11	11	9	5.0
Rotherham	76.8	7.1	21.1	19.0	11	12	10	4.2
Sheffield	76.2	15.4	21.1	25.1	11	11	13	9.7
West Yorkshire (Met. County)	79.2	47.7	21.6	20.0	10	11	10	47.9
Bradford	76.8	12.7	21.0	22.5	12	11	10	10.5
Calderdale	80.7	4.3	22.0	19.3	10	10	9	5.6
Kirklees	79.5	7.9	22.9	16.2	10	11	9	9.6
Leeds	81.2	16.0	20.8	22.0	9	12	13	15.9
Wakefield	76.6	6.8	22.8	15.8	11	12	8	6.3
The Humber	..	24.7	23.2	20.0	11	9	9	20.6
Former county of North Yorkshire	82.0	10.2	26.6	16.3	6	8	8	26.6
East Midlands	79.9	75.9	24.4	19.5	8	11	10	111.2
Derby UA	77.5	5.9	21.9	23.3	10	13	11	4.1
Leicester UA	73.5	8.1	23.8	20.9	14	12	11	7.4
Nottingham UA	71.3	9.8	21.2	27.5	15	12	16	5.5
Rutland UA	78.0	0.2	29.9	5.8	4	9	8	1.3
Derbyshire County	79.0	13.4	23.7	19.4	8	10	9	19.7
Amber Valley	78.9	2.0	25.3	18.4	8	10	9	3.1
Bolsover	75.8	1.6	22.5	16.8	11	11	8	1.3
Chesterfield	74.1	2.9	21.5	23.1	10	13	10	2.2
Derbyshire Dales	82.9	0.8	26.6	17.2	5	7	7	3.5
Erewash	82.8	2.0	25.5	19.5	8	11	9	2.5
High Peak	81.7	1.2	22.6	14.0	7	10	10	2.7
North East Derbyshire	80.1	1.9	24.1	22.3	9	10	9	2.5
South Derbyshire	75.3	0.9	23.4	17.4	7	11	8	2.0

14.6 *(continued)*

| | Economically active 1998-1999[3] (percentages) | Claimant count[4] March 2000 | | | Income Support bene-ficiaries[6] Feb.2000 (percentages) | Businesses registered for VAT 1998 | | Stock of businesses end 1998 (thousands) |
		Total (thousands)	Of which females (percentages)	Of which long-term claimants[5] (percentages)		Registration rates[7] (percentages)	Deregistration rates[7] (percentages)	
Leicestershire County	85.9	6.8	30.3	14.8	5	11	9	18.1
Blaby	84.9	0.8	28.7	14.1	5	10	9	2.3
Charnwood	84.5	2.1	28.8	16.6	5	11	9	4.1
Harborough	89.4	0.5	34.8	12.4	4	11	12	2.9
Hinckley and Bosworth	90.3	1.2	34.2	12.7	5	10	7	3.3
Melton	86.0	0.4	36.0	8.9	5	8	8	1.7
North West Leicestershire	81.4	1.0	27.1	15.1	6	13	9	2.6
Oadby and Wigston	85.9	0.7	28.6	17.9	5	9	8	1.3
Lincolnshire	79.1	10.1	25.2	14.0	8	9	8	19.6
Boston	76.4	0.8	23.9	6.9	8	8	8	1.8
East Lindsey	76.4	2.5	25.5	9.8	10	8	8	4.4
Lincoln	75.2	2.3	21.2	19.6	12	12	9	1.7
North Kesteven	82.1	1.0	28.9	11.7	6	8	10	2.4
South Holland	78.5	0.7	31.0	11.7	7	8	7	2.9
South Kesteven	82.9	1.4	26.5	11.8	6	10	9	3.8
West Lindsey	78.5	1.5	25.1	20.9	8	8	7	2.6
Northamptonshire	83.7	7.9	26.3	16.3	7	12	10	18.9
Corby	80.6	1.0	24.0	13.2	11	10	9	1.0
Daventry	84.8	0.5	33.6	13.1	4	11	9	2.9
East Northamptonshire	83.0	0.7	28.5	15.2	5	9	10	2.3
Kettering	82.5	1.0	26.4	12.7	8	13	10	2.0
Northampton	81.9	3.2	24.5	19.4	8	14	13	4.6
South Northamptonshire	89.5	0.4	33.4	17.6	4	10	8	3.2
Wellingborough	84.0	1.1	26.1	15.5	8	17	13	2.7
Nottinghamshire County	79.8	13.8	24.3	19.6	7	11	9	16.6
Ashfield	80.6	2.7	22.7	23.5	10	11	8	2.0
Bassetlaw	80.0	2.4	25.2	23.6	8	10	8	2.7
Broxtowe	81.3	1.7	26.6	17.3	6	13	8	2.1
Gedling	78.5	1.8	25.8	21.2	6	11	11	2.3
Mansfield	75.1	2.3	21.1	13.5	10	11	9	1.8
Newark and Sherwood	79.2	1.7	24.6	15.0	7	10	9	3.0
Rushcliffe	83.2	1.1	26.3	22.2	4	12	11	2.8
Former county of Derbyshire	78.7	19.3	23.1	20.6	9	11	9	23.8
Former county of Leicestershire	81.8	15.0	26.8	18.0	8	11	10	26.8
Former county of Nottinghamshire	77.4	23.5	23.0	22.9	9	11	11	22.1
West Midlands	79.4	113.5	23.7	25.3	10	11	10	136.3
County of Herefordshire UA	82.1	2.2	27.0	14.9	6	8	8	8.0
Stoke-on-Trent UA	73.4	5.8	24.2	16.9	12	11	11	4.5
Telford and Wrekin UA	80.1	2.5	24.1	12.5	12	13	10	3.3
Shropshire County	79.2	3.4	26.2	16.3	7	8	7	11.1
Bridgnorth	83.2	0.6	26.9	14.4	5	9	7	2.2
North Shropshire	82.8	0.6	25.7	13.3	7	8	6	2.4
Oswestry	87.5	0.6	30.9	20.7	9	8	8	1.2
Shrewsbury and Atcham	74.5	1.2	24.8	17.5	7	10	8	3.1
South Shropshire	73.3	0.4	23.6	14.2	6	7	7	2.2
Staffordshire County	82.8	11.8	26.4	17.0	7	11	9	21.9
Cannock Chase	74.8	1.5	28.0	15.9	9	11	9	2.3
East Staffordshire	87.7	2.0	23.3	18.4	7	11	8	3.2
Lichfield	79.5	1.0	27.5	16.3	7	11	9	3.0
Newcastle-under-Lyme	77.5	1.7	22.9	13.4	7	12	15	2.3
South Staffordshire	87.7	1.4	27.3	20.1	8	11	9	2.8
Stafford	83.9	1.6	24.8	17.4	6	11	8	3.6
Staffordshire Moorlands	86.4	1.1	30.9	11.5	6	7	7	3.2
Tamworth	85.5	1.5	29.6	22.0	8	14	9	1.6
Warwickshire	83.2	5.9	27.2	19.8	6	11	9	16.2
North Warwickshire	83.5	0.8	30.4	15.8	5	11	8	2.0
Nuneaton and Bedworth	81.4	1.8	27.5	16.6	8	12	10	2.3
Rugby	83.0	1.0	27.7	20.1	6	10	9	2.6
Stratford-on-Avon	84.8	0.8	28.9	20.9	5	10	7	5.2
Warwick	83.4	1.5	23.7	25.0	6	13	9	4.1

14.6 *(continued)*

| | Economically active 1998-1999[3] (percentages) | Claimant count[4] March 2000 | | | Income Support beneficiaries[6] Feb.2000 (percentages) | Businesses registered for VAT 1998 | | Stock of businesses end 1998 (thousands) |
		Total (thousands)	Of which females (percentages)	Of which long-term claimants[5] (percentages)		Registration rates[7] (percentages)	Deregistration rates[7] (percentages)	
West Midlands (Met. County)	76.9	74.7	22.4	29.7	13	12	12	54.4
Birmingham	74.3	35.8	22.0	33.6	15	13	14	20.2
Coventry	78.3	6.5	21.2	24.4	11	13	9	5.4
Dudley	82.6	6.9	23.8	29.0	9	11	10	7.6
Sandwell	76.5	9.1	22.5	30.0	14	11	9	6.3
Solihull	82.5	2.9	24.8	24.5	6	13	11	4.5
Walsall	76.7	6.3	23.5	22.7	12	11	11	5.7
Wolverhampton	74.4	7.2	22.7	23.8	13	13	11	4.8
Worcestershire County	84.7	7.1	26.0	15.7	6	11	9	16.9
Bromsgrove	80.2	1.2	24.9	21.6	5	11	10	2.8
Malvern Hills	87.0	0.7	23.2	13.8	6	9	9	2.5
Redditch	87.7	1.4	28.2	16.7	9	12	8	2.1
Worcester	84.8	1.4	22.5	15.8	7	13	11	2.2
Wychavon	82.4	1.1	26.8	9.4	5	9	7	4.6
Wyre Forest	86.4	1.4	29.0	15.4	8	11	10	2.8
Herefordshire and Worcestershire	84.2	9.3	26.2	15.5	6	10	9	24.9
Former county of Shropshire	79.5	5.9	25.3	14.7	8	10	8	14.4
Former county of Staffordshire	80.6	17.7	25.7	17.0	8	11	9	26.4
East	81.4	73.1	25.8	20.2	7	11	10	162.7
Luton UA	80.3	3.9	23.5	23.5	10	12	12	3.9
Peterborough UA	79.2	2.8	23.0	15.5	10	13	10	3.6
Southend-on-Sea UA	76.8	3.8	21.7	30.3	10	15	21	3.5
Thurrock UA	78.3	2.4	26.7	19.8	9	14	9	2.8
Bedfordshire County	83.9	4.3	27.6	14.6	6	11	8	12.1
Bedford	81.7	2.2	24.7	16.6	4	11	9	4.0
Mid Bedfordshire	86.4	0.9	34.0	11.8	9	10	8	4.5
South Bedfordshire	83.6	1.1	28.3	12.9	6	11	9	3.6
Cambridgeshire County	82.1	5.7	26.0	17.7	5	10	9	18.9
Cambridge	71.8	1.6	23.2	20.9	6	14	13	3.0
East Cambridgeshire	88.1	0.7	26.6	19.4	5	10	9	2.8
Fenland	84.8	1.2	26.1	15.0	9	8	9	2.6
Huntingdonshire	83.1	1.5	28.3	15.4	5	11	9	5.2
South Cambridgeshire	85.9	0.8	26.4	18.8	3	9	7	5.2
Essex County	80.2	15.9	27.9	18.6	7	12	9	39.0
Basildon	81.0	2.6	28.4	17.9	10	13	9	4.2
Braintree	79.0	1.4	29.7	13.7	7	11	9	4.3
Brentwood	81.7	0.5	31.1	15.7	5	14	11	2.5
Castle Point	82.2	1.0	28.8	23.1	8	13	9	2.2
Chelmsford	85.3	1.7	29.8	21.3	5	13	12	4.4
Colchester	75.6	1.8	28.9	12.2	6	11	8	4.4
Epping Forest	78.8	1.4	28.0	22.4	6	12	8	4.3
Harlow	83.6	1.3	26.5	24.7	9	15	11	1.5
Maldon	78.1	0.6	27.5	23.6	5	10	8	2.4
Rochford	81.7	0.8	26.7	21.6	5	12	9	2.2
Tendring	75.4	2.4	24.5	16.0	10	9	9	3.1
Uttlesford	81.5	0.4	29.0	15.8	4	9	8	3.5
Hertfordshire	83.8	9.0	25.1	18.9	6	13	10	35.9
Broxbourne	84.7	1.0	29.3	24.2	7	11	10	2.4
Dacorum	84.8	1.1	24.6	15.4	6	14	8	5.1
East Hertfordshire	86.6	0.7	24.6	24.2	5	11	8	5.2
Hertsmere	83.6	0.8	26.5	16.0	7	14	10	3.4
North Hertfordshire	81.9	1.1	25.2	16.4	6	14	9	4.6
St Albans	82.2	0.7	21.3	14.9	5	16	14	5.1
Stevenage	86.9	1.1	24.2	20.1	8	13	10	1.7
Three Rivers	81.3	0.6	23.8	18.9	5	13	9	2.8
Watford	81.8	1.0	24.1	23.2	6	13	10	2.8
Welwyn Hatfield	84.2	0.7	26.4	15.0	6	11	8	2.9

14.6 (continued)

	Economically active 1998-1999[3] (percentages)	Claimant count[4] March 2000 Total (thousands)	Of which females (percentages)	Of which long-term claimants[5] (percentages)	Income Support beneficiaries[6] Feb.2000 (percentages)	Businesses registered for VAT 1998 Registration rates[7] (percentages)	Deregistration rates[7] (percentages)	Stock of businesses end 1998 (thousands)
Norfolk	80.3	15.4	25.8	21.8	8	8	9	22.9
Breckland	82.1	1.6	28.9	15.0	7	8	7	3.6
Broadland	85.2	1.2	27.8	22.8	6	9	9	3.2
Great Yarmouth	73.6	4.1	24.8	24.8	11	8	7	2.3
Kings Lynn and West Norfolk	81.2	2.1	30.2	15.4	9	8	8	4.1
North Norfolk	75.4	1.5	26.0	18.4	8	8	8	3.2
Norwich	79.7	3.8	21.5	26.7	11	11	14	2.9
South Norfolk	80.8	1.1	28.8	19.5	6	8	7	3.8
Suffolk	82.1	9.8	25.0	21.5	7	9	9	20.1
Babergh	81.1	0.9	24.4	18.2	7	9	9	3.1
Forest Heath	86.8	0.5	33.1	15.9	4	12	10	2.0
Ipswich	82.1	2.5	21.8	27.0	10	14	17	2.2
Mid Suffolk	80.8	0.7	31.3	16.1	5	8	8	3.4
St Edmundsbury	85.0	1.0	25.8	13.1	6	9	7	3.2
Suffolk Coastal	80.2	1.2	23.8	18.2	6	9	8	3.6
Waveney	80.0	3.1	25.2	23.9	9	6	9	2.6
Former county of Bedfordshire	82.7	8.2	25.7	18.8	7	11	9	16.0
Former county of Cambridgeshire	81.5	8.6	25.0	17.0	6	11	9	22.5
Former county of Essex	79.6	22.1	26.7	20.7	7	12	10	45.3
London	76.5	187.6	25.9	27.8	11	15	11	270.0
Inner London	..	..	..	..	..	..	..	..
Inner London - West	..	..	..	..	..	..	..	..
Camden	72.9	7.1	28.3	30.7	15	14	18	16.6
City of London	..	0.1	30.6	23.5	7	11	9	11.9
Hammersmith and Fulham	74.1	5.1	26.8	32.9	12	17	8	8.1
Kensington and Chelsea	72.0	3.6	31.6	32.6	8	13	8	11.0
Wandsworth	77.4	6.0	27.1	25.2	9	16	9	9.7
Westminster	70.4	5.4	27.4	31.7	9	16	10	36.8
Inner London - East	..	..	..	..	..	..	..	..
Hackney	66.1	10.2	25.5	27.0	21	19	10	7.1
Haringey	69.1	10.4	24.0	32.6	17	18	11	6.3
Islington	72.1	8.2	28.5	33.0	18	16	11	10.1
Lambeth	76.7	12.0	26.5	30.6	15	17	9	6.8
Lewisham	79.2	9.1	25.0	31.4	13	17	10	4.6
Newham	67.1	9.4	22.6	25.9	22	18	11	4.0
Southwark	72.4	10.4	26.3	30.3	16	17	10	7.5
Tower Hamlets	59.7	8.9	20.7	26.7	21	17	9	7.7
Outer London	..	..	..	..	..	..	..	..
Outer London - East and North East	..	..	..	..	..	..	..	..
Barking and Dagenham	74.0	3.4	25.3	24.9	15	18	18	2.4
Bexley	79.4	3.2	28.8	21.3	7	13	11	4.7
Enfield	79.3	6.6	26.6	29.0	12	15	9	6.5
Greenwich	76.8	7.1	25.7	28.8	13	16	10	4.1
Havering	85.1	3.2	27.9	24.9	7	14	11	5.4
Redbridge	76.8	4.5	25.9	24.0	10	17	12	5.4
Waltham Forest	71.7	6.4	24.2	28.3	14	17	11	4.5
Outer London - South	..	..	..	..	..	..	..	..
Bromley	81.0	3.9	25.8	25.4	6	14	10	8.3
Croydon	81.3	7.0	25.9	28.9	9	14	13	8.2
Kingston upon Thames	79.9	1.5	27.6	18.1	5	15	9	5.1
Merton	85.1	2.9	26.1	20.7	7	16	10	5.5
Sutton	86.7	1.8	27.6	21.7	6	13	10	4.9
Outer London - West and North West	..	..	..	..	..	..	..	..
Barnet	80.4	5.3	26.6	22.8	8	16	19	10.2
Brent	77.3	8.2	24.8	29.5	14	16	11	8.4
Ealing	74.4	6.2	26.0	22.6	11	16	9	9.6
Harrow	81.4	3.0	27.7	24.0	9	16	13	6.3
Hillingdon	81.7	2.8	25.5	17.6	8	14	10	7.4
Hounslow	78.8	2.9	27.7	13.8	11	14	9	7.1
Richmond-upon-Thames	77.8	1.7	27.3	25.0	5	14	8	8.2

14.6 *(continued)*

| | Economically active 1998-1999[3] (percentages) | Claimant count[4] March 2000 | | | Income Support bene-ficiaries[6] Feb.2000 (percentages) | Businesses registered for VAT 1998 | | Stock of businesses end 1998 (thousands) |
		Total (thousands)	Of which females (percentages)	Of which long-term claimants[5] (percentages)		Registration rates[7] (percentages)	Deregistration rates[7] (percentages)	
South East	82.7	88.7	24.0	19.6	6	12	9	253.0
Bracknell Forest UA	86.9	0.7	23.3	15.8	4	16	10	3.3
Brighton and Hove UA	80.0	7.6	25.8	31.3	10	17	15	6.5
Isle of Wight UA	77.2	3.3	24.6	25.6	10	11	8	3.2
Medway UA	81.8	4.4	25.9	20.3	8	14	13	4.8
Milton Keynes UA	82.4	2.4	26.2	11.3	7	16	9	6.1
Portsmouth UA	79.2	3.7	22.6	23.0	9	14	9	3.6
Reading UA	81.1	1.9	21.1	17.0	7	14	16	3.6
Slough UA	77.2	2.0	23.0	22.3	9	16	11	3.0
Southampton UA	77.7	4.4	20.8	18.7	10	14	19	3.8
West Berkshire UA (Newbury)	86.7	0.8	24.6	13.6	4	12	8	6.2
Windsor and Maidenhead UA	86.1	1.0	25.3	20.1	4	13	8	6.7
Wokingham UA	86.7	0.6	25.2	10.1	3	12	8	5.4
Buckinghamshire County	84.4	3.6	23.0	20.4	4	12	9	21.2
Aylesbury Vale	87.0	1.2	26.1	19.4	4	12	8	6.2
Chiltern	84.3	0.5	21.6	14.8	3	11	9	4.1
South Buckinghamshire	80.3	0.4	25.3	24.3	3	11	9	3.5
Wycombe	83.6	1.5	20.3	22.1	5	12	8	7.4
East Sussex County	81.0	6.7	22.6	18.9	8	11	9	14.7
Eastbourne	81.2	1.5	21.9	16.1	9	11	11	1.8
Hastings	79.6	2.3	20.5	19.7	12	11	10	1.7
Lewes	79.2	1.1	23.2	23.1	7	12	11	2.7
Rother	75.5	1.0	25.8	19.1	7	11	9	2.8
Wealden	86.4	0.8	25.4	16.1	5	10	8	5.6
Hampshire County	83.3	9.9	24.8	14.7	5	12	8	37.7
Basingstoke and Deane	83.3	0.9	25.9	14.4	4	13	7	5.0
East Hampshire	85.5	0.7	23.3	20.6	4	11	8	4.4
Eastleigh	86.4	0.8	24.1	11.6	5	14	9	3.1
Fareham	86.2	0.8	25.9	10.8	4	12	10	2.7
Gosport	82.9	1.1	27.3	17.5	6	13	10	1.0
Hart	87.9	0.3	24.2	9.4	3	12	7	3.3
Havant	79.2	1.9	22.3	17.8	8	13	10	2.4
New Forest	78.2	1.4	26.8	12.4	5	10	9	5.3
Rushmoor	83.3	0.7	24.9	12.4	5	13	7	2.3
Test Valley	88.2	0.7	26.5	13.7	4	10	9	4.0
Winchester	78.3	0.6	21.6	14.8	4	12	9	4.2
Kent County	81.1	20.2	23.6	20.6	8	12	9	36.9
Ashford	83.4	1.2	24.8	15.9	7	12	11	3.3
Canterbury	80.5	2.1	22.6	20.0	8	11	13	3.0
Dartford	85.3	1.1	28.3	21.1	6	14	10	2.1
Dover	77.9	2.2	23.2	18.1	8	11	8	2.3
Gravesham	85.6	1.8	23.7	22.2	8	14	9	2.0
Maidstone	78.0	1.4	24.4	15.9	7	11	8	4.6
Sevenoaks	82.5	0.9	26.1	17.9	5	11	8	4.5
Shepway	78.9	2.0	23.2	22.6	10	10	8	2.3
Swale	81.1	2.3	23.1	21.5	9	11	9	3.0
Thanet	75.1	3.7	21.8	25.5	13	12	10	2.2
Tonbridge and Malling	82.7	0.8	26.1	17.1	5	11	9	3.4
Tunbridge Wells	85.8	0.9	23.5	15.7	6	11	9	4.2
Oxfordshire	83.3	4.1	25.6	18.0	5	11	8	20.7
Cherwell	86.0	0.7	27.0	12.0	5	11	8	4.4
Oxford	75.9	1.8	23.0	21.9	7	12	9	2.9
South Oxfordshire	86.8	0.6	27.2	16.8	4	11	8	5.7
Vale of White Horse	81.4	0.6	26.4	18.3	4	11	9	3.8
West Oxfordshire	88.7	0.4	30.6	13.6	4	10	8	3.9

14.6 *(continued)*

| | Economically active 1998-1999[3] (percentages) | Claimant count[4] March 2000 | | | Income Support beneficiaries[6] Feb.2000 (percentages) | Businesses registered for VAT 1998 | | Stock of businesses end 1998 (thousands) |
		Total (thousands)	Of which females (percentages)	Of which long-term claimants[5] (percentages)		Registration rates[7] (percentages)	Deregistration rates[7] (percentages)	
Surrey	84.9	5.6	25.3	13.5	4	12	9	42.7
Elmbridge	83.1	0.7	26.3	13.3	4	12	10	5.6
Epsom and Ewell	85.8	0.4	32.5	12.6	3	11	9	2.1
Guildford	89.6	0.7	25.1	17.8	4	12	9	5.0
Mole Valley	89.2	0.3	22.4	9.9	5	11	7	3.8
Reigate and Banstead	78.0	0.6	23.6	9.4	6	12	8	4.4
Runnymede	83.4	0.5	25.3	14.0	4	10	10	2.8
Spelthorne	83.8	0.7	25.3	16.5	5	11	9	3.5
Surrey Heath	85.4	0.3	22.1	8.5	3	13	8	3.6
Tandridge	90.5	0.4	25.3	12.8	4	11	8	3.3
Waverley	83.6	0.6	25.0	19.7	4	11	9	5.3
Woking	84.1	0.4	23.5	5.6	4	12	9	3.3
West Sussex	83.7	5.9	23.8	16.0	6	12	10	22.9
Adur	79.6	0.6	25.9	16.0	8	11	10	1.4
Arun	83.5	1.2	23.8	14.3	8	12	9	3.6
Chichester	79.6	0.9	25.4	19.8	5	11	8	4.5
Crawley	84.9	1.0	22.8	16.0	7	15	15	1.9
Horsham	85.1	0.7	25.0	14.2	4	11	7	4.8
Mid Sussex	85.8	0.6	23.7	13.6	5	13	9	4.5
Worthing	84.5	1.0	21.0	17.4	8	14	13	2.3
Former county of Berkshire	84.3	7.0	23.2	17.8	5	13	10	28.4
Former county of Buckinghamshire	83.8	6.0	24.2	16.8	5	12	9	27.2
Former county of East Sussex	80.7	14.3	24.3	25.5	9	13	11	21.2
Former county of Hampshire	82.1	17.9	23.4	17.4	6	12	9	45.1
Former county of Kent	81.2	24.6	24.0	20.5	8	12	10	41.7
South West	81.9	70.6	26.5	17.3	8	11	9	149.7
Bath and North East Somerset UA	80.4	1.7	25.6	10.5	6	12	9	5.5
Bournemouth UA	81.2	3.4	22.3	25.2	10	14	16	3.9
City of Bristol UA	78.4	8.3	24.0	19.7	11	15	12	11.0
North Somerset UA	80.2	1.9	26.0	11.6	7	11	9	5.1
Plymouth UA	75.2	5.8	24.2	19.2	10	12	15	3.4
Poole UA	85.4	1.4	24.4	13.9	6	12	10	3.7
South Gloucestershire UA	88.2	2.2	22.5	17.2	7	13	9	5.9
Swindon UA	86.1	1.9	28.6	12.6	9	17	10	4.1
Torbay UA	81.6	3.0	24.9	20.4	13	12	12	2.9
Cornwall and the Isles of Scilly	76.7	11.2	30.3	16.2	9	8	8	16.4
Caradon	82.7	1.3	32.4	14.8	7	7	8	2.7
Carrick	78.7	1.9	26.9	16.6	9	9	10	3.0
Kerrier	72.6	2.3	28.9	19.0	11	8	8	2.5
North Cornwall	83.7	1.6	32.8	16.0	9	7	7	3.6
Penwith	72.0	1.9	30.8	16.7	13	7	10	1.8
Restormel	70.1	2.3	31.2	13.9	9	9	8	2.6
Isles of Scilly	..	..	..	..	-	3	3	0.1
Devon County	83.8	9.7	28.1	17.2	8	9	9	25.7
East Devon	87.1	1.2	28.7	13.0	6	9	9	4.1
Exeter	80.9	1.9	24.3	17.7	8	14	15	2.3
Mid Devon	85.3	0.8	29.5	14.1	8	7	7	3.3
North Devon	83.9	1.7	28.5	18.3	9	7	7	3.6
South Hams	79.5	0.9	31.5	15.3	7	9	9	3.5
Teignbridge	86.5	1.5	27.9	16.8	9	10	9	3.8
Torridge	83.7	1.2	30.1	23.1	8	7	7	2.7
West Devon	81.5	0.5	26.6	18.3	7	6	7	2.4
Dorset County	83.3	3.6	26.9	11.0	6	10	8	12.6
Christchurch	79.4	0.4	23.0	11.5	6	12	11	1.3
East Dorset	79.8	0.5	26.5	8.5	5	11	8	2.9
North Dorset	86.8	0.4	26.8	8.2	5	9	7	2.3
Purbeck	86.1	0.4	29.4	13.6	5	9	10	1.4
West Dorset	82.9	0.8	28.0	11.3	6	7	8	3.6
Weymouth and Portland	85.8	1.2	27.1	11.7	8	14	11	1.1

14.6 *(continued)*

| | Economically active 1998-1999[3] (percentages) | Claimant count[4] March 2000 | | | Income Support bene-ficiaries[6] Feb.2000 (percentages) | Businesses registered for VAT 1998 | | Stock of businesses end 1998 (thousands) |
		Total (thousands)	Of which females (percentages)	Of which long-term claimants[5] (percentages)		Registration rates[7] (percentages)	Deregistration rates[7] (percentages)	
Gloucestershire	83.4	7.6	25.7	21.3	6	11	9	18.7
Cheltenham	82.7	1.7	22.8	21.2	7	14	8	3.4
Cotswold	85.4	0.4	25.7	17.0	4	10	7	4.1
Forest of Dean	79.4	1.1	31.3	21.1	7	9	8	2.9
Gloucester	85.2	2.3	23.4	24.8	9	14	21	1.9
Stroud	84.2	1.2	26.8	16.9	6	10	8	4.0
Tewkesbury	82.8	0.8	29.0	20.7	5	11	8	2.5
Somerset	82.4	5.8	27.4	16.6	7	9	8	17.0
Mendip	86.3	1.3	28.2	16.8	7	8	8	3.8
Sedgemoor	75.4	1.5	27.9	17.4	8	9	7	3.5
South Somerset	84.0	1.2	27.4	15.8	6	9	8	5.2
Taunton Deane	84.8	1.2	24.1	15.9	7	9	9	3.1
West Somerset	78.1	0.6	31.0	17.2	7	9	9	1.4
Wiltshire County	83.2	3.2	27.0	10.8	5	11	8	13.9
Kennet	83.0	0.6	27.7	14.1	4	11	9	2.7
North Wiltshire	83.9	0.8	28.3	6.8	5	11	8	4.4
Salisbury	81.9	0.8	22.4	13.1	5	11	10	3.5
West Wiltshire	83.8	0.9	29.6	10.1	6	11	8	3.3
Bristol/Bath area	..	13.7	25.1	16.5	8	13	10	27.5
Former county of Devon	81.4	18.4	26.4	18.3	9	9	10	32.0
Former county of Dorset	83.3	8.5	24.6	17.2	7	11	10	20.3
Former county of Wiltshire	84.8	5.4	25.2	13.4	6	12	9	18.0

1 See Notes and Definitions to the Labour market chapter.
2 Local government structure as at 1 April 1998. See Notes and Definitions.
3 Based on the population of working age. Data are from the Labour Force Survey and relate to the period March 1998 to February 1999.
4 Count of claimants of unemployment-related benefit, ie. Jobseeker's allowance.
5 Persons who have been claiming for more than 12 months (computerised claims only) as a percentage of all claimants.
6 Claimants and their partners aged 16 or over as a percentage of the population aged 16 or over. Data are from the Income Support Quarterly Statistical Enquiry.
7 Registrations/deregistrations during 1998 as a percentage of the stock at the end of 1997.

Source: Office for National Statistics; Department of Social Security; Department of the Environment, Transport and the Regions; Department of Trade and Industry

NUTS levels 1, 2 and 3 in England[1], 1998

NUTS level 3 areas

1   South Teeside
2   Hartlepool & Stockton
3   Darlington
4   Sunderland
5   Tyneside
6   Halton & Warrington
7   Gt Manchester North
8   Gt Manchester South
9   Blackburn with Darwen
10 Blackpool
11 Sefton
12 Wirral
13 East Merseyside
14 Liverpool
15 East Riding of Yorkshire
16 City of Kingston upon Hull
17 North & North East Lincolnshire
18 York
19 Leeds
20 Bradford
21 Calderdale, Kirklees & Wakefield
22 Sheffield
23 Barnsley, Doncaster & Rotherham
24 Derby
25 South & West Derbyshire
26 East Derbyshire
27 Leicester City
28 Leicestershire CC & Rutland
29 Northamptonshire
30 Nottingham
31 North Nottinghamshire
32 South Nottinghamshire
33 The Wrekin
34 Stoke-on-Trent
35 Staffordshire CC
36 Walsall & Wolverhampton
37 Birmingham
38 Coventry
39 Solihull
40 Dudley & Sandwell
41 Luton
42 Bedfordshire CC
43 Peterborough
44 Southend-on-Sea
45 Thurrock
46 Hertfordshire
47 Inner London - East
48 Inner London - West
49 Outer London - E & NE
50 Outer London - South
51 Outer London - W & NW
52 Milton Keynes
53 Buckinghamshire CC
54 Brighton & Hove

55 Portsmouth
56 Southampton
57 Medway
58 Kent CC
59 N & NE Somerset, South Gloucestershire
60 City of Bristol
61 Plymouth
62 Torbay
63 Bournemouth & Poole
64 Swindon

NUTS level 1 (= GOR)
NUTS level 2
NUTS level 3 (CC = County Council)

1 NUTS (Nomenclature of Units for Territorial Statistics) is a hierarchical classification of areas that provides a breakdown of the EU's economic territory. See Notes and Definitions.

14.7 Gross domestic product: by NUTS 1, 2 and 3 areas[1] at factor cost: current prices

	£ million			£ per head			£ per head (UK=100)		
	1994	1995	1996	1994	1995	1996	1994	1995	1996
United Kingdom[2]	570,944	597,741	629,839	9,777	10,199	10,711	100	100	100
England	484,357	506,246	534,942	9,944	10,352	10,897	102	101	102
North East	21,701	22,612	23,590	8,316	8,680	9,071	85	85	85
Tees Valley and Durham	9,510	9,962	10,343	8,143	8,537	8,872	83	84	83
Hartlepool and Stockton-on-Tees	2,356	2,490	2,623	8,712	9,215	9,674	89	90	90
South Teeside	2,529	2,662	2,746	8,729	9,214	9,582	89	90	90
Darlington	923	987	959	9,171	9,814	9,467	94	96	88
Durham County Council	3,703	3,822	4,015	7,300	7,538	7,922	75	74	74
Northumberland and Tyne and Wear	12,191	12,650	13,246	8,456	8,795	9,233	87	86	86
Northumberland	2,379	2,614	2,666	7,733	8,506	8,672	79	83	81
Tyneside	7,503	7,643	8,097	8,966	9,151	9,720	92	90	91
Sunderland	2,309	2,393	2,484	7,768	8,090	8,441	80	79	79
North West	61,779	64,023	66,978	8,951	9,279	9,719	92	91	91
Cumbria	5,010	5,232	5,329	10,222	10,671	10,862	105	105	101
West Cumbria	2,496	2,581	2,688	10,431	10,824	11,293	107	106	105
East Cumbria	2,514	2,650	2,641	10,022	10,525	10,456	103	103	98
Cheshire	10,645	11,298	11,904	10,911	11,551	12,147	112	113	113
Halton and Warrington	3,558	3,776	4,029	11,459	12,134	12,911	117	119	121
Cheshire County Council	7,087	7,522	7,876	10,655	11,279	11,790	109	111	110
Greater Manchester	23,123	24,105	25,203	8,970	9,349	9,785	92	92	91
Greater Manchester South	14,783	15,336	16,009	10,612	11,005	11,512	109	108	108
Greater Manchester North	8,340	8,769	9,193	7,038	7,401	7,759	72	73	72
Lancashire	12,455	12,667	13,415	8,747	8,883	9,416	90	87	88
Blackburn With Darwen	1,286	1,336	1,357	9,177	9,524	9,730	94	93	91
Blackpool	1,076	1,052	1,126	6,989	6,849	7,383	72	67	69
Lancashire County Council	10,093	10,279	10,932	8,933	9,079	9,651	91	89	90
Merseyside	10,545	10,722	11,128	7,352	7,512	7,834	75	74	73
East Merseyside	2,414	2,402	2,553	7,209	7,195	7,654	74	71	72
Liverpool	4,316	4,379	4,377	9,106	9,302	9,352	93	91	87
Sefton	1,762	1,823	1,962	6,026	6,264	6,772	62	61	63
Wirral	2,053	2,117	2,236	6,162	6,387	6,793	63	63	63
Yorkshire and the Humber	44,021	46,775	48,268	8,760	9,300	9,585	90	91	90
East Riding and North Lincolnshire	8,071	8,689	9,018	9,075	9,771	10,170	93	96	95
Kingston Upon Hull, City of	2,373	2,551	2,612	8,815	9,499	9,789	90	93	91
East Riding of Yorkshire	2,410	2,596	2,754	7,882	8,418	8,920	81	83	83
North and North East Lincolnshire	3,289	3,541	3,653	10,455	11,343	11,734	107	111	110
North Yorkshire	7,150	7,623	7,917	9,846	10,433	10,776	101	102	101
York	2,083	2,260	2,357	11,951	12,958	13,462	122	127	126
North Yorkshire County Council	5,066	5,363	5,559	9,182	9,642	9,935	94	95	93
South Yorkshire	9,662	10,019	10,409	7,402	7,684	7,978	76	75	75
Barnsley, Doncaster and Rotherham	4,932	5,110	5,456	6,362	6,590	7,046	65	65	66
Sheffield	4,730	4,910	4,953	8,923	9,290	9,338	91	91	87
West Yorkshire	19,138	20,444	20,924	9,096	9,709	9,920	93	95	93
Bradford	3,985	4,213	4,417	8,272	8,729	9,138	85	86	85
Leeds	7,724	8,362	8,507	10,662	11,534	11,702	109	113	109
Calderdale, Kirklees and Wakefield	7,428	7,869	8,000	8,274	8,762	8,899	85	86	83
East Midlands	38,757	40,377	41,813	9,448	9,791	10,096	97	96	94
Derbyshire and Nottinghamshire	17,737	18,325	19,202	8,936	9,209	9,631	91	90	90
Derby	2,474	2,492	2,627	10,736	10,745	11,242	110	105	105
East Derbyshire	1,927	2,019	2,016	7,109	7,449	7,448	73	73	70
South and West Derbyshire	3,943	4,205	4,286	8,712	9,242	9,366	89	91	87
Nottingham	3,920	3,986	4,417	13,881	14,043	15,554	142	138	145
North Nottinghamshire	3,315	3,354	3,506	7,864	7,957	8,335	80	78	78
South Nottinghamshire	2,157	2,270	2,349	6,600	6,949	7,181	68	68	67
Leicestershire, Rutland and Northamptonshire	15,586	16,344	16,687	10,311	10,737	10,893	106	105	102
Leicester	3,562	3,690	3,727	12,140	12,481	12,641	124	122	118
Leicestershire County Council and Rutland	5,884	6,288	6,270	9,438	10,024	9,911	97	98	93
Northamptonshire	6,140	6,366	6,689	10,323	10,622	11,068	106	104	103
Lincolnshire[3]	5,434	5,708	5,925	8,973	9,330	9,620	92	92	90

14.7 *(continued)*

	£ million			£ per head			£ per head (UK=100)		
	1994	1995	1996	1994	1995	1996	1994	1995	1996
West Midlands	48,269	50,766	53,249	9,116	9,567	10,016	93	94	94
Herefordshire, Worcestershire and Warwickshire	11,051	12,078	12,833	9,238	10,125	10,719	95	99	100
County of Herefordshire	1,479	1,617	1,646	9,021	9,820	9,984	92	96	93
Worcestershire	4,775	5,192	5,514	8,933	9,730	10,371	90	94	95
Warwickshire	4,796	5,269	5,673	9,664	10,565	11,332	99	104	106
Shropshire and Staffordshire	12,194	12,941	13,954	8,290	8,765	9,448	85	86	88
Telford and Wrekin	1,729	1,915	2,064	12,052	13,242	14,316	123	130	134
Shropshire County Council	2,101	2,235	2,319	7,695	8,121	8,371	79	80	78
Stoke-on-Trent	2,360	2,438	2,614	9,286	9,585	10,275	95	94	96
Staffordshire County Council	6,004	6,352	6,956	7,503	7,919	8,682	77	78	81
West Midlands	25,024	25,747	26,462	9,523	9,763	10,014	97	96	94
Birmingham	10,421	10,672	10,946	10,334	10,489	10,725	106	103	100
Solihull	1,873	2,077	2,166	9,272	10,236	10,620	95	100	99
Coventry	3,151	3,223	3,430	10,417	10,618	11,191	107	104	105
Dudley and Sandwell	5,175	5,311	5,445	8,539	8,762	9,009	87	86	84
Walsall and Wolverhampton	4,404	4,464	4,476	8,654	8,802	8,827	89	86	82
East	49,330	51,665	54,850	9,444	9,827	10,363	97	96	97
East Anglia	20,799	21,777	22,891	9,881	10,258	10,688	101	101	100
Peterborough	1,906	2,013	2,170	12,014	12,634	13,673	123	124	128
Cambridgeshire County Council	5,762	6,268	6,648	10,908	11,726	12,207	112	115	114
Norfolk	6,858	6,896	7,215	8,924	8,929	9,287	91	88	87
Suffolk	6,273	6,600	6,859	9,659	10,049	10,366	99	99	97
Bedfordshire and Hertfordshire	15,374	16,014	17,141	9,929	10,286	10,956	102	101	102
Luton	1,904	1,965	2,066	10,531	10,834	11,386	108	106	106
Bedfordshire County Council	3,349	3,532	3,645	9,244	9,696	9,924	95	95	93
Hertfordshire	10,121	10,517	11,430	10,067	10,400	11,252	103	102	105
Essex	13,157	13,874	14,818	8,380	8,795	9,342	86	86	87
Southend-on-Sea	1,262	1,298	1,310	7,427	7,584	7,607	76	74	71
Thurrock	1,567	1,614	1,766	11,932	12,269	13,353	122	120	125
Essex County Council	10,327	10,962	11,741	8,140	8,599	9,161	83	84	86
London	97,043	100,762	106,658	13,928	14,380	15,077	143	141	141
Inner London	59,390	61,979	65,255	22,315	23,152	24,099	228	227	225
Inner London - West	39,380	41,385	43,928	41,367	43,089	44,811	423	423	418
Inner London - East	20,010	20,594	21,327	11,706	11,997	12,346	120	118	115
Outer London	37,653	38,783	41,402	8,744	8,957	9,482	89	88	89
Outer London - East and North East	10,198	10,553	11,241	6,682	6,910	7,350	68	68	69
Outer London - South	9,414	9,811	10,268	8,489	8,775	9,095	87	86	85
Outer London - West and North West	18,041	18,419	19,893	10,796	10,932	11,647	110	107	109
South East	79,006	82,000	90,440	10,149	10,450	11,455	104	103	107
Berkshire, Buckinghamshire and Oxfordshire	24,411	25,008	27,385	12,098	12,214	13,256	124	120	124
Berkshire	10,398	10,484	11,659	13,518	13,385	14,738	138	131	138
Milton Keynes	2,500	2,613	2,875	13,267	13,548	14,585	136	133	136
Buckinghamshire County Council	5,000	5,235	5,762	10,640	11,068	12,141	109	109	113
Oxfordshire	6,513	6,676	7,089	11,035	11,158	11,753	113	109	110
Surrey, East and West Sussex	24,041	25,594	28,245	9,656	10,210	11,212	99	100	105
Brighton and Hove	2,087	2,202	2,431	8,475	8,875	9,742	87	87	91
East Sussex County Council	3,322	3,665	3,847	6,918	7,592	7,927	71	74	74
Surrey	11,375	11,837	13,458	10,925	11,335	12,853	112	111	120
West Sussex	7,256	7,890	8,509	10,049	10,786	11,542	103	106	108
Hampshire and Isle of Wight	16,701	16,797	19,491	9,652	9,643	11,120	99	95	104
Portsmouth	2,278	2,188	2,559	12,034	11,515	13,441	123	113	126
Southampton	2,545	2,660	2,902	12,021	12,457	13,508	123	122	126
Hampshire County Council	11,041	11,079	13,015	9,164	9,133	10,649	94	90	99
Isle Of Wight	838	870	1,015	6,719	6,952	8,088	69	68	76
Kent	13,853	14,601	15,319	8,959	9,413	9,837	92	92	92
Medway	1,854	1,915	1,925	7,671	7,980	8,041	79	78	75
Kent County Council	12,000	12,686	13,394	9,197	9,675	10,163	94	95	95

Regional Trends 35, © Crown copyright 2000

14.7 *(continued)*

	£ million			£ per head			£ per head (UK=100)		
	1994	1995	1996	1994	1995	1996	1994	1995	1996
South West	44,450	47,265	49,097	9,264	9,792	10,141	95	96	95
Gloucestershire, Wiltshire and North Somerset	22,005	23,530	24,887	10,407	11,069	11,660	106	109	109
Bristol, City of	4,734	5,051	5,253	11,857	12,605	13,144	121	124	123
North and North East Somerset, South Gloucestershire	5,313	5,550	6,184	9,170	9,542	10,567	94	94	99
Gloucestershire	5,497	5,972	6,182	10,004	10,804	11,112	102	106	104
Swindon	2,562	2,721	2,861	14,766	15,660	16,384	151	154	153
Wiltshire County Council	3,899	4,236	4,408	9,445	10,163	10,528	97	100	98
Dorset and Somerset	10,115	10,834	10,963	8,789	9,342	9,414	90	92	88
Bournemouth and Poole	2,956	3,159	3,205	9,910	10,537	10,685	101	103	100
Dorset County Council	2,939	3,223	3,416	7,844	8,505	8,943	80	83	84
Somerset	4,220	4,453	4,342	8,830	9,256	8,996	90	91	84
Cornwall and Isles of Scilly[3]	3,392	3,476	3,680	7,073	7,201	7,614	72	71	71
Devon	8,938	9,425	9,568	8,485	8,902	9,032	87	87	84
Plymouth	2,702	2,765	2,723	10,562	10,740	10,645	108	105	99
Torbay	862	931	934	7,010	7,511	7,571	72	74	71
Devon County Council	5,374	5,729	5,910	7,966	8,458	8,691	82	83	81

1 NUTS (Nomenclature of Units for Territorial Statistics) is a hierarchical classification of areas that provides a breakdown of the EU's economic territory. Data are on the old ESA79 and are consistent with the data published in United Kingdom National Accounts 1997. See Notes and Definitions.
2 Excluding the Continental Shelf and the statistical descrepancy of the income based measure.
3 This area represents both NUTS 2 and 3 levels.

Source: Office for National Statistics

15 Sub-regions of Wales

Unitary Authorities in Wales

Isle of Anglesey

Conwy

Flintshire

Denbighshire

Wrexham

Gwynedd

Ceredigion

Powys

1 Merthyr Tydfil
2 Blaenau Gwent
3 Torfaen

Pembrokeshire

Carmarthenshire

Monmouthshire

Neath Port Talbot

2

Swansea

Rhondda, Cynon, Taff

1

Caerphilly

3

Bridgend

Newport

Cardiff

The Vale of Glamorgan

15.1 Area and population, 1998

	Area (sq km)	Persons per sq km	Population (thousands)			Total population percentage change 1981-1998	Total fertility rate (TFR)[1]	Standard-ised mortality ratio (UK=100) (SMR)[2]	Percentage of population aged			
			Males	Females	Total				Under 5	5-15	16 up to pension age[3]	Pension age[3] or over
United Kingdom	242,910	244	29,128	30,108	59,237	5.1	1.71	100	6.2	14.2	61.4	18.1
Wales	20,779	141	1,439	1,495	2,934	4.3	1.79	101	5.9	14.5	59.6	19.9
Blaenau Gwent	109	661	36	36	72	-4.8	1.94	115	6.1	15.4	59.1	19.4
Bridgend	246	534	64	68	131	4.1	1.81	104	6.2	14.4	60.3	19.4
Caerphilly	278	610	83	86	170	-1.3	2.03	117	6.5	15.6	60.1	17.6
Cardiff	140	2,292	159	162	321	11.9	1.78	91	6.4	15.2	61.3	17.1
Carmarthenshire	2,395	71	82	87	169	2.4	1.79	101	5.4	13.3	58.1	23.1
Ceredigion	1,795	39	35	36	71	15.6	1.52	82	4.7	12.6	60.4	21.9
Conwy	1,130	99	53	59	112	13.1	1.78	91	5.3	13.0	55.5	26.2
Denbighshire	844	107	44	47	91	4.5	1.95	99	5.9	13.8	56.5	23.3
Flintshire	438	336	73	74	147	6.0	1.85	94	6.1	14.3	62.2	17.5
Gwynedd	2,548	46	57	60	118	5.0	1.80	96	5.7	13.7	58.2	22.0
Isle of Anglesey	714	92	32	34	65	-3.9	2.07	98	6.0	14.7	57.9	22.1
Merthyr Tydfil	111	513	28	29	57	-5.9	2.06	127	6.4	16.3	58.6	18.6
Monmouthshire	850	101	42	44	86	12.7	1.65	91	5.5	14.2	60.0	20.7
Neath Port Talbot	442	314	68	71	139	-2.8	1.75	105	5.7	14.4	58.8	20.9
Newport	190	733	68	71	139	5.1	2.00	97	6.5	15.5	59.4	18.7
Pembrokeshire	1,590	72	56	58	114	5.9	1.93	106	5.7	14.7	58.4	21.1
Powys	5,196	24	63	63	126	12.3	1.67	99	5.5	13.6	58.9	22.0
Rhondda, Cynon, Taff	424	567	119	121	240	0.8	1.76	114	6.0	15.0	61.1	18.1
Swansea	378	607	113	117	230	0.1	1.71	98	5.5	13.8	59.9	20.6
Torfaen	126	716	44	46	90	-0.6	1.93	107	6.6	15.2	59.9	18.6
The Vale of Glamorgan	335	362	59	62	121	7.1	1.86	99	6.2	15.3	59.4	19.3
Wrexham	498	251	61	64	125	5.0	1.81	104	5.7	14.6	60.9	19.0

1 The total fertility rate (TFR) is the average number of children which would be born to a woman if the current pattern of fertility persisted throughout her child-bearing years. Previously called total period fertility rate (TPFR).

2 Adjusted for the age structure of the population. See Notes and Definitions to the Population chapter.

3 Pension age is 65 for males and 60 for females.

Source: Office for National Statistics; National Assembly for Wales

15.2 Vital[1,2] and social statistics

	Live births per 1,000 population		Deaths per 1,000 population		Perinatal mortality rate[3]	Infant mortality rate[4]	Percent-age of live births under 2.5 kg	Percent-age of live births outside marriage	Children looked after by LAs per 1,000 population aged under 18
	1991	1998	1991	1998	1997-1999	1997-1999	1998	1998	1999[5]
United Kingdom	13.7	12.1	11.2	10.6	8.3	5.8	..	37.6	..
Wales	13.2	11.4	11.8	11.6	7.9	6.0	7.2	44.4	4.9
Blaenau Gwent	14.8	11.4	12.5	12.8	11.2	8.4	7.3	54.7	8.3
Bridgend	13.3	11.4	11.6	11.5	8.4	5.9	6.8	47.2	5.7
Caerphilly	14.2	12.6	10.0	11.3	7.7	4.1	6.7	51.8	7.4
Cardiff	14.8	12.6	10.8	9.2	7.6	7.0	8.1	43.9	6.1
Carmarthenshire	11.3	10.3	13.0	13.3	8.4	6.2	5.4	39.9	4.4
Ceredigion	10.8	9.2	12.6	10.7	7.3	5.2	7.1	37.3	2.7
Conwy	11.8	10.4	15.6	15.0	10.0	8.1	8.3	42.3	4.1
Denbighshire	12.6	11.4	13.9	14.3	7.4	6.7	6.2	45.4	4.0
Flintshire	13.4	12.2	10.5	9.4	7.9	5.0	6.4	35.9	3.3
Gwynedd	12.2	11.1	13.0	12.4	7.8	4.8	5.6	43.5	2.6
Isle of Anglesey	12.2	11.2	11.6	12.4	7.2	6.3	5.9	39.1	3.4
Merthyr Tydfil	14.7	12.4	11.7	13.1	6.2	3.8	8.4	56.3	7.0
Monmouthshire	11.8	10.0	11.1	11.0	6.9	5.8	5.7	31.9	1.8
Neath Port Talbot	12.4	10.4	13.2	12.5	7.4	6.9	7.9	45.7	5.4
Newport	15.3	12.7	10.6	10.1	4.6	4.0	7.4	47.7	7.3
Pembrokeshire	13.0	11.2	11.2	12.3	8.9	7.1	8.5	40.6	5.5
Powys	12.0	10.0	12.4	12.6	8.9	3.8	5.4	34.5	2.2
Rhondda, Cynon, Taff	13.4	11.7	11.6	11.5	8.2	6.5	8.1	50.6	4.7
Swansea	12.5	10.6	11.8	11.6	9.4	7.4	7.5	45.7	4.7
Torfaen	14.3	12.0	11.4	10.9	7.2	6.0	7.7	48.5	5.1
The Vale of Glamorgan	13.4	11.7	11.5	10.7	7.7	4.9	7.2	41.9	5.6
Wrexham	13.1	11.9	11.2	11.5	7.4	5.8	7.3	42.3	2.8

1 Births and deaths data are based on the usual area of residence of the mother/deceased. See Notes and Definitions to the Population chapter for details of the inclusion/exclusion of births to non-resident mothers and deaths of non-resident persons.

2 Births data are on the basis of year of occurrence in England and Wales and year of registration in Scotland and Northern Ireland. All deaths data relate to year of registration.

3 Still births and deaths of infants under 1 week of age per 1,000 live and still births. Figures for some UAs should be treated with caution as the perinatal mortality rate was based on fewer than 20 events.

4 Deaths of infants under 1 year of age per 1,000 live births. Figures for some UAs should be treated with caution as the infant mortality rate was based on fewer than 20 events.

5 At 31 March. Under 18 mid-1998 population estimates used.

Source: Office for National Statistics; National Assembly for Wales

15.3 Education and training

	Day nursery places per 1,000 population aged under 5 years[1] March 1999	Children under 5 in education[2] (percentages) Jan. 1999	Pupil/teacher ratio 1998/99(numbers)[3]		Percentage of pupils in last year of compulsory schooling[4,5] 1998/99 with		Average A/AS level points score[5,6] 1998/99
			Primary schools	Secondary schools	No graded results	5 or more A*-Cs at GCSE	
United Kingdom	..	64	22.9	16.5	5.4	47.3	17.8
Wales	56.9	78	22.3	16.5	8.1	46.6	16.7
Blaenau Gwent	10.0	76	23.0	16.4	14.3	32.1	13.4
Bridgend	38.6	68	24.0	16.9	12.3	43.1	15.5
Caerphilly	11.4	80	23.2	16.3	10.3	40.6	15.0
Cardiff	100.9	71	22.6	17.1	11.3	43.7	16.6
Carmarthenshire	54.6	82	20.1	16.5	5.5	52.9	15.9
Ceredigion	86.6	63	18.8	15.5	6.3	56.6	21.2
Conwy	100.1	84	23.0	17.2	4.4	48.5	16.8
Denbighshire	144.3	79	23.5	17.4	7.8	48.9	17.5
Flintshire	76.2	83	23.6	16.7	5.4	46.5	17.0
Gwynedd	68.3	73	21.1	15.0	4.3	52.5	15.8
Isle of Anglesey	14.9	68	22.1	16.5	7.1	53.4	17.5
Merthyr Tydfil	20.5	92	22.9	16.3	13.5	38.5	13.9
Monmouthshire	41.1	55	24.0	16.9	6.3	54.0	18.0
Neath Port Talbot	23.4	94	20.6	16.2	6.6	46.6	16.3
Newport	63.9	75	23.6	17.0	6.2	46.8	15.6
Pembrokeshire	53.1	77	20.7	16.1	3.1	50.5	17.7
Powys	56.1	57	20.2	14.5	4.8	55.0	18.4
Rhondda, Cynon, Taff	23.9	89	22.8	16.4	9.2	41.6	15.1
Swansea	44.1	92	22.0	16.6	9.7	45.8	16.6
Torfaen	33.7	75	23.3	16.8	6.5	43.6	17.9
The Vale of Glamorgan	55.9	77	22.4	16.8	6.6	52.7	18.3
Wrexham	96.1	87	23.0	17.3	9.8	46.2	16.9

1 Local authority provided and registered day nurseries only. A small number of places provided by facilities exempt from registration are excluded. Population data used are mid-1998 estimates.

2 Figures relate to all pupils as a percentage of the three and four year old population.

3 Public sector schools only.

4 Pupils in their last year of compulsory schooling as a percentage of the school population of the same age.

5 Figures relate to maintained schools only; hence they are not directly comparable with those in Tables 4.4, 16.3 and 17.3 which are for all schools.

6 Figures for United Kingdom relates to England and Wales average.

Source: National Assembly for Wales; Department for Education and Employment

15.4 Housing and households

	Housing completions 1998 (numbers)		Stock of dwellings 31 December 1998 (thousands)	All households 1998 (thousands)	Local authority tenants: average weekly unrebated rent per dwelling (£) April 1999	Council Tax (£)[1] April 1999
	Private enterprise	Registered social landlords, local authorities etc				
Wales	6,386	1,502	1,257	1,189.1	40.20	602
Blaenau Gwent	40	16	32	28.9	40.10	640
Bridgend	180	30	56	53.4	40.20	647
Caerphilly	513	78	71	68.0	42.50	633
Cardiff	1,125	66	126	127.8	45.60	573
Carmarthenshire	413	42	75	67.7	38.30	666
Ceredigion	153	12	30	29.5	42.00	670
Conwy	149	65	52	47.3	37.70	488
Denbighshire	95	44	39	37.8	36.50	647
Flintshire	363	129	61	58.6	37.50	592
Gwynedd	77	26	56	48.0	37.80	618
Isle of Anglesey	126	16	31	28.9	36.70	534
Merthyr Tydfil	126	15	25	23.6	38.30	712
Monmouthshire	215	17	35	35.5	44.20	485
Neath Port Talbot	293	85	65	56.6	38.40	750
Newport	239	87	57	56.6	43.40	531
Pembrokeshire	248	85	53	46.3	37.20	542
Powys	238	71	55	50.5	39.00	552
Rhondda, Cynon, Taff	555	83	102	96.5	39.00	686
Swansea	488	340	94	94.2	40.00	596
Torfaen	140	52	39	36.2	46.00	564
The Vale of Glamorgan	401	129	49	47.6	45.00	532
Wrexham	209	14	54	50.7	35.20	633

1 See Notes and Definitions.

Source: National Assembly for Wales

15.5 Labour market statistics[1]

	Economically active 1998-1999[2] (percentages)	Total in employment[3] as at September 1998 (thousands)	Employment rate 1998-1999[2] (percentages)[4]	ILO unemployment rate 1998-1999[2] (percentages)[4]	Claimant count[5] March 2000			Average gross weekly full-time earnings, all persons[7] April 1999 (£)
					Total (thousands)	Of which females (percentages)	Of which long term claimants[6] (percentages)	
United Kingdom	78.4	26,508	73.5	6.3	1,194.3	23.5	22.1	398.7
Wales	73.1	1,177	67.7	7.4	61.8	22.2	19.6	353.6
Blaenau Gwent	64.6	22	55.4	..	2.1	22.6	25.3	311.4
Bridgend	73.8	54	70.2	..	2.7	22.1	15.3	367.5
Caerphilly	73.3	69	67.3	..	3.4	21.7	18.2	361.8
Cardiff	70.4	127	66.0	..	6.6	18.7	20.2	364.0
Carmarthenshire	74.4	66	68.5	..	3.5	24.3	18.3	337.6
Ceredigion	75.7	30	72.8	..	1.2	28.2	23.5	..
Conwy	76.9	43	71.5	..	2.5	22.8	19.6	300.3
Denbighshire	74.1	35	69.6	..	1.8	23.0	17.2	341.2
Flintshire	77.2	68	73.3	..	2.4	25.7	18.8	366.5
Gwynedd	68.4	41	61.3	..	3.5	24.5	23.4	322.3
Isle of Anglesey	74.2	29	70.2	..	2.0	26.3	30.1	..
Merthyr Tydfil	63.0	18	54.1	..	1.5	19.8	26.7	..
Monmouthshire	80.1	37	73.9	..	1.2	26.1	21.6	351.4
Neath Port Talbot	69.2	55	63.5	..	3.0	21.7	19.5	384.2
Newport	73.1	53	68.0	..	3.1	21.4	20.9	376.9
Pembrokeshire	67.2	40	61.4	..	3.3	24.4	15.2	356.2
Powys	81.2	54	76.9	..	1.9	27.5	17.1	317.5
Rhondda, Cynon, Taff	71.4	100	65.9	..	4.8	20.4	16.9	338.5
Swansea	74.0	96	68.3	..	5.1	18.7	22.1	351.6
Torfaen	74.7	39	69.3	..	1.7	23.6	13.7	352.1
The Vale of Glamorgan	80.1	51	74.6	..	2.4	19.7	18.1	..
Wrexham	70.6	49	63.7	..	2.0	22.3	14.9	356.6

1 See Notes and Definitions to the Labour market chapter. In some cases sample sizes are too small to provide reliable estimates.
2 For those of working age. Data are from the Labour Force Survey and relate to the period March 1998 to February 1999.
3 For those of working age. Includes those on Government employment and training schemes and unpaid family workers.
4 As a percentage of the economically active.
5 Count of claimants of unemployment-related benefit, ie. Jobseeker's allowance.
6 Persons who have been claiming for more than 12 months as a percentage of all claimants.
7 Earning estimates have been derived from the New Earning Survey and relate to full-time employees whose pay for the survey pay-period was not affected by absence.

Source: Office for National Statistics

NUTS levels 1, 2 and 3 in Wales[1], 1998

1 NUTS (Nomenclature of Units for Territorial Statistics) is a hierarchical classification of areas that provides a breakdown of the EU's economic territory. The NUTS level 1 area is the whole country. See Notes and Definitions.

15.6 Gross domestic product: by NUTS 1, 2 and 3 areas[1] at factor cost: current prices

	£ million			£ per head			£ per head (UK=100)		
	1994	1995	1996	1994	1995	1996	1994	1995	1996
United Kingdom[2]	570,944	597,741	629,839	9,777	10,199	10,711	100	100	100
Wales	23,775	25,088	25,998	8,161	8,601	8,900	84	84	83
West Wales and the Valleys	12,973	13,946	14,706	6,918	7,441	7,862	71	73	73
Isle of Anglesey	452	475	491	6,617	7,069	7,329	68	69	68
Gwynedd	865	945	904	7,399	8,011	7,678	76	79	72
Conwy and Denbighshire	1,265	1,440	1,588	6,261	7,100	7,831	64	70	73
South West Wales	2,315	2,525	2,767	6,562	7,149	7,856	67	70	73
Central Valleys	1,933	1,982	2,032	6,478	6,637	6,815	66	65	64
Gwent Valleys	2,210	2,430	2,626	6,599	7,288	7,894	68	72	74
Bridgend and Neath Port Talbot	2,220	2,346	2,500	8,191	8,679	9,275	84	85	87
Swansea	1,713	1,803	1,797	7,417	7,817	7,806	76	77	73
East Wales	10,801	11,142	11,292	10,410	10,688	10,748	107	105	100
Monmouthshire and Newport	2,341	2,467	2,470	10,565	11,070	11,044	108	109	103
Cardiff and Vale of Glamorgan	4,701	4,726	4,655	11,049	11,035	10,715	113	108	100
Flintshire and Wrexham	2,861	2,974	3,110	10,646	11,052	11,593	109	108	108
Powys	898	975	1,058	7,370	7,973	8,505	75	78	79

1 NUTS (Nomenclature of Units for Territorial Statistics) is a hierarchical classification of areas that provides a breakdown of the EU's economic territory. Data are on the old ESA79 and are consistent with the data published in United Kingdom National Accounts 1997. See Notes and Definitions.
2 Excluding the Continental Shelf and the statistical discrepancy of the income based measure.

Source: Office for National Statistics

16 Sub-regions of Scotland

New Councils in Scotland

Orkney Islands

Shetland Islands

Eilean Siar (Western Isles)

Highland

Moray

Aberdeenshire

13

Angus

Perth & Kinross

12

Argyll & Bute

Stirling

Fife

11

2   6   8

1   5   7   9   10   East Lothian

3   Midlothian

4

North Ayrshire

South Lanarkshire

East Ayrshire

The Scottish Borders

South Ayrshire

Dumfries & Galloway

1   Inverclyde
2   West Dunbartonshire
3   Renfrewshire
4   East Renfrewshire
5   Glasgow City
6   East Dunbartonshire
7   North Lanarkshire
8   Falkirk
9   West Lothian
10 Edinburgh, City of
11 Clackmannanshire
12 Dundee City
13 Aberdeen City

16.1 Area and population, 1998

	Area (sq km)	Persons per sq km	Population (thousands)			Total population percentage change 1981-1998	Total fertility rate (TFR)[1]	Standard-ised mortality ratio (UK=100) (SMR)[2]	Percentage of population aged			
			Males	Females	Total				Under 5	5-15	16 up to pension age[3]	Pension age[3] or over
United Kingdom	242,910	244	29,128	30,108	59,237	5.1	1.71	100	6.2	14.2	61.4	18.1
Scotland	78,133	66	2,484	2,636	5,120	-1.2	1.55	116	5.9	13.9	62.2	18.0
Aberdeen City	186	1,147	105	109	213	0.3	1.43	106	5.5	12.7	64.4	17.4
Aberdeenshire	6,318	36	113	114	226	19.8	1.69	101	6.1	15.2	63.0	15.7
Angus	2,181	50	54	56	110	4.2	1.65	107	5.8	14.1	60.6	19.5
Argyll and Bute	6,930	13	44	46	90	-1.1	1.55	117	5.2	13.3	60.4	21.1
Clackmannanshire	157	310	24	25	49	0.7	1.66	109	6.2	14.8	61.8	17.1
Dumfries and Galloway	6,439	23	72	76	147	1.2	1.74	105	5.7	13.8	58.9	21.6
Dundee City	65	2,252	70	77	147	-13.5	1.64	120	5.8	13.4	60.5	20.3
East Ayrshire	1,252	97	59	63	121	-4.8	1.67	118	6.0	14.6	60.9	18.6
East Dunbartonshire	172	638	54	56	110	-0.1	1.52	97	5.5	13.8	63.3	17.4
East Lothian	678	132	44	46	90	11.0	1.64	109	6.1	13.8	60.7	19.3
East Renfrewshire	173	509	43	45	88	9.6	1.72	97	6.0	14.5	61.9	17.6
Edinburgh, City of	262	1,716	218	232	450	0.9	1.29	106	5.5	11.8	65.1	17.7
Eilean Siar (Western Isles)	3,134	9	14	14	28	-11.4	1.70	113	5.2	14.8	58.7	21.3
Falkirk	299	482	70	74	144	-0.7	1.65	120	6.0	13.8	62.8	17.5
Fife	1,323	264	169	180	349	2.1	1.62	111	5.8	14.3	61.4	18.5
Glasgow City	175	3,540	296	324	620	-13.0	1.37	139	6.1	13.5	63.0	17.5
Highland	25,784	8	102	106	208	6.9	1.91	112	6.0	14.6	60.8	18.6
Inverclyde	162	528	41	44	85	-15.6	1.57	134	5.9	14.6	60.8	18.7
Midlothian	356	227	40	41	81	-3.3	1.71	123	5.9	14.1	63.4	16.6
Moray	2,238	38	43	43	86	2.9	1.86	105	6.4	14.7	60.5	18.4
North Ayrshire	884	158	67	72	140	1.7	1.64	116	5.9	14.8	61.4	18.0
North Lanarkshire	474	690	159	168	327	-4.4	1.69	121	6.2	14.6	63.0	16.2
Orkney Islands	992	20	10	10	20	1.9	1.72	115	5.7	15.2	60.2	18.9
Perth and Kinross	5,311	25	64	69	133	9.2	1.61	103	5.4	13.9	59.6	21.1
Renfrewshire	261	680	86	92	178	-3.9	1.57	126	6.1	14.3	62.4	17.3
Scottish Borders, The	4,734	22	51	55	106	5.0	1.60	101	5.6	13.4	59.2	21.9
Shetland Islands	1,438	16	12	11	23	-13.0	1.74	118	6.5	16.2	62.4	14.9
South Ayrshire	1,202	95	55	60	114	1.1	1.60	108	5.2	13.6	60.0	21.1
South Lanarkshire	1,771	173	149	158	307	-1.0	1.50	125	5.9	14.5	62.9	16.7
Stirling	2,196	38	40	43	83	3.6	1.59	105	5.7	13.6	62.6	18.1
West Dunbartonshire	162	585	45	49	95	-10.3	1.66	130	6.2	15.3	60.8	17.8
West Lothian	425	360	75	78	153	10.0	1.65	131	6.7	14.9	65.1	13.3

1 The total fertility rate (TFR) is the average number of children which would be born to a woman if the current pattern of fertility persisted throughout her child-bearing years. Previously called total period fertility rate (TPFR).

2 Adjusted for the age structure of the population. See Notes and Definitions to the Population chapter.

3 Pension age is 65 for males and 60 for females.

Source: Office for National Statistics; General Register Office for Scotland

16.2 Vital[1,2] and social statistics

	Live births per 1,000 population		Deaths per 1,000 population		Perinatal mortality rate[3]	Infant mortality rate[4]	Percentage of live births outside marriage
	1991	1998	1991	1998	1997-1999	1997-1999	1998
United Kingdom	13.7	12.1	11.2	10.6	8.3	5.8	37.6
Scotland	13.1	11.2	12.0	11.6	8.1	5.3	38.9
Aberdeen City	12.5	11.0	10.8	10.4	8.6	5.1	39.2
Aberdeenshire	13.5	11.5	9.6	9.3	7.5	4.3	25.9
Angus	12.3	10.7	12.9	12.1	5.7	3.7	36.1
Argyll and Bute	13.0	9.4	12.7	14.5	6.8	6.5	35.3
Clackmannanshire	13.7	11.1	10.1	10.6	6.0	4.2	45.4
Dumfries and Galloway	12.1	10.3	13.1	12.3	9.5	3.8	35.7
Dundee City	12.9	11.3	12.4	13.2	8.9	5.3	52.0
East Ayrshire	13.9	11.1	11.9	11.8	8.4	4.7	45.2
East Dunbartonshire	12.4	10.2	8.7	8.8	7.4	3.9	22.5
East Lothian	13.1	11.3	12.6	11.8	4.9	2.6	31.7
East Renfrewshire	13.0	11.5	9.5	9.6	4.0	2.7	20.0
Edinburgh, City of	12.9	10.8	12.5	10.9	7.9	6.4	35.9
Eilean Siar (Western Isles)	11.2	9.8	14.9	14.1	8.8	7.6	24.5
Falkirk	13.5	11.8	11.5	11.2	7.1	5.7	38.8
Fife	12.6	11.0	11.9	11.5	8.2	4.9	38.9
Glasgow City	14.3	11.5	14.4	13.5	8.9	6.6	52.1
Highland	13.0	11.8	11.5	11.5	7.5	4.6	36.8
Inverclyde	12.9	10.7	13.6	13.7	8.4	3.9	46.2
Midlothian	13.5	11.9	10.6	10.8	8.7	7.0	33.7
Moray	13.8	12.0	11.2	10.6	8.0	4.9	28.6
North Ayrshire	13.9	11.0	11.7	11.7	7.8	5.4	45.7
North Lanarkshire	13.6	12.5	11.1	10.6	9.7	5.1	41.7
Orkney Islands	12.1	9.9	11.8	12.5	6.6	3.3	32.0
Perth and Kinross	12.2	10.3	13.0	12.4	8.1	4.8	33.4
Renfrewshire	12.6	11.3	11.5	11.6	6.3	5.8	40.4
Scottish Borders, The	12.1	9.8	13.8	12.6	8.3	5.3	30.2
Shetland Islands	14.4	11.6	10.4	10.6	8.3	3.6	34.0
South Ayrshire	10.8	10.2	13.3	12.8	8.6	6.3	39.7
South Lanarkshire	13.3	10.8	10.5	11.0	7.5	4.3	35.5
Stirling	11.9	11.1	11.8	10.9	5.4	5.4	33.9
West Dunbartonshire	13.1	11.8	12.5	12.2	10.2	8.1	47.5
West Lothian	14.3	12.9	9.2	9.6	9.8	6.2	37.0

1 Births and deaths data are based on the usual area of residence of the mother/deceased. See Notes and Definitions to the Population chapter for details of the inclusion/exclusion of births to non-resident mothers and deaths of non-resident persons.

2 Births data are on the basis of year of occurrence in England and Wales and year of registration in Scotland and Northern Ireland. All deaths data relate to year of registration.

3 Still births and deaths of infants under 1 week of age per 1,000 live and still births. Figures for some UAs should be treated with caution as the perinatal mortality rate was based on fewer than 20 events.

4 Death of infants under 1 year of age per 1,000 live births. Figures for some UAs should be treated with caution as the infant mortality rate was based on fewer than 20 events.

Source: Office for National Statistics; General Register Office for Scotland

16.3 Education and training

	Children under 5 in education[1] (percentages) Feb. 1999[2]	Pupil/teacher ratio 1998/99 (numbers)		Pupils and students participating in post-compulsory education[3] (percentages) 1998/99	Percentage of pupils in last year of compulsory schooling[4,5] 1998/99 with	
		Primary schools	Secondary schools		No graded results	5 or more Grades 1-3 SCE Standard Grade (or equivalent)
United Kingdom	64	22.9	16.5	79	5.9	49.1
Scotland	65	19.4	13.0	95	4.0	57.8
Aberdeen City	74	19.0	12.7	119	4.2	57.4
Aberdeenshire	62	18.6	13.4	72	1.2	69.9
Angus	61	18.5	12.7	85	4.6	54.8
Argyll and Bute	80	17.6	12.6	77	1.7	63.8
Clackmannanshire	70	20.8	12.9	159	4.7	59.4
Dumfries and Galloway	49	18.4	11.7	93	3.3	62.8
Dundee City	67	18.8	12.4	107	7.1	45.8
East Ayrshire	64	20.9	13.8	95	5.1	55.8
East Dunbartonshire	85	22.1	13.9	85	1.6	69.4
East Lothian	76	19.8	13.3	62	7.9	51.1
East Renfrewshire	66	22.1	14.2	95	1.1	76.8
Edinburgh, City of	76	20.1	13.2	132	5.7	59.8
Eilean Siar (Western Isles)	81	12.6	9.2	104	13.2	63.8
Falkirk	58	21.2	12.8	99	0.5	53.2
Fife	56	18.5	13.1	101	4.5	55.6
Glasgow City	68	19.7	13.3	102	7.6	47.5
Highland	62	17.3	11.8	94	2.5	66.1
Inverclyde	75	21.0	13.2	112	2.7	58.5
Midlothian	72	19.0	13.4	62	1.7	55.2
Moray	50	18.8	12.3	102	8.6	57.2
North Ayrshire	63	20.8	13.7	54	3.6	53.7
North Lanarkshire	66	20.0	13.4	90	3.9	52.0
Orkney Islands	59	14.8	10.0	105	-	72.8
Perth and Kinross	79	18.7	12.5	94	9.3	58.3
Renfrewshire	44	21.0	14.0	94	0.1	59.2
Scottish Borders, The	53	17.9	12.0	91	3.1	67.5
Shetland Islands	97	12.7	7.6	75	1.6	73.0
South Ayrshire	53	21.0	13.6	92	0.9	60.6
South Lanarkshire	67	20.4	13.7	79	3.0	58.9
Stirling	66	18.7	12.8	61	3.5	65.1
West Dunbartonshire	69	19.9	13.8	107	5.6	49.2
West Lothian	61	19.8	13.6	100	1.0	53.2

1 Figures relate to all pupils as a percentage of the three and four year old population.
2 January 1999 for the United Kingdom.
3 In Scotland, pupils in S5 at September 1997. The figure for the United Kingdom relates to 16 year olds in education at the beginning of the academic year. Some students in Scotland participate on short courses. They are counted for each course; hence there is double counting which results in some percentages being greater than 100.
4 Pupils in their last year of compulsory schooling as a percentage of the school population of the same age.
5 Figures relate to all schools; hence they are not directly comparable with those in Tables 14.3 and 15.3 which are for maintained schools only.

Source: Scottish Executive; Department for Education and Employment

16.4 Housing and households

	Housing completions 1998 (numbers)		Stock of dwellings[3] 1998 (thousands)	Households[4] 1998 (thousands)	Local authority tenants: average weekly unrebated rent per dwelling (£) April 1999	Council Tax (£)[5] April 1999
	Private enterprise[1]	Housing associations local authorities etc[2]				
Scotland	18,372	2,053	2,286	2,169.9	36.40	849
Aberdeen City	741	21	103	99.4	33.10	824
Aberdeenshire	1,123	43	94	89.0	31.80	719
Angus	213	111	49	46.4	28.70	734
Argyll and Bute	132	199	44	38.1	36.50	881
Clackmannanshire	60	18	21	20.3	31.60	872
Dumfries and Galloway	231	14	67	63.0	34.40	766
Dundee City	432	70	71	66.8	38.70	1,034
East Ayrshire	469	60	52	50.7	30.50	849
East Dunbartonshire	111	22	43	41.6	34.30	830
East Lothian	361	19	39	37.2	31.70	789
East Renfrewshire	402	-	35	33.8	32.60	765
Edinburgh, City of	2,562	132	212	201.8	45.70	889
Eilean Siar (Western Isles)	87	8	13	11.7	38.30	689
Falkirk	632	28	63	60.8	32.10	719
Fife	1,042	119	154	147.9	32.80	809
Glasgow City	1,338	414	289	273.9	44.10	1,074
Highland	992	65	98	87.9	40.10	799
Inverclyde	210	136	39	37.7	27.20	888
Midlothian	238	48	33	31.4	29.50	936
Moray	103	17	38	35.5	29.50	724
North Ayrshire	426	26	62	58.5	31.10	788
North Lanarkshire	1,641	80	134	130.6	34.70	844
Orkney Islands	..	-	10	8.2	34.20	624
Perth and Kinross	672	119	61	56.3	30.80	758
Renfrewshire	779	-	79	76.4	35.90	783
Scottish Borders, The	261	123	50	45.7	31.70	670
Shetland Islands	122	-	10	9.0	42.90	621
South Ayrshire	375	73	50	48.2	32.80	792
South Lanarkshire[6]	1,087	24	128	124.2	38.10	880
Stirling	313	-	36	33.9	34.90	819
West Dunbartonshire[6]	179	-	43	41.8	34.10	981
West Lothian	1,038	64	64	62.4	34.30	858

1 Provisional figures including estimates for outstanding returns.
2 Provisional figures based on incomplete returns.
3 Figures based on the number of residential dwellings on the Council Tax Register at 6 September 1999. The figures for housing stock at local authority level shown in this table are derived using different methods from the regional stock figures shown in Table 6.1. This has led to small discrepancies between the two sets of figures. The figures in Table 6.1 provide the definitive regional estimates.
4 Household estimates at June 1998.
5 See Notes and Definitions.
6 Private sector figures are estimates since return is outstanding.

Source: Scottish Executive

16.5 Labour market statistics[1]

	Economically active 1998-1999[2] (percentages)	Total in employment[3] 1998-1999[2] (thousands)	Employment rate 1998-1999[2] (percentages)[4]	ILO unemploy- ment rate 1998-1999[2] (percentages)[4]	Claimant count[5] March 2000			Average gross weekly full-time earnings, all persons[7] April 1999 (£)
					Total (thousands)	Of which females (percentages)	Of which long term claimants[6] (percentages)	
United Kingdom	78.4	26,508	73.5	6.3	1,194.3	23.5	22.1	398.7
Scotland	77.4	2,259	71.5	7.6	130.6	22.7	19.7	364.9
Aberdeen City	83.5	109	79.6	..	4.3	20.2	11.7	423.2
Aberdeenshire	81.4	109	78.5	..	3.0	24.4	10.4	342.9
Angus	81.7	49	75.1	..	2.7	28.2	17.4	332.5
Argyll and Bute	82.8	38	76.8	..	2.4	25.6	23.7	353.0
Clackmannanshire	81.0	21	72.8	..	1.5	25.6	16.8	..
Dumfries and Galloway	76.8	63	72.4	..	4.0	25.1	21.1	335.4
Dundee City	73.7	63	68.5	..	5.4	20.6	27.3	354.2
East Ayrshire	69.5	43	57.9	16.8	4.3	22.9	25.0	328.6
East Dunbartonshire	79.9	52	77.6	..	1.6	22.7	16.9	..
East Lothian	81.2	42	76.4	..	1.2	20.5	13.9	338.7
East Renfrewshire	82.5	40	80.5	..	1.3	22.5	15.7	..
Edinburgh, City of	78.0	213	73.9	5.3	8.7	20.6	20.5	396.1
Eilean Siar (Western Isles)	83.2	9	68.4	..	1.0	19.9	21.0	..
Falkirk	78.7	65	73.2	..	3.8	23.9	20.6	358.0
Fife	81.4	158	74.1	8.9	9.1	23.5	21.8	337.6
Glasgow City	64.9	217	57.0	12.1	22.5	20.1	25.1	373.1
Highland	79.8	95	75.5	..	5.6	25.8	16.1	335.9
Inverclyde	77.5	37	69.9	..	2.3	20.2	11.6	349.5
Midlothian	82.1	39	78.1	..	1.1	21.2	13.8	360.2
Moray	83.8	42	77.3	..	2.2	28.7	9.0	324.9
North Ayrshire	74.6	56	64.5	13.5	5.3	26.2	16.4	354.8
North Lanarkshire	73.0	138	66.5	8.9	9.1	22.2	18.8	343.7
Orkney Islands	87.8	11	82.1	..	0.3	28.8	26.7	..
Perth and Kinross	82.9	59	76.1	..	2.1	26.5	13.3	325.4
Renfrewshire	76.8	79	70.7	7.9	4.7	20.8	20.9	373.1
Scottish Borders, The	81.9	47	76.6	..	1.8	23.7	13.6	315.7
Shetland Islands	85.0	12	83.3	..	0.4	26.0	13.0	..
South Ayrshire	78.7	51	70.6	10.2	3.3	23.6	19.7	387.0
South Lanarkshire	79.1	147	74.9	5.2	7.0	24.1	19.7	365.6
Stirling	79.9	36	74.8	..	1.7	23.8	15.8	332.8
West Dunbartonshire	76.5	41	71.1	..	3.5	20.3	22.9	..
West Lothian	83.7	78	77.1	7.9	3.5	22.1	11.7	354.6

1 See Notes and Definitions to the Labour market chapter. In some cases sample sizes are too small to provide reliable estimates.
2 For those of working age. Data are from the Labour Force Survey and relate to the period March 1998 to February 1999.
3 Includes those on Government employment and training programmes and unpaid family workers.
4 As a percentage of the economically active.
5 Count of claimants of unemployment-related benefit, ie. Jobseeker's allowance.
6 Persons who have been claiming for more than 12 months as a percentage of all claimants.
7 Earning estimates have been derived from the New Earnings Survey and relate to full-time employees whose pay for the survey pay-period was not affected by absence.

Source: Office for National Statistics

NUTS levels 1, 2 and 3 in Scotland[1], 1998

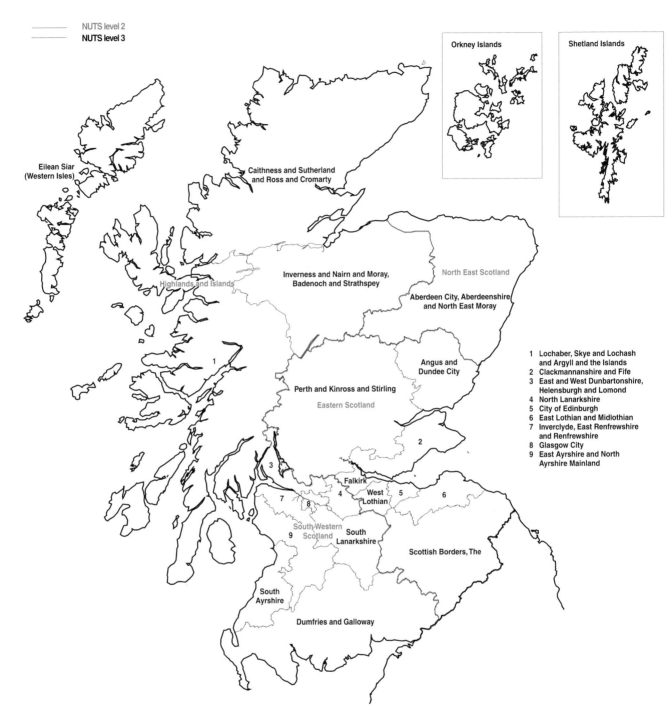

NUTS level 2
NUTS level 3

Orkney Islands

Shetland Islands

Eilean Siar
(Western Isles)

Caithness and Sutherland
and Ross and Cromarty

Inverness and Nairn and Moray,
Badenoch and Strathspey

North East Scotland

Highlands and Islands

Aberdeen City, Aberdeenshire
and North East Moray

Angus and
Dundee City

Perth and Kinross and Stirling

Eastern Scotland

1

2

Falkirk

West
Lothian

5

6

4

7

8

South Western
Scotland

9

South
Lanarkshire

Scottish Borders, The

South
Ayrshire

Dumfries and Galloway

1 Lochaber, Skye and Lochash
   and Argyll and the Islands
2 Clackmannanshire and Fife
3 East and West Dunbartonshire,
   Helensburgh and Lomond
4 North Lanarkshire
5 City of Edinburgh
6 East Lothian and Midlothian
7 Inverclyde, East Renfrewshire
   and Renfrewshire
8 Glasgow City
9 East Ayrshire and North
   Ayrshire Mainland

1 NUTS (Nomenclature of Units for Territorial Statistics) is a hierachical classification of areas that provides a breakdown of the EU's economic territory. The NUTS level 1 area is the whole country.
   See Notes and Definitions.

16.6 Gross domestic product: by NUTS 1,2 and 3 areas[1] at factor cost: current prices

	£ million			£ per head			£ per head (UK=100)		
	1994	1995	1996	1994	1995	1996	1994	1995	1996
United Kingdom[2]	570,944	597,741	629,839	9,777	10,199	10,711	100	100	100
Scotland	49,720	52,518	54,430	9,688	10,224	10,614	99	100	99
North Eastern Scotland[3] (Aberdeen City, Aberdeenshire and North East Moray)	6,900	7,068	7,347	13,545	13,864	14,453	139	136	135
Eastern Scotland	18,778	19,889	21,026	9,963	10,513	11,116	102	103	104
Angus and Dundee City	2,467	2,589	2,642	9,381	9,853	10,123	96	97	95
Clackmannanshire and Fife	3,105	3,222	3,465	7,744	8,047	8,704	79	79	81
East Lothian and Midlothian	911	952	1,009	5,460	5,683	5,999	56	56	56
Scottish Borders	889	948	1,000	8,414	8,929	9,422	86	88	88
Edinburgh, City of	6,630	6,946	7,131	14,946	15,520	15,888	153	152	148
Falkirk	1,298	1,472	1,580	9,110	10,312	11,046	93	101	103
Perth and Kinross and Stirling	1,887	1,970	2,168	8,815	9,157	10,068	90	90	94
West Lothian	1,590	1,789	2,030	10,732	11,965	13,466	110	117	126
South Western Scotland	21,166	22,572	22,964	8,946	9,556	9,747	92	94	91
East and West Dunbartonshire, Helensburgh and Lomond	1,643	1,698	1,690	6,969	7,219	7,211	71	71	67
Dumfries and Galloway	1,351	1,415	1,469	9,142	9,568	9,955	94	94	93
East Ayrshire and North Ayrshire Mainland	1,772	1,925	1,954	6,906	7,502	7,637	71	74	71
Glasgow City	7,369	7,714	7,895	11,819	12,474	12,808	121	122	120
Inverclyde, East Renfrewshire and Renfrewshire	3,147	3,417	3,472	8,895	9,621	9,814	91	94	92
North Lanarkshire	2,314	2,484	2,594	7,083	7,604	7,959	72	75	74
South Ayrshire	1,189	1,314	1,294	10,399	11,466	11,286	106	112	105
South Lanarkshire	2,380	2,604	2,596	7,741	8,470	8,445	79	83	79
Highlands and Islands	2,877	2,990	3,093	7,729	8,020	8,308	79	79	78
Caithness and Sutherland and Ross and Cromarty	693	717	733	7,729	7,974	8,157	79	78	76
Inverness and Nairn and Moray, Badenoch and Strathspey	844	864	897	7,703	7,862	8,152	79	77	76
Lochaber, Skye and Lochalsh and Argyll and the Islands	709	740	759	7,017	7,331	7,540	72	72	70
Eilean Siar (Western Isles)	216	221	233	7,370	7,611	8,072	75	75	75
Orkney Islands	157	167	177	7,942	8,381	8,949	81	82	84
Shetland Islands	258	281	293	11,278	12,161	12,748	115	119	119

1 NUTS (Nomenclature of Units for Territorial Statistics) is a hierarchical classification of areas that provides a breakdown of the EU's economic territory. Data are on the old ESA79 and are consistent with the data published in United Kingdom National Accounts 1997. See Notes and Definitions.
2 Excluding the Continental Shelf and the statistical descrepancy of the income based measure.
3 This area represents both NUTS 2 and 3 levels.

Source: Office for National Statistics

17 Sub-regions of Northern Ireland

Boards and Travel-to-work areas in Northern Ireland

Health and Social Services Boards

Education and Library Boards

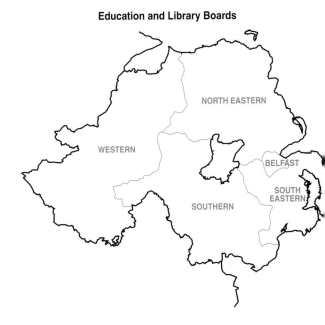

Travel-to-work areas

17.1 Area and population: by Board[1] and district, 1998

	Area (sq km)	Persons per sq km	Population (thousands)			Total population percentage change 1981-1998	Total fertility rate (TFR)[2]	Standardised mortality ratio (UK=100) (SMR)[3]	Percentage of population aged			
			Males	Females	Total				Under 5	5-15	16 up to pension age[4]	Pension age[4] or over
United Kingdom	242,910	244	29,128	30,108	59,237	5.1	1.71	100	6.2	14.2	61.4	18.1
Northern Ireland	13,576	124	827	861	1,689	9.4	1.89	101	7.2	17.3	60.3	15.2
Eastern	1,751	385	325	350	675	5.3	1.82	100	6.8	16.3	59.8	17.0
Ards	380	186	34	36	71	21.8	1.82	99	6.5	15.4	62.4	15.7
Belfast	110	2,623	136	152	287	-9.1	1.78	105	6.8	16.6	58.5	18.1
Castlereagh	85	784	32	34	67	9.4	1.86	93	6.8	13.8	60.3	19.1
Down	649	97	31	32	63	16.9	2.02	109	7.6	18.5	59.3	14.5
Lisburn	447	250	55	57	112	30.9	1.86	92	7.5	17.5	61.4	13.7
North Down	81	938	37	39	76	13.3	1.68	90	5.6	14.8	60.1	19.4
Northern	4,093	104	209	216	425	12.9	1.84	99	7.0	16.6	61.4	15.0
Antrim	421	119	26	25	50	9.0	2.04	106	7.8	16.0	64.3	11.9
Ballymena	630	94	29	30	59	7.4	1.88	106	6.7	16.2	61.4	15.6
Ballymoney	416	61	13	13	26	11.2	1.98	99	7.3	17.5	60.1	15.2
Carrickfergus	81	466	18	19	38	31.2	1.70	100	7.0	16.5	61.5	15.0
Coleraine	486	114	27	29	55	18.4	1.72	93	6.7	15.5	61.5	16.4
Cookstown	514	62	16	16	32	12.5	1.70	86	6.5	19.0	60.0	14.5
Larne	336	91	15	16	31	5.5	1.78	99	6.2	16.0	60.5	17.3
Magherafelt	564	68	19	19	39	17.8	2.34	105	8.4	19.6	59.0	12.9
Moyle	494	31	8	8	15	6.1	1.71	89	5.9	17.5	59.5	17.2
Newtownabbey	151	537	39	42	81	11.6	1.71	100	6.8	15.6	62.0	15.6
Southern	3,075	100	153	155	308	12.2	2.09	103	7.7	18.4	60.1	13.8
Armagh	671	82	28	27	55	11.2	1.89	101	6.9	18.4	60.3	14.4
Banbridge	451	87	20	20	39	30.8	1.75	99	6.8	16.3	62.6	14.3
Craigavon	282	281	39	40	79	7.7	2.10	96	7.7	17.7	60.3	14.3
Dungannon	772	62	24	24	48	9.0	2.15	113	8.0	19.0	59.7	13.4
Newry and Mourne	898	96	43	44	87	11.6	2.35	109	8.5	19.6	59.0	12.9
Western	4,658	60	140	140	280	11.9	2.14	107	7.9	19.5	60.1	12.6
Derry	381	278	52	54	106	17.3	2.19	117	8.4	20.5	60.3	10.8
Fermanagh	1,699	34	29	28	57	9.8	2.17	108	7.4	18.4	58.5	15.7
Limavady	586	55	17	16	32	18.5	2.01	91	7.8	18.7	62.0	11.4
Omagh	1,130	42	24	24	48	6.7	2.00	102	7.4	19.7	59.8	13.1
Strabane	862	43	19	19	37	2.7	2.25	103	7.7	18.6	60.3	13.5

1 Health and Social Services Board areas.
2 The total fertility rate (TFR) is the average number of children which would be born to a woman if the current pattern of fertility persisted throughout her child-bearing years. Previously called total period fertility rate (TPFR). Figures for Northern Ireland are based on births and population data for the previous three years.
3 Averaged for the years 1996, 1997 and 1998 and adjusted for the age structure of the population. See Notes and Definitions to the Population chapter.
4 Pension age is 65 for males and 60 for females.

Source: Office for National Statistics; Northern Ireland Statistics and Research Agency

17.2 Vital[1,2] and social statistics: by Board[3]

	Live births per 1,000 population		Deaths per 1,000 population		Perinatal mortality rate[4] 1998	Infant mortality rate[5] 1998	Percentage live births outside marriage 1998	Children looked after by LAs per 1,000 population aged under 18 1998[6]
	1991	1998	1991	1998				
United Kingdom	13.7	12.1	11.2	10.6	8.2	5.7	37.6	..
Northern Ireland	16.3	14.0	9.4	8.9	8.1	5.6	28.5	5.0
Eastern	15.8	13.2	10.3	9.7	6.1	4.2	34.5	5.9
Northern	15.2	13.5	8.7	8.6	8.5	6.3	25.5	4.9
Southern	18.2	15.1	8.9	8.3	8.8	5.1	20.5	3.2
Western	17.9	15.4	8.7	7.9	11.0	8.1	28.5	5.1

1 Births and deaths data are based on the usual area of residence of the mother/deceased. See Notes and Definitions to the Population chapter for details of the inclusion/exclusion of births to non-resident mothers and deaths of non-resident persons.
2 Births data are on the basis of year of occurrence in England and Wales and year of registration in Scotland and Northern Ireland. All deaths data relate to year of registration.
3 Health and Social Service Board Areas.
4 Still births and deaths of infants under 1 week of age per 1,000 live and still births.
5 Death of infants under 1 year of age per 1,000 live births.
6 At 31 March. Figures are not directly comparable with similar data in the rest of the United Kingdom as Children Order legislation in Northern Ireland is not identical.

Source: Northern Ireland Statistics and Research Agency; Department of Health, Social Services and Public Safety, Northern Ireland

17.3 Education and training: by Board[1]

	Children under 5 in education (percentages) Jan. 1999	Pupil/teacher ratio 1998/99 (numbers)		Pupils and students participating in post-compulsory education (percentages) 1998/99[3]	Percentage of pupils in last year of compulsory schooling[4,5] 1998/99 with	
		Primary schools[2]	Secondary schools		No graded results	5 or more Grades 1-3 SCE Standard Grade (or equivalent)
United Kingdom	64	22.9	16.5	79	5.9	49.1
Northern Ireland	53	19.9	14.6	74	3.5	56.0
Belfast	..	18.8	14.7	..	4.6	58.6
South Eastern[6]	..	20.5	14.7	..	3.2	52.2
Southern	..	19.3	14.6	..	3.5	56.7
North Eastern	..	20.5	14.6	..	2.3	58.0
Western	..	20.3	14.5	..	4.0	54.9

1 Education and Library Boards.
2 In Northern Ireland the primary PTR includes preparatory departments of grammar schools.
3 Pupils and students aged 16 at 1st July. Figures for Northern Ireland exclude those in part-time further education.
4 Pupils in their last year of compulsory schooling as a percentage of the school population of the same age.
5 Figures relate to all schools; hence they are not directly comparable with those in Tables 14.3 and 15.3 which are for maintained schools only.
6 South Eastern figure for day nursery places includes Belfast.

Source: Department for Education and Employment; Northern Ireland Department of Education

17.4 Labour market[1] and benefit statistics: by district

	Economically active 1998-99[2] (percentages)[3]	Employment rate 1998-1999[2] (percentages)[4]	Claimant count[5] March 2000				Income Support beneficiaries[7] March 2000 (percentages)
			Total (thousands)	Of which females (percentages)	Of which long term claimants[6] (percentages)		
United Kingdom	78.4	73.5	1,194.3	23.5	22.1		10
Northern Ireland	72.7	67.0	43.2	22.5	32.6		15
Eastern	72.7	67.3	17.8	20.8	32.5		14
Ards	79.2	76.2	1.5	26.2	26.8		10
Belfast	67.6	61.6	10.5	18.3	35.8		21
Castlereagh	78.4	73.2	0.9	22.4	25.4		6
Down	74.5	67.8	1.5	23.8	29.0		12
Lisburn	74.1	69.1	1.9	22.1	29.4		12
North Down	76.7	71.5	1.4	28.0	26.8		8
Northern	75.5	70.9	9.1	26.2	28.6		12
Antrim	79.2	75.6	0.9	27.1	24.1		10
Ballymena	72.8	66.3	1.2	27.7	29.7		10
Ballymoney	72.5	67.4	0.6	22.4	31.0		14
Carrickfergus	76.5	72.2	0.8	25.7	32.3		9
Coleraine	79.5	73.4	1.6	27.0	28.3		13
Cookstown	68.8	64.9	0.6	26.0	29.3		19
Larne	78.1	74.1	0.6	26.3	23.4		11
Magherafelt	74.1	71.1	0.8	30.3	32.2		15
Moyle	68.0	63.3	0.6	24.9	30.5		16
Newtownabbey	77.8	73.8	1.4	23.8	27.1		10
Southern	72.5	66.2	6.4	24.1	31.9		16
Armagh	74.7	69.5	1.1	27.6	31.6		13
Banbridge	75.2	70.6	0.5	24.7	24.9		10
Craigavon	73.0	69.4	1.5	22.8	29.4		15
Dungannon	74.0	65.8	0.8	27.9	27.5		19
Newry and Mourne	68.1	58.8	2.4	22.1	36.6		20
Western	68.4	60.9	9.8	21.0	37.1		19
Derry	65.7	58.8	4.5	19.3	36.7		21
Fermanagh	71.8	65.1	1.8	22.0	43.2		15
Limavady	74.9	73.4	0.9	24.2	24.4		16
Omagh	72.0	61.9	1.3	25.6	39.0		17
Strabane	61.5	48.6	1.3	18.2	36.9		23

1 See Notes and Definitions to the Labour market chapter.
2 Based on the population of working age.
3 Data are from the Labour Force Survey and relate to the period March 1998 to February 1999.
4 As a percentage of the economically active.
5 Count of claimants of unemployment-related benefit, ie. Jobseeker's allowance.
6 Persons who have been claiming for more than 12 months as a percentage of all claimants.
7 Claimants and their partners aged 16 or over as a percentage of the population aged 16 or over. The figure for Northern Ireland includes those who could not be assigned to a council. The figure for the UK relates to February.

Source: Office for National Statistics; Department of Enterprise, Trade and Investment, and Department of Health, Social Services and Public Safety, Northern Ireland

NUTS levels 1, 2 and 3 in Northern Ireland[1], 1998

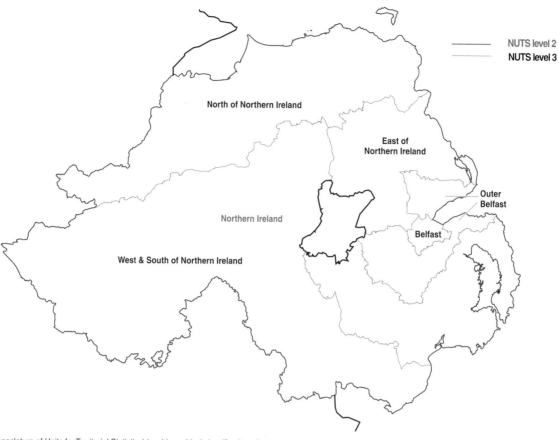

1 NUTS (Nomenclature of Units for Territorial Statistics) is a hierarchical classification of areas that provides a breakdown of the EU's economic territory. The NUTS level 1 area is the whole country. See Notes and Definitions.

17.5   Gross domestic product: by NUTS 1,2 and 3 areas[1] at factor cost: current prices

	£ million			£ per head			£ per head (UK=100)		
	1994	1995	1996	1994	1995	1996	1994	1995	1996
United Kingdom[2]	570,944	597,741	629,839	9,777	10,199	10,711	100	100	100
Northern Ireland[3]	13,092	13,889	14,469	7,973	8,423	8,699	82	83	81
Belfast	3,777	3,902	4,075	12,712	13,150	13,706	130	129	128
Outer Belfast	2,433	2,553	2,666	6,859	7,153	7,374	70	70	69
East of Northern Ireland	2,720	2,966	3,086	7,224	7,831	8,074	74	77	75
North of Northern Ireland	1,889	2,041	2,131	7,228	7,751	8,003	74	76	75
West and South of Northern Ireland	2,272	2,427	2,511	6,454	6,871	7,053	66	67	66

1 NUTS (Nomenclature of Units for Territorial Statistics) is a hierarchical classification of areas that provides a breakdown of the EU's economic territory. Data are on the old ESA79 and are consistent with the data published in United Kingdom National Accounts 1997. See Notes and Definitions.
2 Excluding the Continental Shelf and the statistical descrepancy of the income based measure.
3 This area represents both NUTS 1 and 2 levels.

Source: Office for National Statistics

Notes and Definitions

Government Office Regions within England

Most of the statistics in *Regional Trends* are presented on the Government Office Regions (GORs) of England, together with Wales, Scotland and Northern Ireland. The Government Office for the North West merged with the Government Office for Merseyside in August 1998, so figures for Merseyside are no longer shown separately. In tables, the Government Office for the East of England (formerly the Eastern Region) is referred to as East. Maps of the GORs are on pages 14 to 36 and 183.

Standard Statistical Regions

Prior to the introduction of the GORs, regional statistics were presented on the basis of the Standard Statistical Regions (SSRs) of the United Kingdom. A few tables in *Regional Trends 35* continue to be presented on this classification. The SSRs are shown in a map on page 239.

Sub-regions of England

The implementation of local government reorganisation in England, which took place in four phases on 1 April in each year between 1995 and 1998, is summarised below. The reorganisation involved only the non-metropolitan counties. Unitary Authorities (UA) have replaced the two-tier system of County Councils and Local Authority District Councils in parts of some shire counties and, in some instances, across the whole county. For statistical purposes grouping UAs by geography can be helpful. In Chapter 14 the following groupings are also included:

> *Tees Valley less Darlington* relates to the abolished administrative county of Cleveland (*Tees Valley* relates to the area covered by five UAs; Darlington, Hartlepool, Middlesborough, Redcar and Cleveland, and Stockton-on-Tees);

> *The Humber* relates to the abolished administrative county of Humberside;

> *Herefordshire and Worcestershire* relates to the former administrative county of Hereford and Worcestershire;

> *Bristol/Bath area* relates to the abolished administrative county of Avon.

By legal definition all Unitary Authorities in England are counties. However, some of the responsibilities of UAs are the same as those of districts and so they are often presented together. For the majority of UAs their establishment has been achieved without geographical change. However, for a few Unitary Authorities, there are some boundary changes at District and Ward levels, most notably, the County of Herefordshire UA in the West Midlands and Peterborough UA in the East of England. Full details of these are given in the *Gazetteer of the old and new geographies of the United Kingdom* available from National Statistics Direct Tel. 01633 812078.

The local government structure at 1 April 1998 is used in Chapter 14 and throughout the rest of the book unless otherwise specified. A map showing the Counties and Unitary Authorities is given on page 183.

Counties, Districts and Unitary Authorities in England

	Non-metropolitan areas			Metropolitan areas	
Year	Counties	Districts	Unitary Authorities	London boroughs	Metropolitan boroughs
1994	39	296	0	33	36
1995	38	294	1	33	36
1996	35	274	14	33	36
1997	36	260	27	33	36
1998	34	238	46	33	36

Unitary Authorities of Wales

On 1 April 1996, the 8 counties and 37 districts of Wales were replaced by 22 Unitary Authorities. A map is given on page 218. In Chapter 15, the Unitary Authorities are presented in the tables in alphabetical order.

New Councils of Scotland

On 1 April 1996, the 10 Local Authority regions and 56 districts of Scotland were replaced by 32 Unitary Councils. A map is given on page 226. In Chapter 16, the New Councils are presented in the tables in alphabetical order.

Standard Statistical Regions

SSR boundary

Environment Agency regions

ENGLAND and WALES

Environment Agency region boundary

NHS Regional Office areas
(from April 1996)

ENGLAND and WALES

Health Authority boundary

NHS Regional Office areas
(from April 1999)

ENGLAND and WALES

Health Authority boundary

Police Force areas

ENGLAND and WALES

—— Police Force area boundary

Prison Service regions

ENGLAND and WALES

—— Prison Service region boundary

Department of Trade and Industry regions

ENGLAND

—— DTI region boundary

Tourist Board regions

UNITED KINGDOM

—— Tourist Board region boundary

Northern Ireland

The 26 districts of Northern Ireland are listed in Chapter 17. For some topics, they have been grouped into either the five Education and Library Boards or the four Health and Social Services Boards. The districts comprising the Education and Library Boards are as follows:

Board	Districts
Belfast	Belfast
South	Eastern Ards, Castlereagh, Down, Lisburn, North Down.
Southern	Armagh, Banbridge, Cookstown, Craigavon, Dungannon, Newry and Mourne.
North Eastern	Antrim, Ballymena, Ballymoney, Carrickfergus, Coleraine, Larne, Magherafelt, Moyle, Newtownabbey.
Western	Derry, Fermanagh, Limavady, Omagh, Strabane.

Health and Social Services Boards are as follows:

Northern	as North Eastern Education and Library Board but including Cookstown.
Eastern	as South Eastern Education and Library Board but including Belfast.
Southern	as Southern Education and Library Board but excluding Cookstown.
Western	as Western Education and Library Board.

Maps of the Northern Ireland Boards and Travel-to-work areas are on page 234.

NUTS (Nomenclature of Territorial Statistics) area classification

Data are presented using this classification in Chapter 2, Map 12.5, Tables 14.7, 15.6, 16.6 and 17.5. In Tables 14.1 and 14.4, data for London are also presented on the NUTS classification, which provides additional levels of geographic aggregation between the London as a whole and individual London boroughs.

NUTS is a hierarchical classification of areas that provide a breakdown of the European Union's economic territory for producing regional statistics that are comparable across the Union. It has been used since 1988 in EU legislation for determining the distribution of the Structural Funds.

The NUTS five-tier structure for the UK – reviewed during 1998 as a consequence of the move to using Government Office Regions as the principal classification for English Regions and the local government reorganisation – comprises current national administrative areas, except in Scotland where some NUTS areas comprise whole and/or part local enterprise company areas.

Maps showing the NUTS levels 1, 2 and 3 areas for England, Wales, Scotland and Northern Ireland are on pages 214, 224, 232 and 237 respectively.

Other Regional Classifications

The UK Continental Shelf, now referred to as Extra-Regio, is treated as a separate region in Tables 12.1, 12.3 and 12.6 (see the Notes and Definitions to Chapter 12 Regional accounts).

Maps of non-standard regions used in Regional Trends are shown on pages 239 and 240.

United Kingdom NUTS levels 1 to 5

	Level 1		Level 2		Level 3		Level 4		Level 5	
		Numbers		Numbers		Numbers		Numbers		Numbers
England	Government Office Regions	9	Individual countiies or groups of counties/ London boroughs/ metropolitan counties/ counties/unitary authorities	30	Individual counties/ unitary authorities or groups of counties/ London boroughs/ metropolitan counties/ unitary authorities/ local authority districts	93	Individual London boroughs/metro-politan districts/ unitary authorities/ local authority districts	354	Wards	8,442
Wales	Country	1	Groups of unitary authorities	2	Groups of unitary authorities	12	Individual unitary authorities	22	Wards	865
Scotland	Country	1	Groups of whole/part unitary authorities (councils) and/or local enterprise companies	4	Groups of whole/part unitary authorities (councils) and/or local enterprise companies	23	Individual or groups of whole/part unitary authorities (councils) and/or local enterprise companies	41	Wards	1,247
Northern Ireland	Country	1	Country	1	Groups of district council areas	5	Individual district council areas	26	Wards	582
Total		12		37		133		443		11,136

CHAPTER 2: EUROPEAN UNION REGIONAL STATISTICS

The data appearing in this chapter are based on information in the statistical database REGIO produced by the Statistical Office of the European Communities (EUROSTAT) which uses the Nomenclature of Territorial Units for Statistics (NUTS) classification, described earlier. Data relate to the NUTS level 1 areas for countries in the European Union.

Table 2.3 Economic statistics

Employment statistics are derived from the annual Community Labour Force Survey (CLFS), which uses national Labour Force Survey (LFS) data although there may be minor differences in interpretation compared with the national LFS. Since the survey is conducted on a sample basis, results relating to small regions should be treated with caution. One of the main statistical objectives of the CLFS is to divide the population of working age into three groups: persons in employment, unemployed persons and inactive persons (those not classified as employed or unemployed). The groups are used to derive the following measures:
a) activity rates the labour force as a percentage of the population of working-age;
b) employment/population ratios persons in employment as a percentage of the population of working-age; and
c) unemployment rates unemployed persons as a percentage of the labour force.

The definitions of employment and unemployment used in the CLFS closely follow those adopted by the 13th International Conference of Labour Statisticians and promulgated by the International Labour Organisation (ILO) and are as follows (further detail is available in the EUROSTAT publication *Labour Force Survey, Methods and Definitions, 1992*):

Employment: the employed comprise all persons above a specified age who during a specified brief period either one week or one day were in the following categories:
a) *)paid employment at work or with a job but not at work ie temporarily absent but in receipt of a wage or salary;
b) *)self-employment at work ie persons who during the reference period performed some work for profit or family gain, in cash or kind, or with an enterprise but not at work ie temporarily absent. (An 'enterprise' may be a business enterprise, a farm or a service undertaking.)

Unemployment: the unemployed comprise all persons above a specified age who, during the reference period, were:
a) without work ie were not in paid employment or self-employment;
b) currently available for work ie were available for paid employment or self-employment during the reference period;
c) seeking work ie had taken specific steps in a specified recent period to seek paid employment or self-employment.

Long-term unemployment: persons who have been unemployed for 12 or more consecutive months.

Table 2.3 and Map 2.5 Purchasing Power Standard

The Purchasing Power Standard (PPS) is a unit of measurement calculated by scaling Purchasing Power Parities (PPPs) so that the aggregate for the EU-15 as a whole is the same whether expressed in EUROs (ECUs) or in PPS. Purchasing Power Parities are conversion factors, which make it possible to eliminate the combined effect of price level differences and other factors from a comparison of economic aggregates and thereby obtain a real volume comparison between countries.

Table 2.4 Agricultural statistics

The 'gross margin' of an agricultural enterprise is defined as the monetary value of gross production from which corresponding specific costs are deducted. The 'Standard Gross Margin' (SGM) is the value of gross margin corresponding to the average situation in a given region for each agricultural characteristic eg crop production, livestock production. 'Gross production' is the sum of the value of the principal product(s) and of any secondary product(s). The values are calculated by multiplying production per unit (less any losses) by the farm-gate price, excluding VAT. Gross production also includes subsidies linked to products, to area and/or to livestock.

Basic data are collected in Member States from farm accounts, specific surveys or compiled from appropriate calculations for a reference period which covers three successive years or agricultural production years. The reference period is the same for all Member States. SGMs are first calculated in Member States national currencies and then converted into European currency units (ECUs) using the average exchange rates for the reference period.

CHAPTER 3: POPULATION AND HOUSEHOLDS
Tables 3.1, 3.2, 3.11 and Maps 3.3, 3.4, 3.5 and 3.8 Resident Population

The estimated population of an area includes all those usually resident in the area, whatever their nationality. HM Forces stationed outside the United Kingdom are excluded but foreign forces stationed here are included. Students are taken to be resident at their term-time address. The population estimates for mid-1998 are based on the 1991 Census (with allowance for Census under-enumeration) and take account of births, deaths and net migration between 1991 and mid-1998.

Table 3.6 Social class

Based on the Labour Force Survey (see Notes and Definitions to the Labour market chapter), the table gives percentages of working age people in each social class based on occupations. The method used is designed to group together as far as possible people with similar levels of occupational skills. The basis of the groupings is given in *Volume 3, Standard Occupational Classification*, (HMSO, 1991).

The six occupational social classes in the classification are as follows:
I Professional occupations (including doctors, solicitors, chemists, university professors and clergymen);
II Managerial and technical occupations (including school teachers, computer programmers, personnel managers, nurses, actors and laboratory technicians);

III Skilled occupations

    (N) Non-manual (including typists, clerical workers, photographers, sales representatives and shop assistants);

    (M) Manual (including cooks, bus drivers, railway guards, plasterers, bricklayers, hairdressers and carpenters);

IV Partly skilled occupations (including bar staff, waitresses, gardeners and caretakers);

V Unskilled occupations (including refuse collectors, messengers, lift attendants, cleaners and labourers).

For those in employment in the reference week of the survey, the occupation was that of their main job, and for those not in employment, their last occupation if they had done any paid work in the previous eight years.

Table 3.7 Ethnic Group

The information on the ethnic group of each respondent to the Labour Force Survey is collected using the categories first used in the 1991 Census. Those classified as 'mixed/other' include Chinese, other Asians whose origin is not Indian, Pakistani or Bangladeshi, and those of mixed origin.

Map 3.9 Projected population

The projected population figures for Wales and for England, Scotland and Northern Ireland are not directly comparable with previous editions of Regional Trends as more recent projections are used. There are changes to the assumptions for fertility/mortality/migration in the preparation of each set of projections. The projections used in this edition for Wales and Northern Ireland are 1998-based national projections, while those for England and Scotland are 1996-based sub-national projections.

Table 3.10, 3.13 and 3.14 Births and deaths

Within England and Wales, births are assigned to areas according to the usual residence of the mother at the date of birth, as stated at registration. If the address of usual residence is outside England and Wales, the birth included in any aggregate for England and Wales as a whole (and hence the UK total, but excluded from the figures for any individual region or area. In 1998 there were 352 live births to non-resident mothers.

Birth figures for Scotland include births to both resident and non-resident mothers. Where sub-national data are given (Table 16.2), births have been allocated to the usual residence of the mother if this was in Scotland and to the area of occurrence if the mother's usual residence was outside Scotland. There were 250 births to non-resident mothers in Scotland in 1998.

All figures given for Northern Ireland (including the sub-regional figures in Table 17.2) exclude births to mothers not usually resident in Northern Ireland. However, the UK total includes such births. There were 193 births to non-resident mothers in Northern Ireland in 1998.

As with births, within England and Wales, a death is normally assigned to the area of usual residence of the deceased. If this is outside England and Wales, the death is included in any aggregate for England and Wales as a whole (and hence the UK total), but excluded from the figures for any individual region or area. There were 1,441 deaths to non-residents in 1998.

Death figures for Scotland and Northern Ireland include deaths to both residents and non-residents. Where sub-national data are given (Tables 16.2 and 17.2), deaths of Scottish or Northern Irish residents have been allocated to the usual area of residence, while deaths of non-residents have been allocated to the area of occurrence. In 1998 there were 301 deaths to non-residents in Scotland and 100 to non-residents in Northern Ireland.

Table 3.10 Birth and death rates and rate of natural change.

Unlike Table 3.11 which relates to population change from mid-year to mid-year, the numbers shown in this table relate to calendar years.

Crude birth/death rates and natural change are affected by the age and sex structure of the population. For example, for any given levels of fertility and mortality, a population with a relatively high proportion of persons in the younger age groups will have a higher crude birth rate and consequently a higher rate of natural change than a population with a higher proportion of elderly people.

Table 3.12 Conceptions

The date of conception is estimated using recorded gestation for abortion and stillbirths, and assuming 38 weeks gestation for live births. A woman's age at conception is calculated as the interval in complete years between her date of birth and the estimated age of conception. The postcode of the woman's address is used to determine the region she was living in at the time of the conception.

Table 3.13 Total Fertility Rate

The total fertility rate (TFR) is the average number of children which would be born to a woman if she experiences the current age-specific fertility rates throughout her child-bearing years. It is sometimes called the total period fertility rate (TPFR).

Table 3.14 Standardised mortality ratio

The standardised mortality ratio (SMR) compares overall mortality in a region with that for the United Kingdom. The ratio expresses the number of deaths in a region as a percentage of the hypothetical number that would have occurred if the region's population had experienced the sex/age specific rates of the United Kingdom that year.

Tables 3.15 & 3.16 Inter-regional movements

Estimates for internal population movements are based on the movement of NHS doctors' patients between Family Health Services Authority Areas (FHSAs) in England and Wales and Area Health Boards (AHBs) in Scotland and Northern Ireland. These transfers are recorded at the NHS Central Registers (NHSCRs), Southport and Edinburgh, and at the Central Services Agency, Belfast. The figures have been adjusted to take account of differences in recorded cross-border flows between England and Wales, Scotland, and Northern Ireland.

The figures provide a detailed indicator of population movement within the United Kingdom. However, they should not be regarded as a perfect measure of migration as there is variation in the delay between a person moving and registering with a new doctor. Additionally, some moves may not result in a re-registration, ie individuals may migrate again before registering with a doctor. Conversely, there may be others who move and re-register several times in a year.

The NHSCR at Southport was computerised in 1991. Before 1991, the time lag was assumed to be three months between a person moving and the re-registration with an NHS doctor being processed onto the NHSCR. (It was estimated that processing at NHSCR took two months.) Since computerisation, estimates of internal migration derived from the NHSCR are based on the date of acceptance of the new patient by the FHSA (not previously available), and a one-month time lag assumed.

Table 3.15 Migration

International migration data are derived from the International Passenger Survey (IPS), a continuous voluntary sample survey. The IPS provides information on passengers entering and leaving the United Kingdom by the principal air, sea and tunnel routes. Routes between the United Kingdom and the Irish Republic, and those between the Channel Islands, Isle of Man and the rest of the world are excluded. The IPS has been running since 1961 and in 1998 it covered 263 thousand travellers. The IPS is also used to collect information on the travel account of the Balance of Payments, and for tourism policy. It shows how many people travelled, where they went and why, and gives a picture of how long they stayed and what they spent. It currently samples between 0.1 and 5 per cent of passengers depending on route and time of year.

It is believed that IPS migration figures exclude most people seeking asylum after entering the country and 'visitor switchers': persons admitted as short-term visitors who are subsequently granted an extension of stay for a year or more, for example as students or on the basis of marriage. It is estimated that there were 56 thousand such persons in 1998, after taking account of persons leaving the United Kingdom for a short-term period who stay overseas for longer than originally intended. It also excludes migration between the United Kingdom and the Irish Republic. For 1998, this was estimated as a net outflow of people.

For demographic purposes, a migrant into the United Kingdom is defined as a person who has resided abroad for a year or more and states the intention to stay in the United Kingdom for a year or more, and vice versa for a migrant from the United Kingdom. Migrants defined in this way were asked an additional group of questions which form the basis of these statistics.

In view of the small number of migrants in the sample, it should be noted that the estimates of migration, in particular the differences between inflow and outflow, are subject to large sampling errors. As a rough guide, the standard error for an estimate of one thousand migrants is around 40 per cent, whilst that for an estimate of 40 thousand migrants reduces to about 10 per cent, but on occasions these standard errors can be higher. However, the structure of the sample is such that estimates based on the sampling of passengers on certain routes have much larger standard errors associated with them.

Table 3.19 Household projections

The household projections are trend-based; they illustrate what would happen if past trends in household formation were to continue into the future. They are therefore not policy-based forecasts of what is expected to happen, but provide a starting point for policy decisions. The projections are heavily dependent on the assumptions involved, particularly international and internal migration, the marital status projections (in England and Wales only) and the continuation of past trends in household formation.

CHAPTER 4: EDUCATION AND TRAINING

In England and Wales, only qualified teachers are included for public sector schools (i.e. all other teaching staff at these schools are excluded). In Scotland and Northern Ireland, however, all teachers employed in schools are included other than in independent schools.

Table 4.1 Pupils and teachers by type of school

The pupil-teacher ratio in a school is the ratio of all pupils on the register to all qualified teachers employed within the schools during the census week. Part-time teachers and part-time pupils are included on a full-time equivalent basis. The difference in the age at which pupils transfer from primary to secondary school affects the comparison of pupil-teacher ratios between Scotland and the rest of the United Kingdom.

Table 4.2 Class sizes for all classes

Figures for England, Wales and Scotland include classes where more than one teacher may be present. Figures previously shown in this publication for England related to classes taught by one teacher. In 1999/00, the average Key Stage 1 class, taught by one teacher, had 25.8 pupils, with 8.8 per cent of classes having 31 or more pupils. Further information, including on-teacher class size data for Key Stage 2, primary and secondary school class sizes can be found in *DfEE Statistical First Release 15/2000*.

Table 4.4 and 4.5 Examination achievements

The main examination for pupils at the minimum school-leaving age in England, Wales and Northern Ireland is the General Certificate of Secondary Education (GCSE); in Scotland it is the Scottish Certificate of Education (SCE) S (Standard) Grade. The GCSE is awarded in eight grades, A*-G, while the SCE S Grade is awarded in seven levels, 1 to 7. In Scotland, Standard Grade courses begin in the third year and continue to the end of the fourth year. Each subject has a number of elements, some of which are internally assessed in school. The award for the subject as a whole is given on a 7-point scale at three levels: Credit (1 and 2), General (3 and 4) and Foundation (5 and 6). An award of 7 means that the course has been completed. Pupils who do not complete the course or do not sit all parts of the examination get 'no award'.

GCSE figures relate to achievements by 16-year-olds at the end of the academic year and are shown as percentages of 16-year-olds in school. Standard Grades (in Scotland) relate to achievements by pupils in year S4 at the end of the academic year. That is, the achievements of pupils by the end of their last year of compulsory schooling: some may have been passed a year earlier.

GCE A levels are usually taken after a further two years of post-compulsory education, passes being graded from A-E. The SCE H (Higher) Grade requires only one year of post-compulsory study and for the more able candidates the range of subjects taken may be as wide as at S Grade. GCE A level and equivalent figures for pupils aged between 17 and 19 at the end of the school year are shown as a percentage of the 18-year-old population. This age spread in the examination result figures takes account of those pupils sitting examinations a year early or resitting them. Scottish Higher figures are based on the 17-year-old population as Highers are normally taken one year earlier (in Year S5) than A levels, although they can resit them or take additional subjects in year S6. However the data for Scotland relate only to year S5 pupils' examination results.

Average GCE A/AS level points scores are shown in Table 4.4. Points scores are determined by totalling pupils' individual GCE A/AS results: GCE A-level grades A-E count as 10 to 2 points respectively; and GCE AS grades A-E count as 5 to 1 points respectively.

In Wales, at below GCSE standard, the Certificate of Education examination is also available and is widely used by schools. Many pupils take Welsh as a first language at GCSE. In all countries pupils may sit non-GCE/GCSE examinations such as BTEC (SCOTVEC in Scotland), City and Guilds, RSA and Pitman. A proportion of pupils who are recorded as achieving no GCSE, AS or A level qualification will have passes in one or more of these other examinations.

In Table 4.5, Mathematics figures exclude computing science (England) and computer studies and statistics (Wales) while 'Any science' in England and Wales includes double award, single award and individual science subjects. Double award science was introduced with the GCSEs in 1988. Success in double award science means that the pupil has achieved two GCSEs rather than just one pass with single science or the individual sciences of biology, physics and chemistry. The majority of 15-year-olds now attempt GCSE double award science in preference to the single science subjects, although the individual sciences are still popular in the independent sector. There is no equivalent to double award science in Standard Grade.

Comparisons of examination results for England, Wales and Northern Ireland with those for Scotland are not straightforward because of the different education and examination systems. However, the following should be used as a guideline:

> 5 or more GCSEs at grades A*–C = 5 or more SCE Standard Grades at levels 1–3
> 1–4 GCSEs at grades A*–C = 1-4 SCE Standard Grades at levels 1–3
> GCSEs at grades D–G only = SCE Standard Grades at levels 4–7 only
> 2 or more GCE A levels passes at A–E = 3 or more SCE Higher Grade passes at A–C.

Also see the National Curriculum notes for Table 4.6.

Table 4.6 The National Curriculum: Assessments and Tests

Under the *Education Reform Act (1988)* a National Curriculum has been progressively introduced into primary and secondary schools in England and Wales. This consists of mathematics, English and science (and Welsh in Welsh speaking schools in Wales) as core subjects with history, geography, information technology, music, art, physical education and a modern foreign language (and Welsh in non-Welsh speaking schools in Wales) as non-core subjects. For all subjects measurable local targets have been defined for four key stages, corresponding to ages 7, 11, 14 and 16.

Pupils are assessed formally at the ages of 7, 11 and 14 by their teachers and by national tests in the core subjects of English, mathematics and science (and in Welsh in Welsh speaking schools in Wales). Sixteen year olds are assessed by means of the GCSE examination. Statutory authorities have been set up for England and Wales to advise government on the National Curriculum and promote curriculum development generally. Northern Ireland has its own common curriculum which is similar but not identical to the National Curriculum in England and Wales. Assessment arrangements in Northern Ireland became statutory from September 1996. The National Curriculum does not apply in Scotland, where school curricula are the responsibility of education authorities and individual head teachers, and in practice almost all 14 to 16-year-olds study mathematics, English, science, a modern foreign language, a social subject, physical education, technology and a creative and aesthetic subject.

The Key Stage 1, 2 and 3 figures for England cover all types of school (eg maintained and independent). The Government Office Region figures cover LEA maintained schools only.

Tables 4.7 and 4.8 Further (including adult) education

Further Education (FE) includes home students on courses of further education (FE) in further education institutions. The FE sector includes all provision outside schools that is below higher education (HE) level. This ranges from courses in independent living skills for students with severe learning difficulties up to GCE A level, advanced GNVQ or GSVQ and level 3 NVQ or SVQ courses. The FE sector also includes many students pursuing recreational courses not leading to a formal qualification. Students in England and Wales are counted once only, irrespective of the number of courses for which a student has enrolled. In Scotland and Northern Ireland, students enrolled on more than one course in unrelated subjects are counted for each of these courses with the exception of those on SCE S/GCSE and/or SCE H/GCE courses, who are counted once only irrespective of the number of levels/grades. Most FE students are in FE colleges and (in England) sixth form colleges that were formerly maintained by Local Education Authorities (LEAs), but in April 1993 became independent self-governing institutions receiving funding through the FEFC. There are also a small number of FE students in higher education (HE) institutions, and conversely some HE students in FE institutions.

Students may be of any age from 16 upwards (no minimum age in Scotland), and full or part-time. Full-time students aged under 19 are exempt from tuition fees and fully funded by the Further Education Funding Councils in England, the Further Education Funding Council of Wales, the Scottish Further Education Funding Council and the Department of Higher and Further Education, Training and Employment in Northern Ireland. Students aged 16-18 on FE courses in the Scottish FEIs are exempt from tuition fees, at the discretion of the individual colleges. Students are eligible to apply for support (bursary); the policy for eligibility is at the discretion of the colleges. For other students tuition fees are payable, but may be remitted for students in receipt of certain social security benefits. In some cases discretionary grants may be available from LEAs or the colleges themselves. LEAs continue to make some FE provision (often referred to as 'adult education') exclusively part-time, and predominantly recreational. The majority of LEAs make part or all of this provision directly themselves, but some pay other organisations (usually FE colleges) to do so on their behalf i.e. 'contracted out' provision.

Part-time day courses are mainly those organised for students released by their employers either for one or two days a week (or any part of a week in Scotland), or for a period (or periods) of block release.

Sandwich courses are those where periods of full-time study are broken by a period (or periods) of associated industrial training or experience, and where the total period (or periods) of full-time study over the whole course averages more than 19 weeks per academic year (18 weeks in Scotland). Sandwich course students are classed as full-time students.

National Vocational Qualifications (NVQs) and Scottish Vocational Qualifications (SVQs) are occupational qualifications, available at five levels, and are based on up-to-date standards set by employers.

General National Vocational Qualifications (GNVQs) and General Scottish Vocational Qualifications (GSVQs) combine general and vocational education and are available at three levels:

> *Foundation* – broadly equivalent to four GCSEs at grades D-G or four SCE Standard Grades at levels 4 to 7.
> *Intermediate* – broadly equivalent to five GCSEs at grades A* to C or five SCE Standard Grades at levels 1 to 3.
> *Advanced* – broadly equivalent to two GCE A levels, or three SCE Higher Grade passes; also known as vocational A levels .

There are approximately 368 thousand students in FEIs in England for whom limited information is available and therefore these students are not included in the FE figures.

Since 1996/97 the figures for FE students in England have been extracted from the Individualised Student Record (ISR) which counts those students taking a course in an English further education college on 1 November 1996 each year. Until 1995/96 figures were taken from the Further Education Statistical Record (FESR). Due to differences in data collection and methodology between the two sources, the ISR figures are not directly comparable with figures derived from the FESR.

Tables 4.9 and 4.10 Education expenditure by local education authorities

Continuing education includes expenditure on adult education centres, youth and other community services, (England) teacher and curriculum centres and on awards (fees and maintenance exclusive of parental contributions) to students normally resident within the local authority area prior to going to college or university. (For Scotland, university awards are not included) 'Other educational services' includes school catering services, school welfare, youth service (excluding England) and other facilities such as sports, outdoor activity and residential study centres, and educational research. For Scotland and Northern Ireland, it also includes transport of pupils. Transport of pupils for England and Wales is allocated across the schools sectors. Loan charges are excluded. For Scotland, Administration and Inspection includes all local authority employees in education administration including senior staff.

The proportion of post-compulsory education pupils in schools who attract higher levels of expenditure will vary between regions due to differences in staying-on rates at school as opposed to colleges of further education (including sixth form colleges).

Table 4.11 Higher education

Higher education (HE) students are those on courses that are of a standard that are higher than GCE A level, the Higher Grade of the Scottish Certificate of Education, GNVQ/NVQ level 3 or the BTEC or SCOTVEC National Certificate or Diploma. Higher education in publicly funded institutions is funded by block grants from the three Higher Education Funding Councils (HEFCs) in Great Britain and the Department of Higher and Further Education, Training and Employment in Northern Ireland. Some HE activity takes place in FE sector institutions, some of which is funded by the HEFCs and some by the FEFCs (The Scottish Further Education Funding Council). Most home students on full-time undergraduate courses are eligible for a mandatory award and top-up student loans.

The figures for HE students in English further education colleges for 1998/99 are extracted from the Individualised Student Record (ISR) which counts those students taking a course in an English further education college on the 1 November 1998. Until 1995/96 figures were taken from the Further Education Statistical Record (FESR). Due to differences in data collection and methodology between the two sources the ISR figures are not directly comparable with figures derived from the FESR.

Higher Education institutions figures use the interim December student record, however previous editions of this table were based on the July whole session record, which records a higher total number of students.

Table 4.13 Educational qualifications

Table 4.13 covers all people of working age (16-64 for males, 16-59 for females). Please also see notes to Tables 4.4 and 4.5

Degree or equivalent includes higher and first degrees, NVQ level 5 and other degree level qualifications such as graduate membership of a professional institute.

Higher education qualification below degree level includes NVQ level 4, higher level BTEC/SCOTVEC, HNC/HND, RSA Higher diploma and nursing and teaching qualifications.

GCE A level or equivalent includes NVQ level 3, GNVQ advanced, BTEC/SCOTVEC National Certificate, RSA Advanced diploma, City and Guilds advanced craft, A/AS levels or equivalent, Scottish Highers and Scottish Certificate of Sixth Year Studies and trade apprenticeships.

GCSE grades A*-C or equivalent includes NVQ level 2, GNVQ intermediate, RSA diploma, City and Guilds craft, BTEC/SCOTVEC First or general diploma, GCSE grades A*-C or equivalent, O level and CSE Grade 1.

Other qualifications at NVQ level 1 or below include GNVQ, GSVO foundation level, GCSE grade D-G, CSE below grade 1, BTEC/SCOTVEC First or general certificate, other RSA and City and Guilds qualifications, Youth Training certificate and any other professional, vocational or foreign qualifications for which the level is unknown.

Table 4.14 National Learning Targets for England 2002

Table 4.14 shows the proportions of people meeting the required qualification level for four of the National Learning Targets for England 2002. The Targets shown are for young people and adults and have been set using the competence-based National Vocational Qualifications (NVQs) and their vocational and academic equivalents. It should be noted that the data in Table 4.14 relate to the region in which the person is resident, and not where they obtained the qualifications. This can lead to some distortion of the regional picture of educational standards; although not identifiable from the table, this is particularly relevant in Northern Ireland, as many qualified young people leave home to enter higher education or seek employment in Great Britain.

The four main targets are shown in the table:

Young people
— by the year 2002, 85 per cent of 19 year olds to achieve 5 GCSE passes at grades A*-C, an Intermediate GNVQ or an NVQ level 2.
— by the year 2002, 60 per cent of 21 year olds to achieve 2 GCE A levels, or Advanced GNVQ or NVQ level 3.

Adults
— by the year 2002, 50 per cent of those of working age in employment to be qualified to NVQ level 3, Advanced GNVQ or 2 GCE A level standard.
— by the year 2002, 28 per cent of those of working age in employment to have a vocational, professional, management or academic qualification at NVQ level 4 or above.

In addition to the targets shown, there are also National Learning Targets for 11 year olds, 16 year olds, targets for organisations and a learning participation target for adults.

Tables 4.16 Learning and Training at Work 1999

Learning and Training at Work 1999 (LATW 1999) is a new multi-purpose survey of employers that investigates the provision of learning and training at work. This information was previously collected in the annual Skill Needs in Britain (SNIB) surveys, along with information on recruitment difficulties, skill shortages and skill gaps.

Due to the increasing focus on skills issues on the one hand and employer provided training on the other coupled with the increasing complexity and length of the SNIB questionnaire, DfEE decided to replace the single SNIB study with two separate surveys, one covering skills issues and another training issues. The Learning and Training at Work 1999 report relates to the latter study and includes information about; key indicators of employers' commitment to training, including the volume of off-the-job training provided; and awareness of, and participation in, a number of initiatives relevant to training.

The LATW1999 survey consisted of 4,008 telephone interviews with employers having one or more employees at the specific location sampled, and covered public and private business and all industry sectors. The main stage of interviewing was carried out between 3 November and 21 December 1999. The overall response rate from employers was 63 per cent. Sample design involved setting separate sample targets for each cell on a Government Office region by industry sector by establishment size matrix. In contrast the SNIB surveys covered employers with 25 or more employees in all business sectors, except agriculture, hunting, forestry and fishing, in Great Britain.

Table 4.17 Work-based Learning for Adults (WBLA) and Work-based Training for Young People (WBTYP)

Work-based Learning for Adults (previously known as Work-Based Training for Adults) replaced the former Training for Work (TfW) initiative in April 1998, and is aimed at getting unemployed adults back into work.

Work-based Training for Young People consists of Modern Apprenticeships (MA), National Traineeships (NTr) and Other training for young people (formerly known as Youth Training). These programmes aim to provide the participants with training towards a recognised National Vocational Qualification at levels 2 and 3 or above.

These programmes are delivered through the network of Training & Enterprise Councils (TECs) in England and Wales and local Enterprise Companies (LECs) in Scotland. In England and Wales leavers are followed up six months after they leave WBTA/WBTYP, whereas in Scotland they are followed up three months after completing training.

For Northern Ireland, figures relate to persons on Jobskills. Jobskills is a vocational training programme which over-arches the three distinct but interlinked strands of Access, Traineeship and Modern Apprenticeship and focuses mainly on the attainment of NVQs at Level 2 and above in line with National Learning Targets. Recruitment of adults ceased with effect from June 1998. For statistical purposes Jobskills applicants can join the:
a)   Access and Traineeship strands up to their 18th birthday. However, those with a disability can join up until their 22nd birthday;
b)   Modern Apprenticeships strand up to their 25th birthday.
As trainees can stay on Jobskills for up to three years, outcomes for leavers during '98/'99 may not necessarily be fully representative of outcomes in the longer term.

The reason for the drop in leavers is due to changes to WBLA eligibility which reduced the numbers of starts between 97-98 and 98-99. (Amongst other things the changes excluded 18-24 year olds from the programme. The inception of various New Deal programmes also filtered out further potential WBLA trainees).

CHAPTER 5: LABOUR MARKET

Interpretation of the labour market requires a number of different sources of data to be used. There are five main sources in this chapter: the Labour Force Survey (LFS), the Annual Employment Survey (AES), the Northern Ireland Quarterly Employment Survey (QES), the New Earnings Survey and the claimant count. Problems can arise in drawing together data on the same subject from different sources. For example, the question in the LFS as to whether the respondent is employed produces a measure of employment based on the number of persons, whereas a question addressed to employers asking the number of people they employ, as in AES, produces a measure of the number of jobs. Thus if someone has a second job they will be included twice.

Tables 5.1 and 5.2 Labour force

The labour force includes people aged 16 or over who are either in employment (whether employed, self-employed, on a work-related government-supported employment and training programme or an unpaid family worker) or unemployed. The 'ILO definition' of unemployment counts as unemployed people without a job who were available to start work within two weeks and had either looked for work in the past four weeks or were waiting to start a job they had already obtained in the next two weeks.

Tables 5.4 and 5.5 Annual Employment Survey, Short-term Employment Survey and Quarterly Employment Survey

The Annual Employment Survey (AES) is a sample survey which ran for the first time in 1995 and replaced the Census of Employment which ran until 1993. The AES is the only source of employment statistics for Great Britain analysed by the local area and by detailed industrial classification. The sample was drawn from the Inter-Departmental Business Register (IDBR) and the AES 98 sample comprised 64,000 enterprises. An enterprise is roughly defined as a combination of local units (ie individual workplaces with PAYE schemes or registered for VAT) under common ownership. These enterprises covered 0.5 million local units and 15 million employees (out of a total population of roughly 22 million employees in employment).

The AES results are used to benchmark the monthly/quarterly employment surveys (STTES) which measure 'movements' (by region and industrial group) between the annual survey dates.

The Quarterly Employment Survey (QES) for Northern Ireland is a voluntary survey which covers all employers with at least 25 employees, all public sector employers and a representative sample of smaller firms. Data are collected for both male/

female, full-time/part-time employees. Estimates for Northern Ireland are produced on a quarterly basis with unadjusted figures available at the two-digit or division level of the 1992 Standard Industrial Classification and seasonally adjusted figures available at a broad sector level.

Table 5.10 'Other' occupational group

This covers occupations which require the knowledge and experience necessary to perform mostly routine tasks, often involving the use of simple hand-held tools and, in some cases, requiring a degree of physical effort.

Most occupations in this group do not require formal educational qualifications but will usually have an associated short period of formal experience related training. All non-managerial agricultural occupations are also included in this group, primarily because of the difficulty of distinguishing between those occupations which require only a limited knowledge of agricultural techniques, animal husbandry, etc., from those which require specific training and experience in these areas.

Tables 5.11 and 5.12 New Earnings Survey

These tables contain some of the regional results of the New Earnings Survey 1999, fuller details of which are given for the Government Office Regions in part A and E of the report *New Earnings Survey 1999* (National Statistics Direct), published in October and December 1999. Results for Northern Ireland are published separately by the Department of Enterprise, Trade and Investment, Northern Ireland. The survey measured gross earnings of a 1 per cent sample of employees, most of whom were members of Pay-As-You-Earn (PAYE) schemes for a pay-period which included 14 April 1999. The earnings information collected was converted to a weekly basis where necessary, and to an hourly basis where normal basic hours were reported.

Figures are given where the number of employees reporting in the survey was 30 or more and the standard error of average weekly earnings was 5 per cent or less. Gross earnings are measured before tax, National Insurance or other deductions. They include overtime pay, bonuses and other additions to basic pay but exclude any payments for earlier periods (e.g. back pay), income in kind, tips and gratuities. All the results in this volume relate to full-time male and female employees on adult rates whose pay for the survey pay-period was not affected by absence. Employees were classified to the region in which they worked (or were based if mobile) using postcode information, and to manual or non-manual occupations on the basis of the Standard Occupational Classification 1990 (SOC 90). Part A of the report for Great Britain gives full details of definitions used in the survey.

Full-time employees are defined as those normally expected to work more than 30 hours per week, excluding overtime and main meal breaks (but 25 hours or more in the case of teachers) or, if their normal hours were not specified, as those regarded as full-time by the employer.

Tables 5.14, 5.15 and 5.16 ILO unemployment

The International Labour Organisation (ILO) definition of unemployment is measured through the Labour Force Survey and covers those people who are looking for work and are available for work (see Glossary of terms). The ILO unemployment rate is the percentage of economically active people who are ILO unemployed.

Counts of claimants of unemployment-related benefits are also published. There are advantages and disadvantages with both series, but they are complementary. The ILO unemployment rate is the number of people who are ILO unemployed as a proportion of the resident economically active population of the area concerned. The claimant count rate is the number of people claiming unemploymentrelated benefits as a proportion of claimants and jobs in each area. This explains why the ILO unemployment rate for London, where inward commuting is an important feature of the local labour market, tends to be significantly higher than the equivalent claimant count rate. The differential is much smaller for a region such as the South East where people commute out of the region into London.

A fuller description of ILO unemployment and claimant count, and the way they relate to one another is in the booklet *'How exactly is unemployment measured?'* available from the Office of National Statistics.

Map 5.13, Tables 5.17 and 5.20 Claimant Count statistics

Prior to 7 October 1996, figures in Table 5.17 relate to persons claiming unemployment-related benefits (that is, Unemployment Benefit, Income Support or National Insurance credits) at an Employment Service Office on the day of the monthly count, who on that day were unemployed and satisfied the conditions for claiming benefit. The figures include disabled people, so long as they meet the eligibility criteria and are claiming unemployment-related benefits, but exclude students seeking vacation work and temporarily stopped workers.

From 7 October 1996, a new single benefit, the Jobseeker's Allowance (JSA), replaced Unemployment Benefit and Income Support for unemployed people. People who qualify for JSA through their National Insurance contributions are eligible for a personal allowance (known as contribution-based JSA) for a maximum of six months. People who do not qualify for contribution-based JSA, or whose needs are not met by it, are able to claim a means-tested allowance (known as income-based JSA) for themselves and their dependants for as long as they need it. All those eligible for and claiming for JSA, as well as those claiming National Insurance credits, continue to be included in the monthly claimant count.

National and regional claimant count rates are calculated by expressing the number of claimants as a percentage of the estimated total workforce (the sum of claimants, employee jobs, self-employment jobs, HM Armed Forces and government-supported trainees).

Chart 5.19 Redundancies	Estimates cover the number of people who were not in employment during the reference week and who reported that they had been made redundant in the month of the reference week or in the two calendar months prior to this; plus the number of people who were in employment during the reference week who started their job in the same calendar month as, or the two calendar months prior to, the reference week, and who reported that they had been made redundant in the past three months.
Tables 5.21 and 5.22 The New Deal for Young People	The New Deal for the young unemployed is available to young people aged 18-24 who have been unemployed for more than six months, through four options:

a) a job attracting a wage subsidy of £60 a week, payable to employees for up to six months;
b) a work placement with a voluntary organisation;
c) a six-month work placement with an Environment Task Force; and
d) for those without basic qualifications, a place on a full-time education and training course, which might last for up to one year.

All the options include an element of training. For each young person the programme begins with a 'gateway' period of careers advice and intensive help with work, and with training in the skills needed for the world of work.

Those who are recorded by the Employment Service as having been placed into subsidised employment, plus those who are recorded as having terminated their Jobseeker's Allowance (JSA) claim in order to go into a job. This will undercount the total number going into a job: some who go into a job will not, for whatever reason, record this as the reason for termination of their JSA claim. These will be counted as 'not known'. Past research indicates that the destinations of those who do not give a reason for termination follow a similar pattern to those who do give a reason. Where a young person returns to JSA within 13 weeks of starting an unsubsidised job, the job is discounted.

Tables 5.23 and 5.24 Economic activity rates	The economic activity rate is the percentage of the population in a given age group which is in the labour force.

The economic activity households with at least one member of working age includes households where at least one person of pensionable age lives with at least one person of working age. It also includes households made up entirely of students.

Table 5.26 Labour Disputes	The table shows rates per 1,000 employees of working days lost for all industries and services. The statistics relate only to disputes connected with terms and conditions of employment. Stoppages involving fewer than ten workers or lasting less than one day are excluded except where the aggregate of working days lost is 100 or more. When interpreting the figures the following points should be borne in mind:

a) geographical variations in industrial structure affect overall regional comparisons;
b) a few large stoppages affecting a small number of firms may have a significant effect;
c) the number of working days lost and workers involved relate to persons both directly and indirectly involved at the establishments where the disputes occurred;
d) the regional figures involve a greater degree of estimation than the national figures as some large national stoppages cannot be disaggregated to a regional level and are only shown in the figure for the United Kingdom.

CHAPTER 6: HOUSING Tables 6.1 and 6.4 Dwellings	In the 1981 Census, a dwelling was defined as structurally separate accommodation whose rooms, excluding bathrooms and WCs, are self-contained. In the 1991 Census the definition changed to structurally separate accommodation whose rooms, including bath or shower, WC, and kitchen facilities, are self-contained. The figures in Table 6.1 include vacant dwellings and temporary dwellings occupied as a normal place of residence. Estimates of the stock in England are based on data from the 1981 and 1991 Censuses. In Wales and Scotland data from the Census is supplemented by local authority and other public sector landlords' figures. Northern Ireland stock figures are based on rating lists, Northern Ireland Housing Executive and Housing Association figures. Estimates of the tenure distribution in Table 6.4 are based on the above estimates and certain assumptions regarding the tenure distribution of gains and losses in the housing stock.
Table 6.2 New Dwellings completed	A dwelling is defined for the purposes of this table as a building or any part of a building which forms a separate and self-contained set of premises designed to be occupied by a single family. The figures relate to new permanent dwellings only, i.e. dwellings with a life expectancy of 60 years or more. A dwelling is counted as completed when it becomes ready for occupation, whether actually occupied or not. The figures for private sector completions in Northern Ireland have been statistically adjusted to correct, as far as possible, the proven under-recording of private sector completions in Northern Ireland.
Table 6.9 Average weekly rents: by tenure	*Private sector rents*: average rents for 1998-99, excluding tenants who were living rent-free. Figures include any Housing Benefit but exclude any water and other charges paid as part of rent which would not be eligible for Housing Benefit. Data for England are combined averages from the DSS Family Resources Survey and the DETR Survey of English Housing.

Local authority rents: average unrebated rents at April 1999.

Registered Social Landlord (formerly Housing Association) rents: these figures cover the whole stock at 31 March 1999, and are derived from Housing Corporation returns.

Table 6.10 Selected housing costs of owner occupied	*Mortgage payments*: mortgage interest plus any premiums on mortgage protection policies for loans used to purchase the property. For repayment mortgages, interest is calculated using the amount of loan outstanding and the standard interest rate at time of interview. *Endowment policies*: premium on endowment policies covering the repayment of mortgages and loans used to purchase the property. *Structural insurance*: includes cases where insurance also covers furniture and contents and structural element cannot be separately identified. *Services*: includes payments of ground rent, feu duties (applies in Scotland), chief rent, service charges, compulsory or regular maintenance charges, site rent (caravans), factoring (payments to a land steward) and any other regular payments in connection with the accommodation.
Table 6.11 Average dwelling prices	Average prices in this table are calculated from data collected by the Land Registry. Because of the time lag between the completion of a house purchase and its subsequent lodgement with the Land Registry, data for the final quarter of 1999 are not as complete as those for the final quarter of 1998. The table includes all sales registered up to 31 March 2000.
Table 6.12 Mortgage advances, income for mortgage purchases	Figures in this table are taken from The Survey of Mortgage Lenders, a five per cent sample survey of mortgages at completion stage. Full details of the survey are given in *The New Survey of Mortgage Lenders* by Bob Pannell and David Champion (Department of the Environment) in Housing Finance No.16 November 1992 published by the Council of Mortgage Lenders. First-time buyers include sitting tenant purchases.
Table 6.13 County Court actions for mortgage possessions	The figures do not indicate how many houses have been repossessed through the courts; not all the orders will have resulted in the issue and execution of warrants of possession. The regional breakdown relates to the location of the court rather than the address of the property. *Actions entered*: a claimant begins an action for an order for possession of residential property by way of a summons in a county court. *Orders made*: the court, following a judicial hearing, may grant an order for possession immediately. This entitles the claimant to apply for a warrant to have the defendant evicted. However, even where a warrant for possession is issued, the parties can still negotiate a compromise to prevent eviction. *Suspended orders*: frequently, the court grants the mortgage lender possession but suspends the operation of the order. Provided the defendant complies with the terms of the suspension, which usually require them to pay the current mortgage instalments plus some of the accrued arrears, the possession order cannot be enforced.
Table 6.14 Homeless households by reason	In England and Wales the basis for these figures is households accepted for re-housing by local authorities under the homelessness provisions of Part III of the *Housing Act 1985*, and Part VII of the *Housing Act 1996*. In Scotland the basis of these figures is households assessed by the local authorities as homeless or potentially homeless and in priority need, as defined in Section 24 of the *Housing (Scotland) Act 1987*. In Northern Ireland, the *Housing (Northern Ireland) Order 1988 (Part II)* defines the basis under which households (including one-person households) are classified as homeless. The figures relate to priority cases only.
CHAPTER 7: HEALTH Chart 7.3, Tables 7.6 and 7.7 General Household Survey and Continuous Household Survey	The General Household Survey (GHS) and Continuous Household Survey (CHS) are continuous surveys which have been running since 1971 for the GHS and 1983 for the CHS, and are based each year on samples of the general population resident in private (that is, non-institutional) households in Great Britain and Northern Ireland respectively. They are multi-purpose surveys, providing information on aspects of housing, employment, education, health and social services, health related behaviour, transport, population and social security. Since the 1988 GHS the fieldwork has been based on a financial rather than calendar year and due to this data was not collected for the first quarter of 1988.
Chart 7.3 Limiting long-standing illness	'Long-standing illness' is measured by asking respondents if they have a long-standing illness, disability, or infirmity. Long-standing means anything that has troubled the respondent over a period of time that is likely to affect the respondent over time. A limiting long-standing illness/infirmity is one which limits the respondent's activity in any way.
Tables 7.4 and 7.9 and Chart 7.10 Age Standardisation	Age standardisation re-weights the sample in each region so as to give it the same age profile as the whole population (mid-1997 estimates), thereby removing the effects of age difference from the comparisons of cardiovascular conditions.

| Table 7.4 The Health Survey for England | The Health Survey for England is an annual survey which has been running since 1991. It is usually based on a sample of the general population living in private households, but does include an institutional sample when necessary (e.g. in the 2000 survey the focus was on older people and so residential care homes were included in the sample). The survey has an annually repeating core and each year special focus topics are included. The 1998 survey focused on cardiovascular disease. As well as questionnaire information, the survey includes objective measurements such as height, weight, blood pressure, ECG and lung function measurements, and blood and saliva samples are analysed. |

Table 7.5 National Food Survey

The National Food Survey (NFS) is a continuous sample survey in which about 6,700 households per year in the UK keep a record of the type, quantity and costs of foods entering the home during a one week period. Nutritional intakes are estimated from the survey data. The survey included Northern Ireland from 1996 onwards. From 1994, data are also available on food eaten out in Great Britain (but not Northern Ireland).

Detailed survey results and definitions are published by The Stationary Office in an annual report *National Food Survey*.

Table 7.7 Alcohol consumption

A unit of alcohol is 8 grams of pure alcohol, approximately equivalent to half a pint of ordinary strength beer, a glass of wine, or a pub measure of spirits.

Sensible Drinking, the 1995 report of an inter-departmental review of the scientific and medical evidence of the effects of drinking alcohol, concluded that daily benchmarks were more appropriate than previously recommended weekly levels since they could help individuals decide how much to drink on single occasions and to avoid episodes of intoxication with their attendant health and social risks. The report concluded that regular consumption of between three and four units a day for men and two to three units for women does not carry a significant health risk. However, consistently drinking more than four units a day for men, or more than three for women, is not advised as a sensible drinking level because of the progressive health risk it carries. The government's advice on sensible drinking is now based on these daily benchmarks.

Chart 7.8 Drug use

Results can be found in *Drug Misuse Declared in 1998: results from the British Crime Survey*: Ramsay & Partridge (1999), Home Office Research Study 197.

Tables 7.9 and 7.11 Age-adjusted mortality rates

Mortality rates vary with age so the rates for different areas can be affected by the age structure of their populations. The figures in Tables 7.9 and 7.11 have been adjusted to take into account these differences in age structure. The rates have been standardised to the mid-1991 UK population for males and females separately: this means it is permissible to compare rates across areas for each gender, but not to compare males and females.

The causes of death included in Table 7.9 correspond to International Classification of Diseases (9th Revision) codes as follows:

> all circulatory diseases -- 390–459;
> ischaemic disease – 410–414;
> cerebrovascular disease – 430-438;
> all respiratory diseases – 460–519;
> bronchitis et al – 490-493+496;
> cancer (malignant neoplasms) – 140–208;
> all injuries and poisoning – 800–999;
> road accidents – E810–E819;
> suicides and open verdicts – E950-E959 and E980–E989.

The data in these tables relate to registrations in the reference year.

7.10 Cancer: comparative incidence ratios

The directly age-standardised rates in each country and region of the United Kingdom have been calculated using the European standard population. This is done by multiplying the age specific incidence rates in each area by the number of people in the corresponding age groups in the standard population and summing to give the overall rate per 100,000 population. This gives comparable overall rates for areas which have different population structures. The standardised incidence of selected cancer sites in each area have been compared with the United Kingdom as a whole (expressed as the ratio of the rates multiplied by 100) - the comparative incidence ratio.

Directly age-standardised registration rates per 100,000 population for the United Kingdom, 1996 are:

Selected site	Males	Females
Lung	79.6	37.4
Colorectal	55	37.5
Breast	..	104.3
Prostate	64.4	..

7.11 Cervical and breast cancer screening

Figures for the two cancer screening programmes are snapshots of the coverage of the target population for each programme at 31 March 1999.

Figures for the Scottish Breast Screening Programme are an estimate of the coverage of the target population over the three year period 1 April 1996 to 31 March 1999. These figures are derived from the number of women in the 50-54, 55-59 and 60-64 year age groups who have attended a routine screening appointment or a self/GP referral appointment during this period and a mid-year estimate of the female population in Scotland aged 50-64 in 1998. Medically ineligible women are not excluded from the target population. Northern Ireland figures for breast screening may include a small number of women who have been counted more than once due to early recall for screening during the relevant three year period. The maximum extent of any such double count can be calculated as less than 0.4 per cent.

All population data for Scotland were obtained from the General Register Office for Scotland.

Table 7.14 and Chart 7.15 NHS hospital waiting lists

The waiting list figures are Health Authority (HA) population based. That is, they are based on figures received from Health Authority based returns and include all patients resident within the HA boundary plus all patients registered with GPs who are members of a Primary Care Group (PCG) for which the HA is responsible, but are resident in another HA, and excludes any patient resident in the HA, but registered with a GP who is a member of a PCG responsible to a different HA. Other exclusions are all patients living outside England and all privately funded patients waiting for treatment in NHS hospitals. However they do include NHS funded patients, living in England, who are waiting for treatment in Scotland, Wales, Northern Ireland, abroad, and at private hospitals, which are not included in the corresponding provider based return.

In Scotland data is collected by trusts for each individual patient waiting for NHS in-patient or day care treatment - information on Scottish residents waiting outside Scotland is not collected centrally. Average waiting times are calculated from the waiting time associated with each individual patient record.

Figures from Northern Ireland are provider based. They include all patients waiting for treatment at NI Trusts including private patients and patients from outside Northern Ireland.

Mean waiting time: This is calculated approximately for any category as the total waiting times for patients still on the list for that category divided by the corresponding number of people waiting in that category.

Median waiting time: The waiting time of 50 per cent of those patients will be less than the median length. This is a better indicator of the 'average' case since it is generally unaffected by abnormally long or short waiting times at the end of the distribution.

Table 7.16 NHS hospital activity

Data for England and Wales are based on Finished Consultant Episodes (FCEs). An FCE is a completed period of care of a patient using a NHS hospital bed, under one consultant within one NHS Trust. If a patient is transferred from one consultant to another, even if this is within the same NHS Trust, the episode ends and another one begins. The transfer of a patient from one hospital to another with the same consultant and within the same NHS Trust does not end the episode. Data for Scotland and Northern Ireland are based on a system where transfers between consultants do not count as a discharge except in Scotland where figures include patients transferred from one consultant to another within the same hospital, provided there is a change of speciality (or significant facilities e.g. a change of ward). Transfers from one hospital to another, with the same consultant, however, count as a discharge. Newborn babies are included for Northern Ireland but excluded for the other countries. Deaths are included in all four countries.

For Scotland figures include NHS beds/activity in Joint-User and Contractual Hospitals; these hospitals account for a relatively small proportion of total NHS activity. This is a change in presentation from previous years. For outpatient data, the change in recording has meant a slight discontinuity with data for earlier years.

A day case is a person who comes for investigation, treatment or operation under clinical supervision on a planned non-resident basis, who occupies a bed for part or all of that day, and who returns home as planned the same day. Scotland figures will also include day cases that have been transferred to or from in-patient care.

An outpatient is defined as a person seen by a consultant for treatment or advice. A new outpatient is one whose first attendance of a continuous series (or single attendance where relevant) at a clinical outpatient department for the same course of treatment falls within the period under review. Each outpatient attendance of a course or series is included in the year in which the attendance occurred. Persons attending more than one department are counted in each department.

In Northern Ireland, the outpatient figures separated into GP referrals and consultant initiated attendances. It is possible for a first attendance to be initiated by a consultant. The number of attendances in the 'new attendances' refers to GP referrals only, and therefore may not include all new attendances.

Mean duration of stay: this is calculated for any category as the total bed-days for that category divided by the number of ordinary admissions (Finished Consultant Episodes in England, in-patient discharges (including transfers) in Scotland, and deaths and discharges in Northern Ireland) for that category. An ordinary admission is one where the patient is expected to remain in hospital for at least one night. For Scotland figures exclude learning disabilities and non-psychiatric specialities.

Population figures are based on estimates for 1998 Health Authorities for persons all ages.

It should be noted that where figures are presented to the nearest whole number, this is to facilitate the calculation of rates and the aggregation of age bands.

Cases treated per available bed are for ordinary admissions (in-patient discharges including transfers in Scotland) and do not include day case admissions.

Table 7.18 NHS Hospital and Community Health Service directly employed staff

General Medical Practitioners (ie family GPs), General Dental Practitioners, the staff employed by the practitioners, pharmacists in General Pharmaceutical Services and staff working in other contracted out services are not included in the figures.

Medical and dental staff included are those holding permanent paid (whole-time, part-time, sessional) and/or honorary appointments in NHS hospitals and Community Health Services. Figures include clinical assistants and hospital practitioners. Occasional sessional staff in Community Health Medical and Dental Services for whom no whole-time equivalent is collected are not included. The whole-time equivalent of staff holding appointments with more than one region is included in the appropriate region.

Nursing, midwifery and health visiting staff included health care assistants, and excluded nurse teachers and students on '1992' courses. Scientific, therapeutic and technical staff comprises Scientific and Professional and Technical staff incorporating PAMs. Administration and estates comprises Administration and Clerical, Senior Managers and Works staff. Other staff comprises Ancillary, Trades, Ambulance staff and support staff.

Table 7.19 General Practitioners

The figures for general medical practitioners (GPs) include unrestricted principals, PMS contracted GPs and PMS salaried GPs.

An Unrestricted Principal is a practitioner who is in contract with a HA to provide the full range of general medical services and whose list is not limited to any particular group of persons. In a few cases, he/she may be relieved of the liability for emergency calls out-of-hours from patients other than his/her own. Most people have an Unrestricted Principal as their GP.

A PMS Contracted Doctor is a practitioner who is in a contract with a HA to provide the full range of services through the PMS pilot contract and like Unrestricted Principals they have a patient list.
A PMS Salaried Doctor is a doctor employed to work in a PMS pilot either by the PMS Contracted Doctor, and who provides the full range of services and has a list of registered patients.

Other types of General Medical Practitioners include GP Retainers, Restricted Principals, Assistants, Associates (Scotland only), GP Registrars, Salaried Doctors (para 52 SFA) and PMS Other.

The figures for General Dental Practitioners include principals, assistants and vocational dental practitioners in the general dental service. Salaried dentists are excluded. Neither the Hospital Dental Service nor the Community Dental Service are reflected. All Scottish data are provisional.

Table 7.20 Places Available in Residential Care Homes

The figures for England relate to residential places in homes registered under part III of the *National Assistance Act 1948*, the *Registered Homes Act 1984* and the *Registered Homes (Amendment) Act 1991*. They include residential places in homes registered for both residential and nursing care. Places are displayed by type of registered home at 31 March 1999.

CHAPTER 8: INCOME AND LIFESTYLES

Table 8.1 and 8.2 Household income

The Family Expenditure Survey (FES) is a continuous, random sample survey of private households in the United Kingdom and collects information about incomes as well as detailed information on expenditure. All members of the household aged 16 or more keep individual diaries of all spending for a period of two weeks. To increase the reliability of regional breakdowns, three years of data have been combined, the financial years 1996-97 to 1998-99. The total sample over this period was 19,454 households. Results are weighted for non-response.

See the FES annual report, *Family Spending*, for a description of the concepts used and details of the definitions of expenditure and income.

Tables 8.4, 8.8 and 8.14 Family Resources Survey (FRS)

The Family Resources Survey (FRS) is a continuous survey of around 24 thousand private households in Great Britain and is sponsored by the Department of Social Security. Results are based on weighted survey data, which are adjusted for non-response. The overall response rate was 66 per cent for 1998-99 but varied regionally. In common with other surveys, there is evidence to suggest some problems of misreporting certain types of benefit, such as the under-reporting of Income Support, where respondents have stated that all money received comes from a single benefit eg Retirement Pension.

Table 8.3 Measure of income

The measure of income used in compiling Table 8.3 is that used in the Department of Social Security's Households Below Average Income. The income of a household is the total income of all members of the household after the deduction of income tax, National Insurance contributions, contributions to occupational pension schemes, additional voluntary contributions to personal pensions, maintenance/child support payments and Council Tax. Income includes earnings from employment and self-employment, social security benefits including Housing Benefit, occupational and private pensions, investment income, maintenance payments, educational grants, scholarships and top-up loans and some in-kind benefits such as luncheon vouchers.

No adjustment has been made in Table 8.3 for any differences between regions in cost of living as the necessary data for adjustment are not available. In the analysis of regions this inability to adjust costs implicitly suggests that there is no difference in cost of living between regions. As this is unlikely to be true, statements have been sensitivity tested where possible against alternative cost of living regimes. Results suggest that estimates of income before housing costs are deducted are not sensitive to regional price differentials, but results after deduction of housing costs are. In particular, for London and to a lesser extent the South West, living standards may be overstated, and in Wales, the North East, and in Yorkshire and the Humber living standards may be understated.

Income is adjusted for household size and composition by means of the McClements equivalence scale (see below). This reflects the common sense notion that a household of five will need a higher income than a single person living alone to enjoy a comparable standard of living. The total equivalised income of a household is used to represent the income level of every individual in that household; all individuals are then ranked according to this level.

McClements equivalence scale

	Before housing costs	After housing costs
Household member:		
First adult (head)	0.61	0.55
Spouse of head	0.39	0.45
Other second adult	0.46	0.45
Third adult	0.42	0.45
Subsequent adults	0.36	0.4
Each dependent aged:		
0–1	0.09	0.07
2–4	0.18	0.18
5–7	0.21	0.21
8–10	0.23	0.23
11–12	0.25	0.26
13–15	0.27	0.28
16 or over	0.36	0.38

Table 8.5 and Chart 8.6 Survey of Personal Incomes

The Survey of Personal Incomes uses a sample of around 80 thousand cases drawn from all individuals for whom income tax records are held by the Inland Revenue: not all cases in the sample are taxpayers - about 6 per cent do not pay tax because the operation of personal allowances and reliefs removes them from liability. The data in Table 8.5 relate to individuals whose income over the year amounted to the threshold for operation of Pay-As-You-Earn (£4,045 in 1997-98) or more. Below this threshold, coverage of incomes is incomplete in tax records. A more complete description of the survey appears in *Inland Revenue Statistics*.

Table 8.5 Distribution of income liable to assessment for tax

The income shown is that which is liable to assessment in the tax year. In most cases, this is the amount earned or receivable in that year, but for business profits and professional earnings the assessments are normally based on the amount of income arising in the trading account ending in the previous year. Those types of income that were specifically exempt from tax eg certain social security benefits are excluded.

Incomes are allocated to regions according to the place of residence of the recipient, except for the self-employed, where allocation is according to the business address. For many self-employed people home address and business address are the same, and for the majority the region will correspond.

The table classifies incomes by range of total income. This is defined as gross income, whether earned or unearned, including estimates of employees' superannuation contributions, but after deducting employment expenses, losses, capital allowances, and any expenses allowable as a deduction from gross income from lettings or overseas investment income. Superannuation contributions have been estimated and distributed among earners in the Survey of Personal Incomes consistently with information about numbers contracted in or out of the State Earnings Related Pension Scheme and the proportion of their earnings contribution. The coverage of unearned income also includes estimates of that part of the investment income (whose liability to tax at basic rate has been satisfied at source) not known to tax offices. Sampling errors need to be borne in mind when interpreting small differences in income distributions between regions.

Chart 8.6 Average total income and average income tax payable

Income tax is calculated as the liability for the income tax year, regardless of when the tax may have been paid or how it was collected.

The income tax liability shown here is calculated from the individual's total income, including tax credits on dividends, and interest received after the deduction of tax grossed up at the appropriate rate. Allowable reliefs etc, and personal allowances are deducted from total income in order to calculate the tax liability. However, relief given at source on mortgage interest is not deducted as it cannot be estimated with sufficient reliability at regional level.

The average of total incomes for males and females by Government Office Region are based on all individuals with total income in excess of the single person's allowance, which was £4,045 in 1997-98. This will include some individuals who are not liable to tax because of the operation of their personal allowances and reliefs. The average income tax payable for males and females by Government Office Region are based on those individuals who are liable to tax.

Table 8.8 Households in receipt of benefit

Income Support is a non-contributory benefit payable to people working less than 16 hours a week, whose incomes are below the levels (called 'applicable amounts) laid down by Parliament. The applicable amounts generally consist of personal allowances for members of the family and premiums for families, lone parents, pensioners, the disabled and carers. Amounts for certain housing costs (mainly mortgage interest) are also included.

Housing Benefit is administered by local authorities. People are eligible only if they are liable to pay rent in respect of the dwelling they occupy as their home. Couples are treated as a single benefit unit. The amount of benefit depends on eligible rent, income, deductions in respect of any non-dependants and the applicable amount. 'Eligible rent' is the amount of a tenant's rental liability which can be met by Housing Benefit. Payments made by owner-occupiers do not count. Deductions are made for service charged in rent which relate to personal needs.

Council Tax Benefit is also administered by local authorities. Generally, it mirrors the Housing Benefit scheme in the calculation of the claimant's applicable amount, resources and deductions in respect of any non-dependants.

Jobseeker's Allowance (JSA) replaced Unemployment Benefit and Income Support for unemployed people on 7 October 1996. It is payable to people under pensionable age who are available for, and actively seeking, work of at least 40 hours per week. Certain groups of people, including carers, are able to restrict their availability to less than 40 hours depending on their circumstances. There are contribution-based and income-based routed of entry to JSA. Both types of JSA are included under the Jobseeker's Allowance column of the table.

Retirement Pensions are paid to men aged 65 or over and women aged 60 or over who have paid sufficient National Insurance contributions over their working life. A wife who cannot claim a pension in her own right may qualify on the basis of her husband's contributions. The table excluded non-contributory rate.

Incapacity Benefit replaced Sickness and Invalidity Benefits from 13 April 1995. It is paid to people who are assessed as being incapable of work and who meet the contribution conditions. The figures do not include expenditure for Statutory Sick Pay (SSP).

Industrial injuries includes pensions, gratuities and sundry allowances for disablement and specified deaths arising from industrial causes. Child Benefit is normally paid to children up to the age of 16. Benefit may continue up to age 19 for children in full-time education up to 'A' level standard. 16 and 17 year olds are also eligible for a short period after leaving school.

A brief description of the main features of the various benefits paid in Great Britain is set out in *Social Security Statistics* (published annually by Corporate Document Services). Detailed information on benefits paid in Northern Ireland is contained in *Northern Ireland Annual Abstract of Statistics* and *Northern Ireland Social Security Statistics*.

Chart 8.9 Children's spending

In the Family Expenditure Survey, children aged between 7 and 15 are asked to complete diaries of their daily expenditure. In the period 1996-97, 1997-98 and 1998-99 survey, 5,725 children from 18,607 households in Great Britain completed diaries over a two week period. Some details of the survey are given in the notes to Tables 8.1 and 8.2.

Expenditure covers anything children buy with their own money. The data in the chart do not therefore include money spent on children. They include money spent by children on school dinners, and on fares to and from school. However, money spent direct by the parent on these items is excluded. Spending by the child on behalf of the parent is also excluded, eg where the child is given money to buy a loaf of bread from the local shop.

Chart 8.10 Charitable giving

The figures related to charitable donations and subscriptions (excluding entrance fees to bazaars, jumble sales, etc.). This includes animal charity, Big Issue (charity), Marie Curie Memorial Foundation, missionary box, Mother Union collection, Poppy (charity), Red Cross donation, Rugby Life Line, Salvation Army, school fund, sponsor money, Sunday School collection.

Table 8.11 Household expenditure

This table contains results from the Family Expenditure Survey for the period 1996-97, 1997-98 and 1998-99. Some details of the survey are given in the notes to Tables 8.1 and 8.2.

Expenditure excludes savings or investments (eg life assurance premiums), income tax payments, National Insurance contributions and the part of rent paid by Housing Benefit.

Housing expenditure of households living in owner-occupied dwellings consists of the payments by these households for Council Tax (rates in Northern Ireland), water, ground rent, etc, insurance of the structure and mortgage interest payments. Mortgage capital repayments and amounts paid for the outright purchase of the dwelling or for major structural alterations are not included as housing expenditure.

Estimates of household expenditure on a few items are below those which might be expected by comparison with other sources eg alcoholic drink, tobacco and, to a lesser extent, confectionery and ice cream.

Tables 8.12 and 8.13 National Food Survey

See notes to Table 7.5.

Table 8.17 Local voluntary work

The Survey of English Housing is a continuous random survey of about 20,000 households per year, sponsored by the DETR. It interviews one person in each household - almost always the household's head or their partner. Therefore, for convenience, the respondents in Table 8.17 are referred to as 'householders'.

The table shows the proportions of householders who had been involved in some form of voluntary work to improve their local area or neighbourhood in the 12 months prior to the survey in 1996-97 (this includes people who have volunteered in the last 12 months but has stopped volunteering at the time of interview). This could include: improving local people's quality of life (eg being involved with cultural, sport or health activities) helping with activities for local children improving the local environment tackling crime and improving community safety (such as being involved with Neighbourhood Watch) involvement with local employment initiatives (eg helping the unemployed get back to work, such as unemployment worker centres or work placement groups) addressing local housing issues improving local people education skills (eg as a school governor or involvement in PTA work, adult education or voluntary training scheme).

Table 8.19 National Lottery Grants

In 1996, National Lottery grants included 183 grants worth £116 million made UK-wide or to institutions of national significance. It included 356 grants worth £509 million not allocated to a specific region. No grants were made overseas.

In 1997, National Lottery grants included 368 grants worth £608 million made UK-wide or to institutions of national significance. They included 610 grants worth £747 million not allocated to a specific region. A further 4 grants worth 129,087 were made overseas.

In 1998, National Lottery grants included 425 grants worth £627 million made UK-wide or to institutions of national significance. They included 938 grants worth £817 million not allocated to a specific region. A further 9 grants worth £319 thousands were made overseas.

In 1999, National Lottery grants included 649 grants worth £678 million made UK-wide or to institutions of national significance. They included 1,706 grants worth £1,131 million n1ot allocated to a specific region. A further 13 grants worth £490 thousand have been made overseas.

CHAPTER 9: CRIME AND JUSTICE

The figures are compiled from police returns to the Home Office or directly from court computer systems, from police returns to The Scottish Office Home Department and from statistics supplied by the Royal Ulster Constabulary in Northern Ireland.

Tables 9.1, 9.3, 9.4, 9.6 and 9.7 and Map 9.10 Offences

Recorded crime statistics broadly cover the more serious offences. Up to March 1998 most indictable and triable-either-way offences were included, as well as some summary ones; from April 1998, all indictable and triable-either-way offences were included, plus a few closely related summary ones. Recorded offences are the most readily available measures of the incidence of crime, but do not necessarily indicate the true level of crime. Many less serious offences are not reported to the police and cannot, therefore, be recorded while some offences are not recorded due to lack of evidence. Moreover, the propensity of the public to report offences to the police is influenced by a number of factors and may change over time.

In England and Wales and Northern Ireland, indictable offences cover those offences which must or may be tried by jury in the Crown Court and include the more serious offences. Summary offences are those for which a defendant would normally be tried at a magistrates' court and are generally less serious- the majority of motoring offences fall into this category. In general in Northern Ireland non-indictable offences are dealt with at a magistrates' court. Some indictable offences can also be dealt with there.

In Scotland the term 'crimes' is generally used for the more serious criminal acts (roughly equivalent to indictable offences); the less serious are termed 'offences'. In general, the Procurator Fiscal makes the decision as to which court a case should be tried in or, for lesser offences, whether alternatives to prosecution such as a Fixed Penalty might be considered. Certain crimes, such as rape and murder, must be tried by a jury in the High Court. Cases can also be tried by jury in the Sheriff Court. The majority of cases (97 per cent) are tried summarily (without a jury), either in the Sheriff Court or in the lay District Court.

If a person admits to committing an offence he may be given a formal police caution by, or on the instruction of, a senior police officer as an alternative to court proceedings. The figures exclude informal warnings given by the police, written warnings issued for motoring offences and warnings given by non-police bodies eg a department store in the case of shoplifting. Cautions by the police are not available in Scotland, but warnings may be given by the Procurator Fiscal.

Tables 9.2, 9.13 and 9.14 Crime Surveys

The British Crime Survey (BCS) was conducted by the Home Office in 1982, 1984, 1988, 1992, 1994, 1996 and 1998. Each survey measured crimes experienced in the previous year, including those not reported to the police. The survey also covers other matters of Home Office interest including fear of crime, contacts with the police, and drug misuse. The 1998 survey had a nationally representative sample for 14,947 people aged 16 or over in England and Wales. The sample was drawn from the Small User Postcode Address File - a listing of all postal delivery points. The response rate was 79 per cent. The results of the latest British Crime Survey will be published in October 2000.

Scotland participated in sweeps of the BCS in 1982 and 1988 and ran its own Scottish Crime Surveys in 1993, 1996 and 2000 based on nationally representative samples of around 5,000 respondents aged 16 or over interviewed in their homes. In addition around 400 young people aged between 12 and 15 completed questionnaires in each of the surveys. The sample was drawn from addresses randomly generated from the Postcode Address file. Both the 1993 and 1996 surveys had response rates of 77 per cent and the 2000 survey had a response rate of 72 per cent. The results of the latest Scottish Crime Survey are due to be published in Autumn 2000.

The Northern Ireland Crime Survey was commissioned by the Northern Ireland Office in 1997. The survey was conducted throughout Northern Ireland and fieldwork took place between February 1998 and May1998. More than 3,000 people aged 16 years and above participated in the survey. They were sampled from the Valuation and Lands Agency (VLA) list, which is the most up-to-date listing of private addresses in Northern Ireland. The response rate was 70 per cent.

In each of the surveys, respondents answered questions about offences against their household (such as theft or damage of household property) and about offences against them personally (such as assault or robbery). However, none of the surveys provides a complete count of crime. Many offence types cannot be covered in a household survey (eg. shop lifting, fraud or drug offences). Crime surveys are also prone to various forms of error, mainly to do with the difficulty of ensuring that samples are representative, the frailty of respondents' memories, their reticence to talk about their experiences as victims, and their failure to realise an incident is relevant to the survey.

Table 9.3 Clear Up rates

In England and Wales and Northern Ireland offences recorded by the police as having been cleared up include offences for which persons have been charged, summonsed or cautioned, those admitted and taken into consideration when persons are tried for other offences, and others where the police can take no action for various reasons.

In Scotland a revised definition of cleared up came into effect from 1 April 1996. Under the revised definition a crime or offence is regarded as cleared up where there exists a sufficiency of evidence under Scots Law, to justify consideration of criminal proceedings not withstanding that a report is not submitted to the procurator fiscal because either:
a) by standing agreement with the procurator fiscal, the police warn the accused due to the minor nature of the offence or
b) reporting is inappropriate due to the age of the accused, death of the accused or other similar circumstances.

The clear-up rate is the ratio of offences cleared up in the year to offences recorded in the year. Some offences cleared up may relate to offences recorded in previous years. There is some variation between police forces in the emphasis placed on certain of the methods listed above and, as some methods are more resource intensive than others, this can have a significant effect on a force's overall clear-up rate.

Table 9.5 Seizure of controlled drugs

The figures in this table, which are compiled from returns to the Home Office, relate to seizures made by the police and officials of HM Customs and Excise, and to drugs controlled under the *Misuse of Drugs Act 1971*. The Act divides drugs into three main categories according to their harmfulness. A full list of drugs in each category is given in Schedule 2 to the *Misuse of Drugs Act 1971*, as amended by Orders in Council.

Table 9.8 Persons found guilty of offences

In England, Wales and Northern Ireland the term 'suspended sentence' is known as 'fully suspended sentence' and 'immediate custody' includes unsuspended sentences of imprisonment and sentence to detention in a young offender institution. Fully suspended sentences are not available to Scottish courts.

Table 9.14 Community Attitudes Survey

See earlier entry for crime surveys also included in the publication. The Community Attitudes Survey is a continuous survey of public attitudes and views on crime, law and order and policing issues. The survey was initially commissioned by a number of Northern Ireland government departments and agencies in 1992. The survey field period for this year ran between November 1997 and October 1998. The sample comprised 2,400 addresses sampled from the Valuation and Lands Agency (VLA) list, which is the most up-to-date listing of private addresses in Northern Ireland. Interviews were obtained with over 1,400 selected respondents.

CHAPTER 10: TRANSPORT
Table 10.3 Age of household cars

The main or only car available to the household applies to the vehicle with the greatest annual mileage. In the majority of cases this will be the newest car.

Tables 10.4, 10.6 and Chart 10.5 National Travel Survey	The National Travel Survey (NTS) is the only comprehensive national source of travel information for Great Britain that links different kinds of travel with the characteristics of travellers and their families. Since July 1988, the NTS has been conducted on a small scale continuous basis. The last of the previous ad hoc surveys was carried out in 1985/86.

Information is collected from about 3,000 households in Great Britain each year, every member provides personal information (eg age, gender, working status, driving licence, season ticket) and details of journeys carried out in a sample week, including purpose of journey, method of travel, time of day, length, duration, and cost of any tickets bought.

Travel included in the NTS covers all journeys by GB residents within Great Britain for personal reasons, including travel in the course of work. Travel information is recorded at two levels for multi-stage journeys: journey and stage.

A journey is defined as a one-way course of travel having a single main purpose. It is the basic unit of personal travel in the survey. A round trip is split into two journeys, with the first ending at a convenient point about half way round as a notional stopping point for the outward destination and return origin.

A stage is that portion of a journey defined by the use of a specific method of transport or of a specific ticket (a new stage being defined if either the mode or ticket changes).

The purpose of a journey is normally taken to be the activity at the destination, unless that destination is 'home' in which case the purpose is defined by the origin of the journey. The classification of journeys to 'work' are also dependent on the origin if the journey.

Tables 10.7 and 10.8 Roads

Major roads: motorways and A roads.
Principal roads: important regional or local roads for which local authorities are the Highway Authorities (non-trunk A roads).
A Roads: trunk and principal roads (excluding motorways).
Minor roads: comprise of B, C and unclassified roads.
Built-up roads: all those having a speed limit of 40mph or less (irrespective of whether there are buildings or not).
Non built-up roads: all those with a speed limit in excess of 40 mph.

Table 10.7 Annual average daily flow

Traffic estimates are derived from roadside traffic counts and takes two forms, occasional 12 hour counts at a large number of sites to estimate the absolute level of traffic (the rotating census) and frequent count at a small number of sites (the core census) to estimate changes in the amount of traffic.

Tables 10.9 and 10.10 Road accidents

An accident is one involving personal injury occurring on the public highway (including footways) in which a road vehicle is involved and which becomes known to the police within 30 days. The vehicle need not be moving and it need not be in collision with anything.

Persons killed are those who sustained injuries which caused death less than 30 days after the incident.

A serious injury is one for which a person is detained in hospital as an in-patient, or any of the following injuries whether or not they are detained in hospital: fractures, concussion, internal injuries, crushing, severe cuts and lacerations, severe general shock requiring medical treatment, injuries causing death 30 or more days after the accident.

There are many reasons why accident rates per head of population (for all roads) and per 100 million vehicle kilometres (for major roads) vary by region. They will be influenced by the mix of pedestrian and vehicle traffic within each region, which vary as a result of the considerable differences in vehicle ownership. In addition, an area that 'imports' large numbers of visitors or commuters will have a relatively high proportion of accidents related to vehicles or drivers from outside the area. A rural area of low population density but high road mileage can be expected, other things being equal, to have lower than average accident rates.

Table 10.13 Seaports

The Coastal regions are defined as:
East Coast – Orkneys to Harwich inclusive;
Thames and Kent – Colchester to Folkstone inclusive;
South Coast – Newhaven to Lands End;
West Coast – Lands End to Stornoway.

CHAPTER 11: ENVIRONMENT

Tables 11.6- 11.10 The Environment Agency

The Environment Agency for England and Wales was formally created on 8 August 1995 by the Environment Act 1995. It took up its statutory duties on 1 April 1996. The Agency brings together the functions previously carried out by the National Rivers Authority, Her Majesty's Inspectorate of Pollution, the waste regulatory functions of 83 local authorities and a small number of units from the then Department of the Environment dealing with the aspects of waste regulation and contaminated land. One of the key reasons for setting up the Agency was to promote a more coherent and integrated approach to environmental management.

The Agency's principal aim is to protect and improve the environment. It's business can be grouped under two broad headings:

a) pollution prevention and control which includes regulating the disposal of controlled waste, protecting and improving the quality of rivers estuaries and coastal waters, and regulating major industrial processes, nuclear sites and premises authorised to dispose of radioactive waste; and,

b) water management covering water resources, flood defence, fisheries, recreation, conservation and navigation.

The Agency has a budget of around £550 million per annum. About 30 per cent of this is allocated to the prevention and control of pollution; nearly 50 per cent is spent on flood defence, and the remaining 20 per cent on the Agency's other waste management functions.

Table 11.7 Rivers and Canals: by biological and chemical quality

The chemical quality of rivers and canal waters in the United Kingdom are monitored in a series of separate national surveys in England and Wales, Scotland and Northern Ireland. In England and Wales the National Rivers Authority (now superseded by the Environment Agency) developed and introduced the General Quality Assessment (GQA) Scheme to provide a rigorous and objective method for assessing the basic chemical quality of rivers and canals based on three determinants: dissolved oxygen, biochemical oxygen demand (BOD) and ammoniacal nitrogen. The GQA grades river stretches into six categories (A-F) of chemical quality and these in turn have been grouped into two broader groups - good/fair (classes A, B C and D) and poor/bad (classes E and F).

In Northern Ireland, the grading of the 1991 and 1995 surveys is also based on the GQA scheme. In Scotland, the classification system for chemical quality is not directly comparable with the GQA. The system was changed in 1996. The GQA category ìgood/fairî is assumed to be equivalent to Class 1 prior to 1996 and to Classes A1, A2 and B from 1996 onwards.

To provide a more comprehensive picture of the health of rivers and canals, biological testing has also been carried. Biological grading is based on the monitoring of small animals (ie invertebrates) which live in or on the bed of the river. Research has shown that there is a relationship between species composition and water quality. Using a procedure known as the River Invertebrate Prediction and Classification System (RIVPACS), species groups recorded at a site were compared with those which would be expected to be present in the absence of pollution, allowing for the different environmental characteristics in different parts of the country.

In England and Wales and Northern Ireland, two different summary statistics (known as ecological quality indices (EQI) were calculated and the biological quality assigned to one of six bands (A-F) based on a combination of these two statistics. These six bands have been grouped into two broader groups good/fair (classes A, B, C and D) and poor classes (E and F). In Scotland, a third EQI was also calculated and the grading system based on all threes EQIs and river and canal waters classified as good/moderate or poor/very poor. The results for Scotland are not directly comparable with those of England and Wales and Northern Ireland. In 1995, the Scottish Biological survey covered about 33 per cent of the freshwaters covered by the chemical classification.

Table 11.8 Water pollution incidents

The Environment Agency for England and Wales defines four categories of pollution incidents:

Category 1

A major incident involving one or more of the following:

a) potential or actual persistent effect on water quality or aquatic life;

b) closure of potable water, industrial or agriculture abstraction necessary;

c) major damage to aquatic ecosystem;

e) major damage to agriculture and/or commerce;

d) serious impact on man;

f) major effect on amenity value.

Category 2

A significant pollution which involves one or more of the following:

a) notification to abstractors necessary;

b) significant damage to aquatic ecosystem;

c) significant effect on water quality;

d) damage to agriculture and/or commerce;

c) impact on man;

d) amenity value to the public, owners or users.

Category 3

Minor incident involving one or more of the following:

a) a minimal effect on water quality;

b) minor damage to ecosystem;

c) amenity value only marginally affected;

d) minimal impact on agriculture and/or commerce.

Category 4
An incident where no impact on the environment occurred.

Department of the Environment (Northern Ireland) defines four categories of pollution incidents: High Severity; Medium Severity; Low Severity and Unsubstantiated. They are broadly the same as the categories used by the Environment Agency.

The Scottish Environment Protection Agency (SEPA) presently reports on two categories of pollution incidents:
Category 1
Serious incidents are those which cause a breach of any appropriate environmental quality standard in the receiving water. These incidents are reported as 'significant' in SEPA's annual report and compare broadly with all of Category 1 and a, b, c and d of Category 2 used by the Environment Agency.

Category 2
Minor incidents are those which do not cause a breach of any appropriate environmental quality standard in the receiving water. These incidents are reported as 'routine' in SEPA's annual report and compare broadly with e and f of Category 2 and all of Category 3 used by the Environment Agency.

Map 11.4 Radon affected areas

Radon accounts for half of the average overall dose of radioactivity received by the UK population. The health hazard associated with radon is from its radioactive decay products. These may be inhaled and deposited in the lungs where radiation from them can damage lung tissue, and may increase risk of lung cancer. Parts of the United Kingdom have been designated as Radon Affected Areas, that is areas where more than 1 in 1000 homes are estimated to have radon concentrations above the National Radiological Protection Board (NRPB) recommended action level of 200Bq/m^3. Above this level, NRPB recommend that action be taken to limit exposure of households of high levels of radon.

Map 11.11 and Table 11.12 Designated Protected areas

National Parks, Areas of Outstanding Natural Beauty in England and Wales and Northern Ireland, Defined Heritage Coasts in England and Wales and National Scenic Areas in Scotland are the major areas designated by legislation to protect their landscape importance. Green Belts have been designated in England, Scotland and Northern Ireland to restrict the sprawl of built up areas onto previously undeveloped land and to preserve the character of historic towns. Other areas, such as National Nature Reserves, Special Protection Areas, Marine Nature Reserves, are protected for their value as wildlife habitat, in particular for endangered species, Sites in the United Kingdom are protected by sites of Special Scientific Interest (SSSI) status.

Table 11.13 Land use change statistics

Details of changes in land use are recorded for the Department of the Environment by Ordnance Survey (OS) as part of its map revision work in England. The data recorded by OS in any one year depend on OS resources and how these are deployed on different types of map revision survey. The main consequence of this is that physical development (eg new houses) tends to be recorded relatively sooner than changes between other uses (eg between agriculture and forestry), some of which may not be recorded for some years. The statistics are best suited to analyses of changes to urban uses and of the recycling of land already in urban uses.

Land is classified into 24 categories which are then grouped into 'urban uses' and 'rural uses'. Urban uses include: residential; transport and utilities; industry and commerce; community services; vacant land (classified according to whether it was previously developed or within a built-up area, but not previously developed). Rural uses include; agriculture; forestry; open land and water; minerals and landfill; outdoor recreation; defence.

Maps 11.15 and 11.16 Household waste

Map boundaries relate to Local Authority Districts prior to local government reorganisation; see other Regional Classifications at the beginning of the Notes and Definitions.

CHAPTER 12: REGIONAL ACCOUNTS

Regional GDP estimates for years back to 1989 have been revised reflecting conceptual and developmental changes. Further revisions to 1997 and 1998 estimates will be published when Inland Revenue data for wages and salaries and other new data become available. The sources and methodology used to compile the regional accounts are given in a booklet in the *Studies in Official Statistics series, No 31, Regional Accounts*, (HMSO)and more recently in the Eurostat publication *Methods used to compile regional accounts*. More up-to-date information on the data, including the effects of the new European System of Accounts 1995 (ESA95), is contained in *Economic Trends*, October 1999 and August 2000 editions (TSO).

Tables 12.1, 12.3,12.4, Chart 12.2 and Map 12.5 Gross domestic product (GDP)

Estimates of regional GDP can be presented on either a residence or workplace basis. Residence based estimates allocate the income of commuters to where they live rather than to their place of work. Workplace based estimates allocate these incomes to the region in which commuters work. The estimates of GDP presented in Table 12.1, Chart 12.2 and Tables 12.3 and 12.4 are residence based, while those in Map 12.5 are workplace based estimates.

Regional estimates of GDP are compiled as the sum of factor incomes, ie incomes earned by residents, whether corporate or individual, from the production of goods and services. This approach breaks the total down into two major components: compensation of employees and operating surplus/mixed income. The figures for all regions are adjusted to sum to the national totals as published in *1999 Blue Book (United Kingdom National Accounts)* (TSO).

In order to accommodate within regional accounts activity that cannot be allocated to any particular region, such as the offshore oil and gas extraction industry, a region known as the Extra Regio is included.

Table 12.6 Household income and household disposable income

Regional household income measures the income of the household sector, which currently includes private households and non-profit making institutions serving households. Household disposable income is derived after deducting taxes, social security contributions, net non-life premiums and miscellaneous current transfers. The regional figures are consistent with the national estimates published in the 1999 Blue Book.

Tables 12.7 and 12.8 Individual consumption expenditure

Individual consumption expenditure measures expenditure by households and non-profit making bodies resident in a region. Estimates are based mainly on the Family Expenditure Survey and are subject to sampling error and should be used with caution. The regional figures are consistent with the national estimates published in the 1999 Blue Book.

CHAPTER 13: INDUSTRY AND AGRICULTURE

The industrial breakdown used is in accordance with the Standard Industrial Classification (SIC) Revised 1992. Agriculture, industry and services are broken down as follows:

Map 13.1, Tables 13.2, 13.3, 13.4, 13.5, 13.6, Maps 13.16 and 13.17 Industrial Breakdown

AGRICULTURE:

Section A   Agriculture, hunting and forestry
Section B   Fishing

INDUSTRY:

Section C   Mining and quarrying
Section D   Manufacturing
Section E   Electricity, gas and water supply
Section F   Construction

SERVICES:

Section G   Wholesale and retail trade; repair of motor vehicles, motorcycles and personal and household goods
Section H   Hotels and restaurants
Section I   Transport, storage and communications
Section J   Financial intermediation
Section K   Real estate, renting and business activities
Section L   Public administration and defence; compulsory social security
Section M   Education
Section N   Health and social work
Section O   Other community, social and personal service activities

Tables 13.3 and 13.4 Inter-Departmental Business Register

The IDBR is a structured list of business units for the selection, mailing and grossing of statistical inquiries. Information is provided at both the enterprise and local unit level. The enterprise is usually the business registered for VAT and/or PAYE. The local units are the individual sites (or factories shops etc.) operated by the enterprise. The IDBR covers more than 98 per cent of UK output. All analyses are based on enterprises that are VAT and/or PAYE registered.

Tables 13.5 and 13.6 Annual Business Inquiry

The Annual Business Inquiry (formerly the Annual Inquiry into Production) covers UK businesses engaged in the production and construction industries: Divisions 1-5 of the Standard Industrial Classification (SIC) Revised 1980 and Section C to F of the SIC Revised 1992. Regional information is available only for manufacturing industry: ie Divisions 2-4 of the SIC 1980 and Section D of the SIC 1992.

Businesses often conduct their activities at more than one address (local unit) but it is not usually possible for them to provide the full range of data for each. For this reason, data are usually collected at the enterprise level. Gross value added (GVA) is estimated for each local unit by apportioning the total GVA for the business in proportion to the total employment at each local unit using employment from the IDBR.

Gross value added (GVA) at basic prices is defined as:
The value of total sales and work done, adjusted by any changes during the year in work in progress and goods on hand for sale
less: the value of purchases, adjusted by any changes in the stocks of material, stores and fuel etc.
less: payments for industrial services received
less: net duties and levies etc.
less: the cost of non-industrial services, rates and motor vehicle licences.
It includes taxes on production (like business rates), net of subsidies but excludes taxes less subsidies on production (for example, VAT and excise duty).
GVA per head is derived by dividing the estimated GVA by the total number of people employed.
The data include estimates for businesses not responding, or not required to respond, to the inquiry.

Table 13.7 Export and import trade with EU and non-EU countries

Data is sourced from Customs declarations submitted in respect of trade with countries outside the European Union and 'Supplementary Declarations' submitted under the Intrastat EU statistical reporting system. Whereas all imports and exports outside the EU are recorded, the Intrastat system is based on returns from registered companies that exceed a set annual threshold in their trading with the EU. For 1999 the threshold was £230,000. So, whereas the Intrastat data accounts for 97.5 per cent of the value of the UK's trade with the EU, only a relatively small proportion of the total number of companies that are trading with the EU are counted. The totals of the value of the Regional Trade in Goods Statistics do not equate to the totals already published as the UK-wide Overseas Trade Statistics. Certain goods, such as North Sea crude oil, ships and aircraft stores, and transactions involving overseas companies with no place of business in the UK, cannot be allocated to a UK region.

Table 13.8 Direct inward investment: project successes

Data on projects which have attracted inward investment appear in this table. They are based on information provided to Invest-UK, part of the Department of Trade and Industry by the beneficiary companies at the time of the decision to invest. There is no obligation to notify the department, so the figures relate only to those projects where Invest-UK or its regional partners were involved or have come to their notice. They also take no account of subsequent developments: for example, if a company goes bankrupt several years later.

Table 13.10 Government expenditure on regional preferential assistance to industry

The types of assistance included in this table for Great Britain are: Regional Development Grants prior to 1996/97; Regional Selective Assistance; Regional Enterprise Grants; expenditure on Land and Factories by the English Industrial Estates Corporation (until 1993-94 after which this falls under the province of the Single Regeneration Budget), Scottish Enterprise, the Welsh Development Agency; and expenditure on Land and Factories and Grants by the Development Board for Rural Wales and Highlands and Islands Enterprise.

Northern Ireland has a different range of financial incentives available and so the figures have not been aggregated into a United Kingdom total. The items included are: Industrial Development Board grants and loans; expenditure on land and factories; Standard Capital Grants; and Local Enterprise Development Unit grants and loans.

All figures are gross and include payments to nationalised industries. GB payments relate only to projects situated in the Assisted Areas of Great Britain. A map showing the areas qualifying for preferential assistance to industry was included in Regional Trends 31.

Table 13.11 EU Structural Funds

Funds are allocated in the prices of the year of the European Commission (EC) decision. For the majority of the allocations shown in the table, this was 1999. Those that were allocated in the prices of earlier years have been inflated to 1999 using Treasury GDP deflators.

Regions may be eligible for funding in one of two categories. 'Objective 1' funds promote the development of regions, which are lagging behind the rest of the EU. To be eligible regions need to have a per capita GDP of 75 per cent or less of the EU average In these areas, emphasis is placed on creating a sound infrastructure: modernising transport and communication links, improving energy and water supplies, encouraging research and development, providing training and helping small businesses.

Areas suffering from industrial decline may be designated 'Objective 2'. These areas need help adjusting their economies to new industrial activities; they have high unemployment rates, and a high but declining share of industrial activity. EU grants may be provided to help create jobs, encourage new businesses, renovate land and buildings, promote research and development, and foster links between universities and industry. In addition, rural areas where economic development needs to be encouraged may be designated 'Objective 2'. In these areas the focus is on developing jobs outside agriculture in small businesses and tourism, and improvements to transport and basic services are promoted to prevent rural depopulation.

Grants under Objectives 1 and 2 are disbursed under the terms of Single Programming Documents or their equivalents, which provide a strategic framework relevant to the region concerned. The other objective under which grants are allocated, Objective 3, which covers long-term unemployment, jobs for young people and modernisation of farms, is not defined geographically. In addition the Structural Funds provide support for Community-wide Initiatives. These Initiatives account for 8 per cent of the Structural Funds budget.

Table 13.12 Business registration and deregistrations

Annual estimates of registrations and deregistrations are compiled by the Department of Trade and Industry. They are based on VAT information which the Office for National Statistics (ONS) holds. The estimates are a good indicator of the pattern of business start-ups and closures, although they exclude firms not registered for VAT, either because their main activity is exempt from VAT, or because they have a turnover below the VAT threshold (£49,000 with effect from 1 December 1997) and have not registered voluntarily. Large rises in the VAT threshold in 1991 and 1993 affected the extent to which the VAT system covers the small business population. This means that the estimates are not entirely comparable before and after these years.

Tables 13.17–13.21 Agriculture census	The annual census encompasses the 239 thousand main agricultural holdings in the United Kingdom in 1999. Estimates for minor holdings are included in the national totals for England, Wales, Scotland and Northern Ireland; estimates are not included for the English regions. Generally, minor holdings are characterised by a small agricultural area, low economic activity and a small labour input.
Table 13.19 Less Favoured Areas	Land in the Less Favoured Areas is commonly infertile, unsuitable for cultivation and with limited potential which cannot be increased except at excessive cost. Such land is mainly suitable for extensive livestock farming.
Table 13.20 Areas and yields	The figures for specific crops relate to those in the ground on the date of the June census or for which the land is being prepared for sowing at that date. In England and Wales cereal production is estimated from sample surveys held in September, November and April; oilseed rape production is estimated from a sample survey held in August. In Scotland, cereals and oilseed rape yields are estimated by local office staff in mid-September, followed by sample surveys later in the year. The Department of Agriculture for Northern Ireland estimates cereal and oilseed rape yields from a stratified sample survey of 200 farms carried out in the autumn of each year.
CHAPTERS 14–17: SUB REGIONAL STATISTICS	Sub-regional data complement the data shown regionally in Chapters 3 to 13. A wide range of data are presented, covering population, vital statistics, education, housing and households, labour market, deprivation and economic statistics. The statistics cover Government Office Regions, counties/unitary authorities and, where available, local authority districts in England; Unitary Authorities in Wales; the Council areas in Scotland; Health and Social Service Boards/Education and Library Boards/districts as available in Northern Ireland. Tables 14.7, 15.6, 16.6 and 17.5 present data on the NUTS area classification (see Regional Classifications at the beginning of the Notes and Definitions).
	In the local authority tables for England, where data is often collected at district and Unitary Authority level and can be easily combined, county, regional and national totals are given to make comparison easier. However, for national surveys, local estimates have to be derived by disaggregating and sometimes different sources are used to derive estimates for lower geographical levels. It is not therefore necessarily the case that data in this chapter are strictly comparable with data in other chapters. These data identify local as well as regional trends and because of the level of disaggregation more caution in interpretation is necessary.
	There are specific and known problems in comparing population, employment and unemployment data for small areas. For example, for the claimant count rate the numerator is residence-based while the denominator is largely workplace-based; this should be borne in mind when comparing claimant count rates for small areas.
	Allowing for the difficulties in interpreting such geographically desegregated data, the figures in the relevant sub-regional tables can be used to give a broad picture of a particular local authority and how it compares with others. The tables are intended to take a reasonably broad sweep across a range of subjects. More detailed statistics on specific topics may be readily available. For example:

> *Key population and vital statistics* (local and health authority areas of England and Wales)
> *Local Housing Statistics England* (annual statistics by Local Authority area)
> *Projections of Households in England to 2021* (statistics for counties, metropolitan districts and London boroughs)
> *Labour Market Trends* (unemployment by local authority districts and parliamentary constituency).

Tables 14.1-17.1 Standardised Mortality Ratio	The standardised mortality ratio (SMR) compares overall mortality in a region with that for the United Kingdom. The ratio expresses the number of deaths in a region as a percentage of the hypothetical number that would have occurred if the region's population had experienced the sex/age specific rates of the United Kingdom in that year.
Table 14.3-17.3 Education	Pupils in last year of compulsory schooling with no graded results are those who either did not attempt any GCSE, GSE, CSE or SCE examinations or did not achieve a sufficient standard to be awarded a grade.
Table 14.4 Housing completions	The housebuilding figures are compiled from data provided by local authorities and by the National House-Building Council. If a local authority has not sent back statistical returns for 1999, the table shows that the data are not available. County, regional and England figures, however, include estimated figures that allow for these missing data. It is inappropriate to derive figures for any missing authorities from these estimated totals.
Tables 14.4 16.4 Council Tax	Amounts shown for Council Tax are headline Council Tax for the area of each billing authority for Band D, 2 adults, before transitional relief and benefit. The ratios of other bands are: A 6/9, B 7/9, C 8/9, E 11/9, F 13/9, G 15/9 and F 18/9.
	Averages are calculated by dividing the sum of the tax requirement for each area by the tax base for the area. The taxbase is calculated by weighting each dwelling on the valuation list to take account of exemptions, discounts and disabled relief and the valuation band it falls into. It therefore represents the number of Band D equivalent (fully chargeable) dwellings.

Table 14.5–16.5 Labour markets

This table contains some of the regional results of the New Earnings Survey 1999, fuller details of which are given for the Government Office Regions in part A and E of the report *New Earnings Survey 1999*, (National Statistics Direct) published in October and December 1999. Results for Northern Ireland are published separately by the Department of Enterprise, Trade and Investment, Northern Ireland. The measured gross earnings of a 1 per cent sample of employees, most of whom were members of Pay-As-You-Earn (PAYE) schemes for a pay-period which included 14 April 1999. The earnings information collected was converted to a weekly basis where necessary, and to an hourly basis where normal basic hours were reported.

Figures are given where the number of employees reporting in the survey was thirty or more and the standard error of average weekly earnings was 5 per cent or less. Gross earnings are measured before tax, National Insurance or other deductions. They include overtime pay, bonuses and other additions to basic pay but exclude any payments for earlier periods (e.g. back pay), most income in kind, tips and gratuities. All the results in this volume relate to full-time male and female employees on adult rates whose pay for the survey pay-period was not affected by absence. Employees were classified to the region in which they worked (or were based if mobile) using post code information, and to manual or non-manual occupations on the basis of the Standard Occupational Classification (SOC). Part A of the report for Great Britain and the United Kingdom give full details of definitions used in the survey.

Full-time employees are defined as those normally expected to work more than 30 hours per week, excluding overtime and main meal breaks (but 25 hours or more in the case of teachers) or, if their normal hours were not specified, as those regarded as full-time by the employer.

Table 14.6 Business registrations and deregistrations

Annual estimates of registrations and deregistrations are compiled by the Department of Trade and Industry. They are based on VAT information which the Office for National Statistics holds. The estimates are a good pattern of business start-ups and closures, although they exclude firms not registered for VAT, either because they have a turnover below the VAT threshold (£49,000 with effect from 1 December 1997) and have not registered voluntarily; or because they trade in VAT exempt goods or services. Large rises in the VAT threshold in 1991 and 1993 affected the extent to which the VAT system covers the small business population. This means that the estimates are not entirely comparable before and after these years.

Index

Figures in the index refer to table or chart numbers, page numbers refer to maps. The Notes and Definitions are not indexed.

Printed in the United Kingdom by the Stationery Office
TJ003393 C35 1/01 582164 19585